Craig M. Jensen, Etsuo Akiba and
Hai-Wen Li (Eds.)

Hydrides: Fundamentals and Applications

MDPI

This book is a reprint of the Special Issue that appeared in the online, open access journal, *Energies* (ISSN 1996-1073) in 2015, available at:

http://www.mdpi.com/journal/energies/special_issues/fundamentals-applications

Guest Editors
Craig M. Jensen
Department of Chemistry, University of Hawaii
USA

Etsuo Akiba
Department of Mechanical Engineering, Kyushu University, Fukuoka
Japan

Hai-Wen Li
International Research Center for Hydrogen Energy, Kyushu University, Fukuoka
Japan

Editorial Office	*Publisher*	*Senior Assistant Editor*
MDPI AG	Shu-Kun Lin	Guoping (Terry) Zhang
St. Alban-Anlage 66		
Basel, Switzerland		

1. Edition 2017

MDPI • Basel • Beijing • Wuhan • Barcelona • Belgrade

ISBN 978-3-03842-208-2 (Hbk)
ISBN 978-3-03842-209-9 (PDF)

Table of Contents

List of Contributors

Mohamed F. Aly Aboud Sustainable Energy Technologies Center, College of Engineering, King Saud University, Riyadh 11421, Saudi Arabia; Mining, Metallurgical and Petroleum Engineering Department, Faculty of Engineering, Al-Azhar University, Nasr City, Cairo 11371, Egypt.

Etsuo Akiba International Institute for Carbon-Neutral Energy Research (WPI-I2CNER); International Research Center for Hydrogen Energy, Kyushu University, Fukuoka 819-0395, Japan.

Zeid A. ALOthman Advanced Materials Research Chair, Chemistry Department, College of Science, King Saud University, Riyadh 11451, Saudi Arabia.

Katsutoshi Aoki Institute for Materials Research, Tohoku University, Sendai 980-8577, Japan.

Marcello Baricco Department of Chemistry, Centre for Nanostructured Interfaces and Surfaces (NIS), National Interuniversity Consortium of Materials Science and Technology (INSTM), University of Turin, Turin 10125, Italy.

Paul R. Beaumont Department of Chemistry, University of Hawaii at Manoa, 2545 McCarthy Mall, Honolulu, HI 96822-2275, USA.

Sergio Brutti CNR-ISC, U.O.S. La Sapienza, 00185 Roma, Italy; Department of Science, University of Basilicata, 85100 Potenza, Italy.

Hujun Cao Dalian National Laboratory for Clean Energy, Dalian Institute of Chemical Physics, Chinese Academy of Sciences, Dalian 116023, China.

Radovan Černý Department of Quantum Matter Physics, Laboratory of Crystallography, University of Geneva, CH-1211 Geneva, Switzerland.

Dhanesh Chandra Department of Chemical and Materials Engineering, University of Nevada, Reno, NV 89557, USA.

Ping Chen Dalian National Laboratory for Clean Energy, Dalian Institute of Chemical Physics, Chinese Academy of Sciences, Dalian 116023, China.

Fermin Cuevas ICMPE/CNRS-UPEC, UMR 7182, 94320 Thiais, France.

Stefano Deledda Physics Department, Institute for Energy Technology, Kjeller NO-2027, Norway.

Umit B. Demirci IEM (Institut Europeen des Membranes), UMR 5635 (CNRS-ENSCM-UM2), Universite Montpellier 2, F-34095 Montpellier, France.

Roman V. Denys Institute for Energy Technology, Kjeller NO 2027, Norway.

Viney Dixit Hydrogen Energy Centre, Department of Physics, Banaras Hindu University, Varanasi 221 005, India.

Michael Dolan CSIRO Energy Technology, Pullenvale, Queensland 4069, Australia.

Ruoming Duan School of Materials Science and Engineering and Key Laboratory of Advanced Energy Storage Materials of Guangdong Province, South China University of Technology, Guangzhou 510641, China.

Zhigang Zak Fang Department of Metallurgical Engineering, University of Utah, Salt Lake City, UT 84112, USA.

Yaroslav Filinchuk Institute of Condensed Matter and Nanosciences, Université Catholique de Louvain, 1348 Louvain-la-Neuve, Belgium.

Evan MacA. Gray Queensland Micro- and Nanotechnology Centre, Griffith University, Nathan 4111, Brisbane, Australia.

Duncan H. Gregory WestCHEM, School of Chemistry, University of Glasgow, Joseph Black Building, Glasgow G12 8QQ, UK.

Matylda N. Guzik Physics Department, Institute for Energy Technology, Kjeller NO-2027, Norway.

Mohamed A. Habila Advanced Materials Research Chair, Chemistry Department, College of Science, King Saud University, Riyadh 11451, Saudi Arabia.

Bjørn C. Hauback Physics Department, Institute for Energy Technology, Kjeller NO-2027, Norway.

Liqing He Department of Mechanical Engineering, Faculty of Engineering, Kyushu University, Fukuoka 819-0395, Japan.

Teng He Dalian National Laboratory for Clean Energy, Dalian Institute of Chemical Physics, Chinese Academy of Sciences, Dalian 116023, China.

Minghong Huang School of Materials Science and Engineering and Key Laboratory of Advanced Energy Storage Materials of Guangdong Province, South China University of Technology, Guangzhou 510641, China.

Terry D. Humphries Hydrogen Storage Research Group, Fuels and Energy Technology Institute, Department of Physics, Astronomy and Medical Radiation Sciences, Curtin University, Perth, WA 6845, Australia.

Jacques Huot Hydrogen Research Institute, Université du Québec à Trois-Rivières, Trois-Rivières, QC G9A 5H7, Canada.

Tamio Ikeshoji Institute for Materials Research, Tohoku University, Sendai 980-8577, Japan.

Ankur Jain Institute for Advanced Materials Research, Hiroshima University, Higashi-Hiroshima 739-8530, Japan.

Pragya Jain Hydrogen Energy Centre, Department of Physics, Banaras Hindu University, Varanasi 221 005, India; Hydrogen Research Institute, Université du Québec à Trois-Rivières, Trois-Rivières, QC G9A 5H7, Canada.

Craig M. Jensen Department of Chemistry, University of Hawaii at Manoa, Honolulu, HI 96822-2275, USA.

Torben R. Jensen Center for Materials Crystallography (CMC), Interdisciplinary Nanoscience Center (iNANO) and Department of Chemistry, University of Aarhus, DK-8000 Århus C, Denmark.

Michel Latroche ICMPE/CNRS-UPEC, UMR 7182, 94320 Thiais, France.

Morten B. Ley Center for Materials Crystallography (CMC), Interdisciplinary Nanoscience Center (iNANO) and Department of Chemistry, University of Aarhus, DK-8000 Århus C, Denmark.

Guanqiao Li WPI-Advanced Institute for Materials Research (WPI-AIMR), Tohoku University, Sendai 980-8577, Japan.

Hai-Wen Li International Research Center for Hydrogen Energy, WPI International Institute for Carbon-Neutral Energy Research (WPI-I2CNER), Kyushu University, Fukuoka 819-0395, Japan.

Xingguo Li Beijing National Laboratory for Molecular Sciences (BNLMS), the State Key Laboratory of Rare Earth Materials Chemistry and Applications, College of Chemistry and Molecular Engineering, Peking University, Beijing 100871, China.

Miaolian Ma School of Materials Science and Engineering and Key Laboratory of Advanced Energy Storage Materials of Guangdong Province, South China University of Technology, Guangzhou 510641, China.

Jianfeng Mao WestCHEM, School of Chemistry, University of Glasgow, Joseph Black Building, Glasgow G12 8QQ, UK.

Motoaki Matsuo Institute for Materials Research, Tohoku University, Sendai 980-8577, Japan.

Fabrice Morelle Institute of Condensed Matter and Nanosciences, Université Catholique de Louvain, 1348 Louvain-la-Neuve, Belgium.

Romain Moury Max-Planck-Institut für Kohlenforschung, 45470 Mülheim an der Ruhr, Germany.

Jiri Muller Physics Department, Institute for Energy Technology, Kjeller NO-2027, Norway.

Shin-ichi Orimo WPI-Advanced Institute for Materials Research (WPI-AIMR), Tohoku University, Sendai 980-8577, Japan.

Liuzhang Ouyang China-Australia Joint Laboratory for Energy & Environmental Materials; School of Materials Science and Engineering and Key Laboratory of Advanced Energy Storage Materials of Guangdong Province, South China University of Technology, Guangzhou 510641, China.

Oriele Palumbo CNR-ISC, U.O.S. La Sapienza, 00185 Roma, Italy.

Annalisa Paolone CNR-ISC, U.O.S. La Sapienza, 00185 Roma, Italy.

Michael Powell Pacific Northwest National Laboratory, Richland, WA 99352, USA.

Elsa Roedern Center for Materials Crystallography (CMC), Interdisciplinary Nanoscience Center (iNANO) and Department of Chemistry, University of Aarhus, DK-8000 Århus C, Denmark.

Ewa C. E. Rönnebro Pacific Northwest National Laboratory, Richland, WA 99352, USA.

Yolanda Sadikin Department of Quantum Matter Physics, Laboratory of Crystallography, University of Geneva, CH-1211 Geneva, Switzerland.

Ivan Saldan Physics Department, Institute for Energy Technology, Kjeller NO-2027, Norway; Department of Physical and Colloid Chemistry, I.F. National University of Lviv, Lviv UA-79005, Ukraine.

Suchismita Sarker Department of Chemical and Materials Engineering, University of Nevada, Reno, NV 89557, USA.

Pascal Schouwink Department of Quantum Matter Physics, Laboratory of Crystallography, University of Geneva, CH-1211 Geneva, Switzerland.

Onkar N. Srivastava Hydrogen Energy Centre, Department of Physics, Banaras Hindu University, Varanasi 221 005, India.

Lixian Sun Guangxi Collaborative Innovation Center of Structure and Property for New Energy and Materials, Guilin 541004, China.

Peter M. M. Thygesen Center for Materials Crystallography (CMC), Interdisciplinary Nanoscience Center (iNANO) and Department of Chemistry, University of Aarhus, DK-8000 Århus C, Denmark.

Francesco Trequattrini Department of Physics, Sapienza University of Rome, 00185 Roma, Italy.

Jenny G. Vitillo Department of Chemistry, Centre for Nanostructured Interfaces and Surfaces (NIS), National Interuniversity Consortium of Materials Science and Technology (INSTM), University of Turin, Turin 10125, Italy; Science and Technology Department, University of Insubria, Como 22100, Italy.

John C. Walmsley SINTEF Materials & Chemistry, Trondheim NO-7465, Norway.

Han Wang Dalian National Laboratory for Clean Energy, Dalian Institute of Chemical Physics, Chinese Academy of Sciences, Dalian 116023, China.

Hui Wang China-Australia Joint Laboratory for Energy & Environmental Materials; School of Materials Science and Engineering and Key Laboratory of Advanced Energy Storage Materials of Guangdong Province, South China University of Technology, Guangzhou 510641, China.

Colin J. Webb Queensland Micro- and Nanotechnology Centre, Griffith University, Nathan 4111, Brisbane, Australia.

Matthew Westman Pacific Northwest National Laboratory, Richland, WA 99352, USA.

Greg Whyatt Pacific Northwest National Laboratory, Richland, WA 99352, USA.

Guotao Wu Dalian National Laboratory for Clean Energy, Dalian Institute of Chemical Physics, Chinese Academy of Sciences, Dalian 116023, China.

Junzhi Yang Beijing National Laboratory for Molecular Sciences (BNLMS), the State Key Laboratory of Rare Earth Materials Chemistry and Applications, College of Chemistry and Molecular Engineering, Peking University, Beijing 100871, China.

Volodymyr A. Yartys Institute for Energy Technology, Kjeller NO 2027, Norway; Norwegian University of Science and Technology, Trondheim NO 7491, Norway.

Olena Zavorotynska Physics Department, Institute for Energy Technology, Kjeller NO-2027, Norway.

Feng (Richard) Zheng Pacific Northwest National Laboratory, Richland, WA 99352, USA.

Min Zhu China-Australia Joint Laboratory for Energy & Environmental Material; School of Materials Science and Engineering and Key Laboratory of Advanced Energy Storage Materials of Guangdong Province, South China University of Technology, Guangzhou 510641, China.

Claudia Zlotea ICMPE/CNRS-UPEC, UMR 7182, 94320 Thiais, France.

About the Guest Editors

Etsuo Akiba is the Deputy Director of the International Research Center for Hydrogen Energy, Head of Hydrogen Storage Division of International Institute for Carbon-Neutral Energy Research (I2CNER), and Professor of Faculty of Engineering, Kyushu University. He earned Ph.D. degree from The University of Tokyo in 1979 in physical chemistry. He joined the National Institute for Advanced Industrial Science and Technology (AIST) in 1979. He has worked for the Kyushu University since December 2010. His major field is research and development of hydrogen storage materials. He has published around 250 reviewed articles. He has received several awards including the Herbert C. Brown Award for Innovations in Hydrogen Research, Purdue University, USA in 2008 and The IPHE Technical Achievement Award in 2010.

Craig Jensen is a full professor in the Department of Chemistry of the University of Hawaii. He is an inorganic chemist with broad experience in catalyst development and the synthesis and characterization of novel inorganic and organometallic materials. Prof. Jensen has authored or co-authored 141 peer-reviewed publications and eight U.S. patents and presented over 140 invited seminars and conference talks in 24 different countries. He was named the U.S. Department of Energy Hydrogen program's "1999 Research Success Story" and presented with their "R&D" award in 2004. Since 1997, he has been a member of the expert groups of the International Energy Association's tasks aimed at the development of improved hydrogen storage materials. In 2003, he founded Hawaii Hydrogen Carriers, LLC and has since served as the company president. He was a co-chairman of the 2006 International Symposium on Metal–Hydrogen Systems and the 2007 Hydrogen–Metal Systems Gordon Research Conference.

Hai-Wen Li is currently an associate professor at the International Research Center for Hydrogen Energy and the International Institute for Carbon-Neutral Energy Research (WPI-I2CNER), Kyushu University, Japan. After obtaining his Ph.D. degree in 2005 from the Kitami Institute of Technology under the supervision of Prof. Kiyoshi Aoki, he started to work as a postdoctoral researcher with Prof. Shin-ichi Orimo at the Institute for Materials Research, Tohoku University. He was awarded a JSPS Postdoctoral Research Fellowship for two years (2006-2008). He worked as an assistant professor at the Institute for Materials Research, Tohoku University from 2008 to 2011. His research interests focus on investigating fundamental, physical and chemical properties of interstitial and non-interstitial hydrides, aiming at developing advanced energy storage materials for high-density hydrogen storage and electrochemical applications.

Preface to "Hydrides: Fundamentals and Applications"

Both the Japanese and Hawaiian archipelagos are both completely devoid of petroleum resources. Thus the coauthors of this Special Issue live in societies that feel an urgent need to develop alternative energy sources. The utilization of hydrogen as an energy carrier in the form of metal hydrides has long been proposed as a key component in strategies for the harnessing of renewable energy sources. In the past, these considerations alone were the impetus for our studies of metal hydrides. However, the motivation for research in this area is evolving.

PEM fuel cell powered automobiles have recently been commercialized. This has resulted in efforts to develop metal hydride technologies that will enable rapid expansion of the already existing market that is based on high-pressure hydrogen. The scope of the potential practical applications of metal hydrides has also been extended beyond hydrogen storage to ionic conductors and thermal energy storage. Our goal in assembling this Special Issue of Energies was to provide readers with a sense of the future directions that can be anticipated in metal hydride research in view of these changes. We hoped to accomplish this by providing a sampling of the variety of cutting-edge research efforts on metal hydrides that are currently underway throughout the scientific world. We are now happy to present the 15 outstanding contributions (13 original research papers and 2 reviews) that can be found within this Special Issue. With the inclusion of authors from 16 different countries, the issue truly presents a global view. This volume also covers a wide range of materials (classical metal hydrides, complex hydrides, and metal hydride composites); and applications (onboard hydrogen storage, off-board hydrogen storage, and thermal energy storage). Although none of the papers focus directly on battery applications or ionic conductors, the findings reported here are also relevant to these topics.

We wish to express our deep gratitude to all the contributors to this Special Issue and those that served as reviewers.

Craig Jensen, Hai-Wen Li and Etsuo Akiba
Guest Editors

Metal Hydrides for High-Temperature Power Generation

Ewa C. E. Rönnebro, Greg Whyatt, Michael Powell, Matthew Westman,
Feng (Richard) Zheng and Zhigang Zak Fang

Abstract: Metal hydrides can be utilized for hydrogen storage and for thermal energy storage (TES) applications. By using TES with solar technologies, heat can be stored from sun energy to be used later, which enables continuous power generation. We are developing a TES technology based on a dual-bed metal hydride system, which has a high-temperature (HT) metal hydride operating reversibly at 600–800 °C to generate heat, as well as a low-temperature (LT) hydride near room temperature that is used for hydrogen storage during sun hours until there is the need to produce electricity, such as during night time, a cloudy day or during peak hours. We proceeded from selecting a high-energy density HT-hydride based on performance characterization on gram-sized samples scaled up to kilogram quantities with retained performance. COMSOL Multiphysics was used to make performance predictions for cylindrical hydride beds with varying diameters and thermal conductivities. Based on experimental and modeling results, a ~200-kWh/m^3 bench-scale prototype was designed and fabricated, and we demonstrated the ability to meet or exceed all performance targets.

Reprinted from *Energies*. Cite as: Rönnebro, E.C.E.; Whyatt, G.; Powell, M.; Westman, M.; Zheng, F.(R.); Fang, Z.Z. Metal Hydrides for High-Temperature Power Generation. *Energies* **2015**, *8*, 8406–8430.

1. Introduction

To reduce energy consumption and greenhouse gas emissions, we need more efficient ways to utilize energy. In the International Energy Agency (IEA) technology roadmap from 2014 [1–3], energy storage technologies are categorized by output: electricity and thermal (heat or cold). Broadly speaking, energy storage is a system integration technology that allows for the improved management of energy supply and demand. Energy storage is utilized in various areas and for various applications and includes batteries, flywheels, electrochemical capacitors, superconducting magnetic energy storage (SMES), power electronics and control systems. Thermal energy storage (TES) technologies operate with the goal of storing energy for later use as heating or cooling capacity and are contributing to improved energy efficiency.

Thermal energy storage (TES) is a key technology for implementing renewable energies, enabling grid applications and for efficiently utilizing energy for various applications within the areas of buildings and transportation; however, cost reduction

1

is necessary [4]. It is an emerging technology market, which recently has been identified as a key enabling storage method for more efficient energy use in heating and cooling applications. The key challenge is that the materials currently used are not efficient enough and of too high a cost. Materials development programs are needed to improve performance, but materials research needs to go hand in hand with systems engineering to design the most efficient systems.

Thermal energy storage can be stored as a change in internal energy of a material as sensible heat, latent heat and thermochemical or a combination of these. Sensible heat storage is storage that occurs with a temperature change when heat is added or removed. The storage capacity is fairly low. Latent heat storage is typically phase change materials (PCM) that have a phase change that occurs at constant temperature. The storage capacity is 3–5-times higher than for sensible heat storage. Thermochemical heat storage has the potential for much higher energy density, above 10–20-times sensible heat storage, and energy is reversibly stored in chemical exothermic/endothermic reactions, such as in metal hydrides.

During recent years, solar technologies have been strongly emerging, and photovoltaics (PV) has become significantly cheaper. Furthermore, several concentrating solar power (CSP) plants have been built, in Spain and in the USA. However, both PV and CSP technologies are currently not used with storage (except for two CSP plants), which limits the use to when the Sun is shining. It has recently been emphasized how thermal energy storage can provide continuous usage and dispatchability and enable renewable energies for various applications and grid implementation.

The state-of-the-art thermal energy storage for concentrating solar power (CSP) and solar dishes is molten salt storage (mix of nitrates, also called solar salt), using sensible energy storage. The energy storage density if low (153 kJ/kg), which necessitates enormous storage tanks. The salt is fluid at elevated temperatures and pumped up to the receiver in the solar power tower, where it is heated. It is thereafter pumped into a hot tank at 565 °C for heat generation. When the salt comes back from the power block, it is stored in a cold tank at 285 °C until it is pumped back up to the receiver for the next cycle. The main benefit of using molten salt storage is the low cost of nitrates. The main challenge is the high freezing point of the salt (222 °C), which results in having to re-heat the molten salt, which is energy consuming. There are also issues with the corrosion of containers and pipes. Moreover, this system has expensive parts, *i.e.*, heat exchangers and pumps, which results in a high system cost.

There are currently no advanced materials commercially available for thermal energy storage (TES) of ⩾600 °C. Common shortcomings with TES materials are low energy densities and limited capacity and life cycle due to irreversible side reactions. Metal hydrides have among the highest practical energy densities among known materials, and some metal hydrides have eight-times higher energy density than molten salts [5–8], *i.e.*, >700 kJ/kg and >150 kWh/m^3, so an metal hydride (MH) system can be at least eight-times smaller, far exceeding the U.S. DOE target of 25 kWh/m^3. We can reach higher energy efficiencies by operating at higher temperatures (the molten salts store latent heat at about 400–550 °C). This technology is simple, straight forward, without moving parts, and we believe that we can lower the cost relative to other technologies to meet the DOE cost target of $15/kWh. Metal hydrides do not freeze at the anticipated temperatures, so they will not need any energy to re-heat, like molten salts, and they are known to achieve a long life cycles, so we expect that they can meet the 30-year lifetime target.

We have chosen metal hydrides that are low cost, readily abundant and non-toxic, based on titanium. By using titanium hydride for the high-temperature (HT) bed, we believe we can exceed DOE performance targets as indicated from our results presented here.

Table 1. Comparison of titanium hydride with other heat storage technologies. TES, thermal energy storage.

TES Material	T and P Range	Heat of Reaction	Gravimetric Energy Density **	Materials Features
Titanium hydride	650–700 °C 1–3 bar	150 kJ/mol	778 kJ/kg (practical) 3190 kJ/kg (theo)	▪ High exergetic efficiency ▪ Long life cycle expected ▪ T, P allow simple, stainless steel construction
Magnesium hydride	450–500 °C 40–100 bar	75 kJ/mol	2814 kJ/kg (theo)	▪ Lower exergetic efficiency due to lower temperature ▪ Oxidation of MgH$_2$ to MgO limits the life cycle ▪ High pressures require high-cost tank materials
Calcium hydride	1100–1400 °C 1–5 bar	186 kJ/mol	4426 kJ/kg (theo)	▪ High exergetic efficiencies ▪ Oxidation of CaH$_2$ to CaO limits the life cycle ▪ Expensive tank materials needed at >1100 °C
Solar salt/molten salt (NaNO$_3$/KNO$_3$ 60:40 mixture)	300–550 °C N/A	15 kJ/mol	153 kJ/kg	▪ Lower exergetic efficiency due to lower temperature ▪ Lower volumetric energy densities; 8 times larger space requirements than for TiH

Note: ** Gravimetric energy density for materials (not the system). Theo = theoretical. Titanium hydride is given as the practically-obtained kJ/kg for cycling on the plateau pressure. Theoretical is 3190 kJ/kg. The kJ/kg given for MgH$_2$ and CaH$_2$ is theoretical for full capacities, not practical.

Table 1 compares titanium hydride with a molten salt system and two other metal hydrides that are being considered by other teams. Metal hydrides have been investigated for TES; MgH$_2$ requires operation at a lower temperature and much higher pressure, which results in lower exergetic efficiency while increasing structural requirements [9]. CaH_2 operates at very high temperatures, which necessitates the use of expensive nickel-based alloys for the container. Both MgH_2 and CaH_2 have issues with oxidation that can limit the life cycle. The challenge is to avoid irreversible losses due to chemical side reactions and to maintain capacity during cycling. Titanium is less prone to oxidation after a thin, nanosized oxide shell of titanium oxide has formed, which protects from further oxidation while allowing for hydrogen diffusion. The state-of-the-art molten salt $NaNO_3/KNO_3$ technology operates at a lower temperature, which reduces the exergetic efficiency and has a low heat of reaction that leads to a low energy storage density.

2. Results and Discussion

2.1. Operating Principle of Dual Bed Metal Hydride Thermochemical Energy Storage

To accelerate the development and deployment of high-temperature thermal energy storage for renewable energies, we are developing a thermochemical energy storage (TCES) system with superior energy densities and lower cost than the current state-of-the-art. The system consists of two connected metal hydride (MH) beds: a high-temperature (HT) bed operating at $\geqslant 650\,^\circ C$ and a low temperature (LT) bed operating near ambient temperature (~40 $^\circ$C). When heat is added to the HT-reservoir, such as by solar energy, H_2 is released in an endothermic reaction that absorbs ~150 kJ per mol H_2 [10–12]. The hydrogen moves to the LT-reservoir, where it forms a hydride at near ambient temperature and releases ~25–35 kJ/mol H_2 of heat to the environment. When heat is retrieved from storage, hydrogen returns to the HT hydride bed and undergoes a reaction that releases ~150 kJ/mol H_2, which can be utilized to generate power. Our operation is expected to occur between H:Ti ratios of 0.88 and 1.55 in a range of ~650–700 $^\circ$C. The reaction enthalpy is assumed constant with respect to temperature in extrapolating the reaction enthalpy to the operating temperature range of 650–700 $^\circ$C. The HT hydride isotherm and cycling behavior were demonstrated and will be discussed further below. In the region of operation for the HT bed, significant hydrogen loading changes occur with little change in pressure. As a result, the temperature at which heat is stored is very close to the temperature at which it is returned. This makes it possible to achieve high exergetic efficiency, since, unlike sensible heat storage, the source heat temperature is sustained as heat is withdrawn from the reservoir. Sensible heat associated with the hydrogen released is recovered from waste heat before H_2 storage.

To validate the ability of utilizing the identified candidate for high-temperature TES, we: (1) demonstrate the life cycle at >600 $^\circ$C; (2) measure the thermal diffusivity; (3) scale up with retained performance; and (4) demonstrate a bench-scale unit with ~200 kWh/m^3, <6 h charging time.

2.2. Performance of High-Temperature Metal Hydride

We investigated several titanium-based alloys to optimize the pressure and temperature operation range with maximum capacity; however, the best performing material is titanium, and we will in the following subsections report the characterization of titanium hydride.

2.2.1. Isotherms of Titanium Hydride

We measured absorption and desorption isotherms using our custom-built Sievert's system at various temperatures between 630 and 680 °C. To obtain accurate data, we used the criteria for reaching steady state at a stable temperature and pressure condition below 0.075 °C and 0.075 psia. The pressure-composition isotherms are similar to the literature [13,14]. We compared ITP (International Titanium Powder)-Ti-powder with Ti-sponge, which, after screening, had been selected as the two best candidates based on the plateau pressure and the capacity in the plateau region and found that the plateau pressure of the ITP-Ti powder is significantly sloping, while the Ti-sponge plateau pressure is flat and well defined, which is beneficial for reversibly cycling the Ti-powder between higher and lower hydrogen content. Figure 1 shows the isotherms of Ti-sponge at ~640 °C.

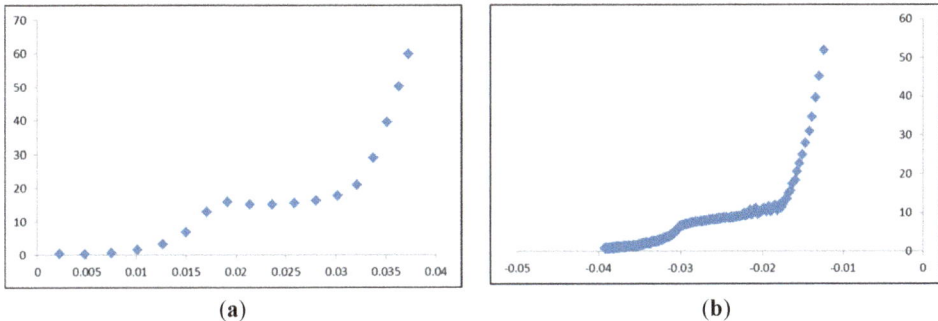

Figure 1. Isotherms of 2.5 grams of Ti-sponge (raw data): (**a**) absorption at 638 °C; plateau pressure = 15 psi (1.03 bar); (**b**) desorption at 638 °C: plateau pressure = 9 psi (0.62 bar). The Y-axis is pressure in psia and the X-axis is bound H_2 in mol.

A summary of the isotherms at various temperatures is shown in Figure 2. It appears that at 638 °C, the plateau pressure is close to one bar. For thermal energy storage applications, we aim at operation near atmospheric pressure to avoid expensive high-pressure tank materials.

These measured data were used to create a model that predicts isotherm behavior. The model predictions are shown in Figure 3. The dashed lines in Figure 3 illustrate that at 2.4 bar, a swing in the H:Ti ratio between 0.88 and 1.55 can be achieved between 700 °C and 650 °C, which corresponds to a swing of 1.4 wt% with respect to titanium mass.

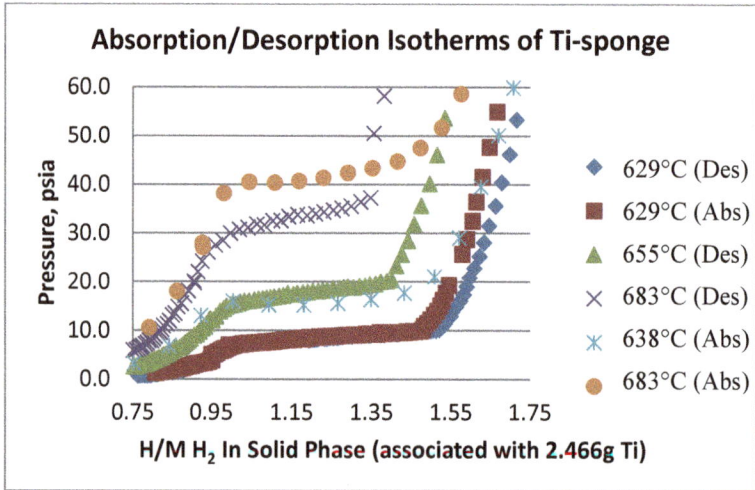

Figure 2. Absorption (Abs) and desorption (Des) isotherms of Ti-sponge at various temperatures. H/M is ratio of moles hydrogen per moles titanium.

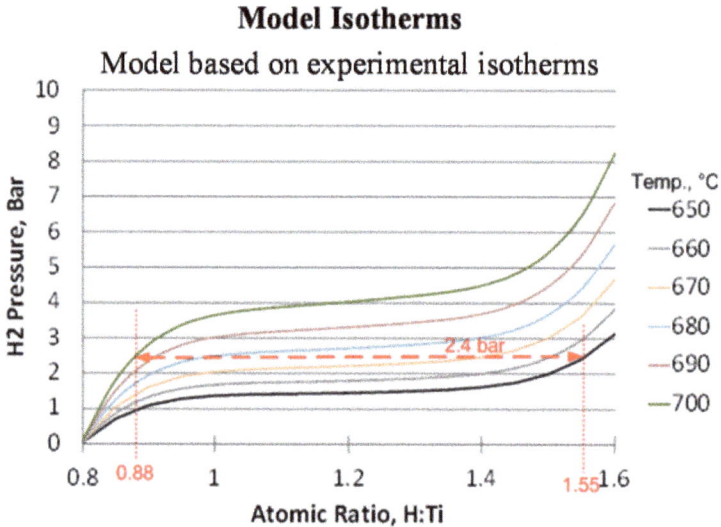

Figure 3. Model isotherms in the region of interest. Equilibrium loading is predicted to swing from 0.88 mol H:mol Ti at 700 °C to 1.55 mol H:mol Ti at 650 °C for a swing of 0.67 mol Hmol Ti. Actual loading swings achieved in the model depend on local temperatures reached during the loading/unloading cycle.

2.2.2. Life Cycle of Ti-Sponge

We performed a life cycle test to show 60 charging and discharging cycles without capacity drop, executed by inserting and releasing hydrogen gas at about 650 °C and one bar H_2 pressure. To our knowledge, the life cycle of titanium hydride has previously not been shown. The life cycle target for TES is ~10,000 cycles (30 years). As can be seen in Figure 4, we showed 60 cycles without capacity drop.

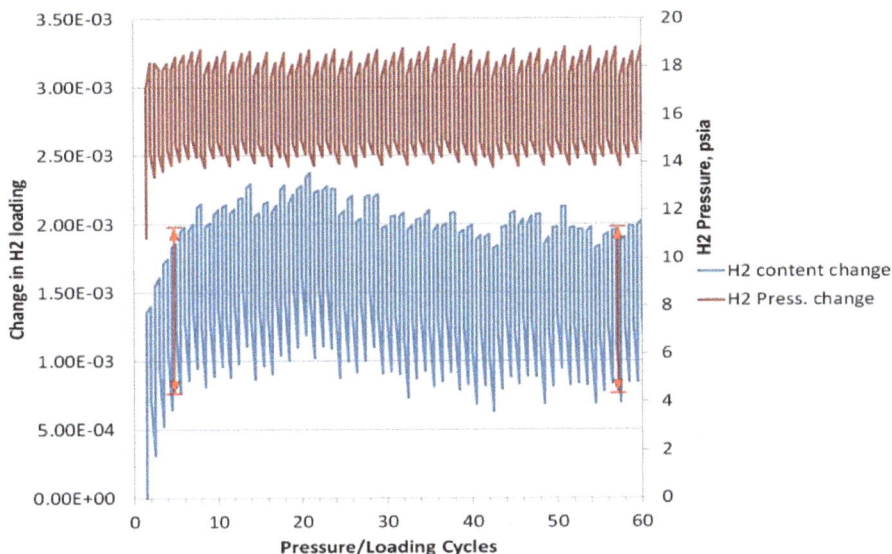

Figure 4. Life cycle of 2.55-gram Ti-sponge at 650 °C showing 60 cycles over 70 h (60 absorption/60 desorption) without a capacity drop in hydrogen pressure (psia).

2.2.3. Thermal Diffusivity of Titanium and Titanium Hydride

To design and build a TES prototype with optimized thermal heat management, it is important to know the thermal diffusivity in the Ti bed with and without hydrogen. It is crucial to have sufficient heat and mass transfer to obtain a long life cycle. To assess the need for thermal enhancement for the bench-scale prototype, we estimated the thermal diffusivity of a series of Ti-based pellets using a commercial instrument to collect data in argon during heating. We chose to focus on key parameters, *i.e.*, temperature dependence, powder compaction, thermal enhancement with graphite and the impact of life cycle, to study trends as a guide to design the first generation prototype. In the future, we will report thermal conductivity measurements using a custom-built device with a larger hydride bed, collecting data during absorption and desorption cycles in a hydrogen atmosphere.

The thermal conductivity as calculated from the measured diffusivity of titanium is about 4–7 W/(mK) at 400 °C, as can be since in Figure 5. The increase after 400 °C is likely because of oxidation, forming titanium oxides on the surface of the pellet. Interestingly, the TiH_2 (−325 mesh) sample had a thermal conductivity of about 12 W/(mK) at 400–500 °C, and the plot is represented in Figure 6. Ito *et al.* [15] studied the electrical and thermal properties of titanium hydrides of $TiHx$ (x = 1.53–1.75). They found that the thermal conductivity of the hydride was the same as that of the metal and increased slightly with increasing temperature. At 323 K, the thermal conductivity was measured to be 21.8 W/(mK) for Ti and 19.7 W/(mK) for $TiH_{1.75}$, which are significantly higher than what we obtained. The reason is likely due to difficulties with measuring thermal diffusivity on porous samples with our currently-employed instrument (described in the Experimental Section). However, we could obtain data to study the trends. In the future, we will design a device for thermal diffusivity measurements during hydrogen absorption and desorption.

It is worthwhile to note that thermal conductivity in argon is lower than in hydrogen atmosphere. The better the thermal diffusivity, the longer the life cycle that can be achieved. We also studied thermal enhancers to optimize the heat management on a materials level.

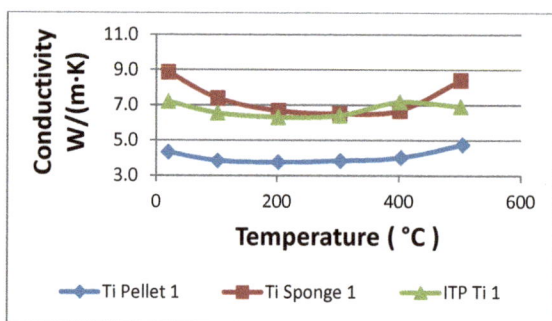

Figure 5. Thermal conductivity up to 500 °C of three different titanium samples; Ti from University of Utah, Ti-sponge and ITP- Ti (International Titanium Powder).

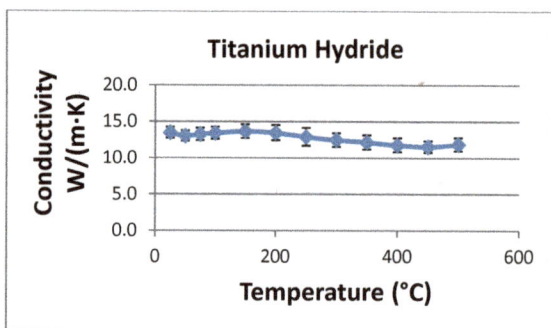

Figure 6. Thermal conductivity of titanium hydride, −325 mesh.

2.2.4. Thermal Enhancement with Graphite

We investigated solutions to enhance thermal conductivity by adding a thermal enhancer, leveraging results from recent studies on other metal hydrides. We prioritized two options: (1) adding a conductive powder as a thermal enhancer, *i.e.*, graphite; or (2) incorporating the Ti-powder in a conductive foam, *i.e.*, copper. Based on our experimental results, we chose the option of using a copper foam as an internal structure of the container for the high-temperature TES prototype bed, as will be discussed below.

A recently-explored metal hydride for TES and hydrogen storage is magnesium hydride. Magnesium has a very low thermal conductivity of less than 1 W/(mK). It has been shown in the literature that if adding 10% expanded natural graphite (ENG), the thermal conductivity is increased to 8 W/(mK) [16]. Therefore, we decided to add graphite to titanium hydride to assess the increase in thermal conductivity. As can be seen in Figure 7, the conductivity of titanium hydride with 10% graphite is two-times the conductivity of titanium hydride, *i.e.*, 20 W/(mK) at 500 °C.

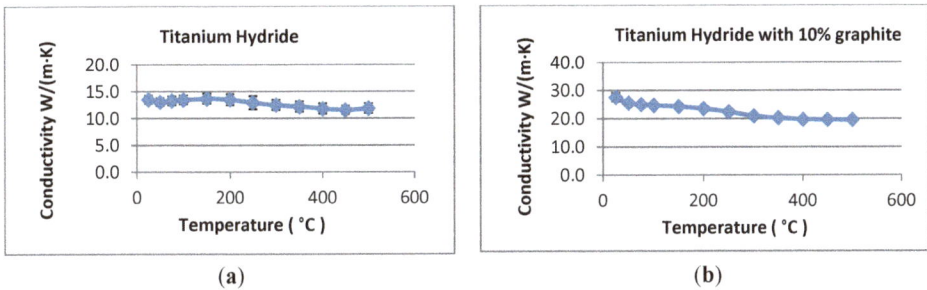

(a)   (b)

Figure 7. Thermal conductivity of (a) titanium hydride and (b) titanium hydride with 10% graphite showing thermal enhancement of two times.

2.2.5. Impact of Compaction on Thermal Diffusivity

We performed a study of the level of compaction of titanium hydride powder and, not surprisingly, found that with higher static pressure applied to increase compaction, the thermal diffusivity is enhanced up to about three times, as can be seen in Figure 8.

2.2.6. Impact of Life Cycle on Thermal Diffusivity

We measured the thermal diffusivity of titanium hydride after multiple cycles of hydrogenation and dehydrogenation had been performed. Interestingly, the thermal diffusivity relative to the original titanium hydride powder remained the same, as can be seen in Figure 9. Most metal hydride materials undergo a decrease in particles size during cycling, but Ti-powder particles appear to sinter together, thus not affecting the thermal diffusivity significantly, at least in the range of dozens of cycles. The effect of hundreds of cycles is beyond the scope of the current project, but will be reported in the future.

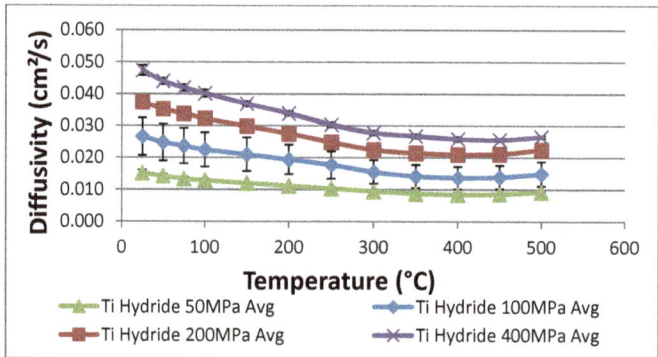

Figure 8. Thermal diffusivity of titanium hydride at various levels of compaction (pressure in MPa), (Avg= average).

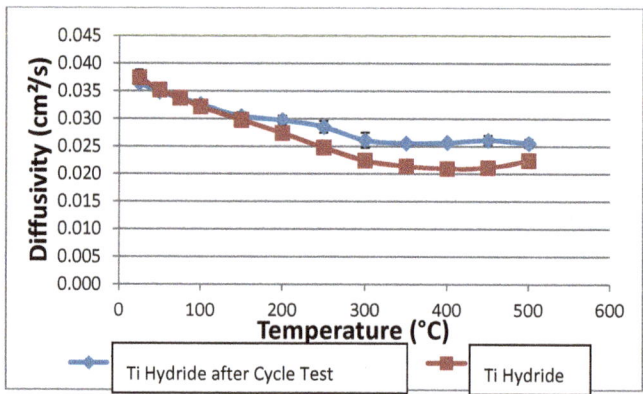

Figure 9. Thermal diffusivity of titanium hydride after multiple cycles compared to original titanium hydride.

2.3. Scale-Up to Kilogram Quantities

The scale-up to kilogram quantities of selected titanium hydride was performed using titanium sponge powders. One of the objectives was to increase both the packing density and specific surface area of the powder to optimize performance. Both the surface area and the packing density of Ti-sponge powders, however, depend on particle sizes. The finer the particle size, the lower the packing density and the higher the specific area. One of the ways to maximize both is to use a custom-designed blend of different particle sizes. "As received" powders were

10

sieved, and the particle size distributions were analyzed. Different blends of powders were mixed and blended using a tumbler mixer. The results showed that a blend of 2:4 powder with 66.67% of −325 mesh powder is the best blend so far with 1.84 g/cm³. About 15 kilograms of Ti-hydride powder with retained performance based on studying isotherms were prepared for inclusion in the container with Cu-foam.

2.4. Numerical Modeling of Hydrogen Uptake and Release for Candidate Prototype Designs

Physical property and hydrogen uptake data for titanium powder were used in a numerical model to predict hydrogen uptake and release rates for several candidate prototype designs. COMSOL Multiphysics was used to make performance predictions for cylindrical hydride beds with varying diameters and thermal conductivities. The modeling results imply that 3 kWh of thermal storage per day can be achieved using two 12.7-cm (5 in) diameter, 29 cm (11.5 in)-long cylinders filled with titanium hydride powder and 8% dense, reticulated copper foam. The basis for this conclusion is provided in the subsections that follow.

2.4.1. Numerical Model Development

The prototype energy storage system is intended to operate at near-constant pressure and to use a temperature swing of ± 10 °C to drive the uptake and subsequent release of hydrogen. As the temperature in the bed changes, the concentration of stored hydrogen responds accordingly. The relationship between stored hydrogen equilibrium pressure, temperature and the heat of reaction can be determined based on a Clausius-Clapeyron expression of the form:

$$Q = -R \left[\frac{\partial \ln P}{\partial \left(\frac{1}{T} \right)} \right]$$

where T is absolute temperature in Kelvin, R is the gas constant (8.314 J/mol K), Q is the heat of adsorption in J/mol and P is pressure. This can be rearranged to yield pressure P_2 at a temperature T_2 if the values P_1 and T_1 are known from an isotherm:

$$P_2 = e^{\left[\frac{-Q}{R} \left(\frac{1}{T_2} - \frac{1}{T_1} \right) + \ln(P_1) \right]}$$

This expression requires an isotherm from which to extrapolate. For our modeling, we measured an isotherm for titanium hydride at 638 °C and then used these data to predict the variation with temperature of the hydrogen concentration in the hydride bed at a constant pressure of 1.10 atm (16.13 psia; 1.1 bar). This relationship was used in the subsequent modeling of the hydride beds. In addition to the uptake/release dependence on temperature, modeling of the hydride beds

11

requires estimates for the hydride bulk density, thermal conductivity, heat capacity and the heat of reaction between hydrogen and titanium.

The powder density was taken to be 3038 kg/m^3 based on the average measured density of the titanium pellets used in the laser flash experiments described previously. Heat capacity was set equal to 600 J/kgK based on literature values for titanium in the temperature range of interest (roughly 638 °C).

The thermal conductivity of the powder was set to 0.2 W/mK. This value is significantly lower than measured in the laser-flash experiments, but it is expected to reflect the reduced conductivity of a hydride bed after many thermal cycles. Repeated cycling may break up the hydride into relatively small (~2–20 microns) particles, and this effect will reduce the bulk thermal conductivity to between 0.1 and 0.4 W/mK. We measured the thermal conductivity after several cycles as described above without seeing a particle size reduction; however, the thermal conductivity reduction may not be pronounced until after several hundreds or thousands of cycles. Rather than rely on the higher thermal conductivities measured in the laser flash experiments, we chose to use a conservatively low bed thermal conductivity to ensure that the prototype system will meet the design objectives. The heat of reaction for hydrogen and titanium hydride varies somewhat with the fraction of hydrogen stored in the hydride. Data for the reaction enthalpy were correlated with stored hydrogen concentration and entered into the numerical model. Over the concentration range of interest, the heat of reaction varies between about 140 and 155 kJ per mole of H$_2$ reacted.

The COMSOL models for the heat-storage prototypes were set up assuming radial symmetry within the device. COMSOL was used to solve the dynamic heat conduction equation with a simultaneous chemical reaction:

$$\partial T/\partial t = k/(\rho Cp)\,((\partial^2 T)/(\partial r^2) + 1/r\,\partial T/\partial r) + 1/(\rho Cp)\,q(r,t)$$

where T is the local temperature, q is the heat generation rate from the hydride reaction, r is radial direction, t is time, k is the hydride powder thermal conductivity, r is the hydride bulk density and Cp is the hydride heat capacity.

The relevant boundary conditions for the model are: (1) no heat loss or gain from the ends of the cylinder; and (2) temperature changes of the outside cylindrical surface are specified numerically as a function of time as described in the text of the next section. The concentration of hydrogen bound in the hydride material was tracked as a function of radial position and time. At time = 0, the hydride was assumed to be fully saturated with hydrogen at the specified starting temperature and pressure. An outside wall temperature change was then applied sigmoidally over a 30-s duration, and the changes in radial temperature and hydrogen concentration profiles were tracked over time thereafter. Reaction kinetics were assumed fast compared to the relatively slow process of heat conduction through the hydride powder, so in the model, the local bound hydrogen concentration was set equal to its equilibrium value based on the local temperature and specified hydrogen gas pressure.

2.4.2. Modeling Results

Initial modeling efforts focused on predicting the rate of hydrogen release from cylindrical beds of hydride powder. For these calculations, the bed was initially assumed to be at 635 °C and loaded to 1.35 moles hydrogen per mole of titanium. At time = 0, the outside wall of the cylinder is rapidly heated to 645 °C, and hydrogen is released as progressively more of the bed is heated. With each mole of hydrogen released, however, approximately 150 kJ of heat is required to drive the reaction, so heating the bed proceeds much more slowly than it would if there were no reaction taking place. Figure 10 shows a series of radial temperature profiles within a one inch-diameter bed starting with the blue line at the bottom for time = 0 (T_{init} = 634.85 °C = 908 K) and then moving from right to left with each successive line representing 20 min of elapsed time. The left-most line (colored red) indicates the temperature profile after 10 h. The r-coordinate in the graph starts at 3.2 mm because the bed geometry is assumed to include a 6.4-mm (0.25-in) outside diameter, porous metal tube that is included to facilitate the addition and removal of hydrogen. Figure 11 shows a similar plot for the hydrogen concentration within the bed.

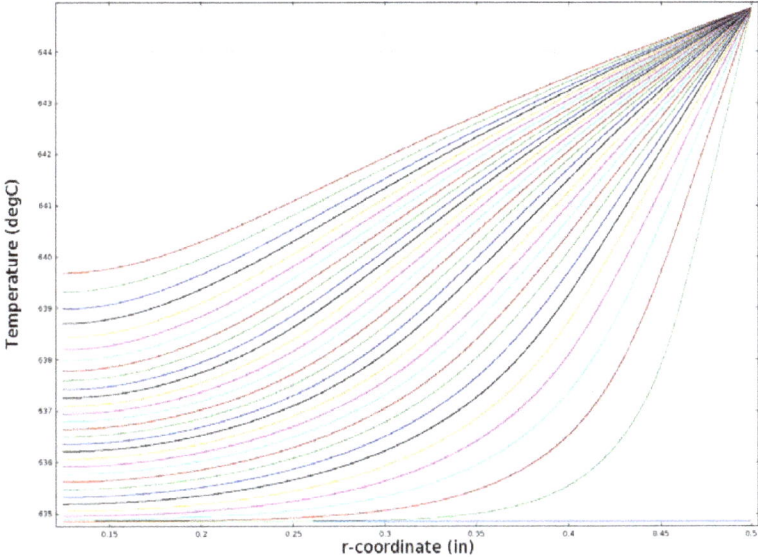

Figure 10. Modeled temperature profile in a one inch-diameter cylindrical bed (k = 0.2 W/mK).

13

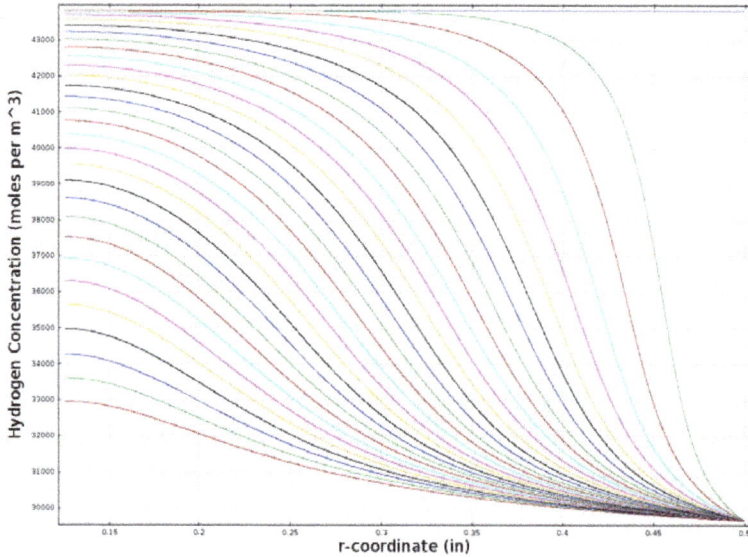

Figure 11. Modeled hydrogen concentration profile in a one inch-diameter cylindrical bed (k = 0.2 W/mK).

The heat-storage prototype is designed with the assumption that the bed is cycled once per day with approximately 6 h of hydrogen loading, 6 h of thermal equilibration, 6 h of unloading and, then, six more hours of thermal equilibration. The one inch-diameter bed still contains a substantial fraction of its loaded hydrogen after 6 h of heating, so the one inch diameter is likely a bit larger than we desire for our prototype system. Figure 12 shows the averaged hydrogen concentration within the bed as a function of time for a bed with and without the central 6.4 mm (0.25 in)-diameter porous metal core. Roughly 10 h is required to reduce the hydrogen concentration to about 10% of its initial value. Similar simulations were run for alternative tube diameters, and the hydrogen-release time was found to scale approximately with the square of the tube diameter. A bed diameter of approximately 1.9 cm (0.75 in) yields a hydrogen release time of about 6 h, which is the nominal target for the demonstration system.

Multiple load/unload cycles were simulated for the 1.9 cm (0.75 in)-diameter hydride bed with a 0.63 cm (0.25 in)-diameter porous tube along its axis. The hydride was assumed to be loaded to 1.35 moles of hydrogen atoms per mole of titanium atoms. This condition represents the maximum expected loading at 635 °C and 16.13 psia. At time = 0, the exterior wall of the cylinder is assumed to be heated to 645 °C and held constant for the next 6 h, while hydrogen is released and the gas pressure in the bed is held constant at 1.1 atm (16.13 psia). At time = 6 h, the heat flow into the bed is stopped, and the bed is allowed to come to thermal equilibrium

14

until time = 12 h. At time = 12 h, the exterior wall of the cylinder is decreased back to 635 °C. Between time = 12 h and time = 18 h, the hydrogen concentration in the bed increases as the bed is cooled as it delivers stored heat and hydrogen gas is supplied from an external source at 1.1 atm. Finally, between time = 18 h and time = 24 h, the heat flow out of the bed is stopped, and the bed is allowed to come to thermal equilibrium. The cycle is then repeated starting at time = 24 h, when the exterior wall is again heated to 645 °C.

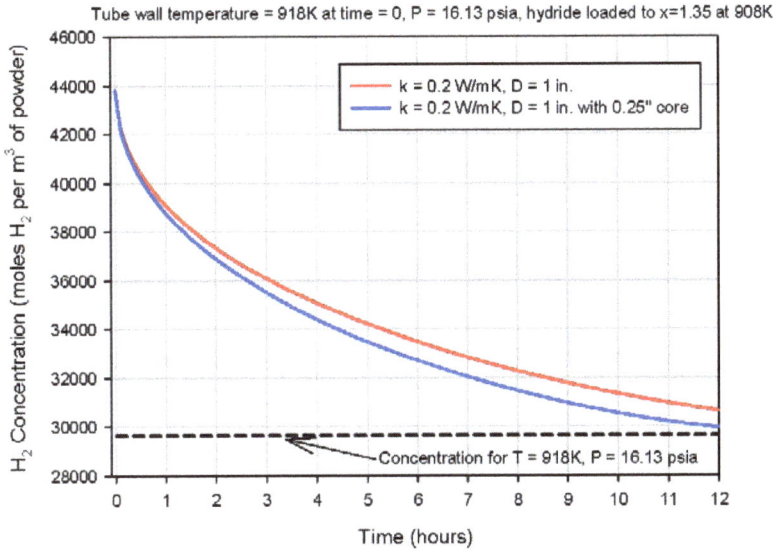

Figure 12. Modeled average hydrogen concentration in the hydride bed *vs.* time for a one-inch diameter.

During each 24-h cycle, the applied exterior-wall temperature swings between 635 °C and 645 °C. This 10-degree temperature swing is only half of the 20-degree swing available for the prototype system. We limited the temperature swing in the model to 10 degrees to ensure that there is adequate temperature driving force available between the heating/cooling fluid and the exterior surface of the hydride bed. The 10-degree driving force should be sufficient for gas-phase convective heating/cooling of the hydride bed. Figure 11 shows the results for cylinders ranging in size from 10 cm (4 in)–15 cm (6 in). In all cases, we assumed that there is a 1.3-cm (0.5 in) porous-metal tube in the center of the bed to facilitate the addition and removal of hydrogen gas.

The 12.7 cm (5 in)-diameter cylinder is predicted to yield roughly 90% hydrogen release after the targeted 6 h of heating, so the performance of a cylinder with this diameter was simulated for four complete 24-h cycles. Figure 12 shows the predicted

15

volume-averaged hydrogen concentration inside the bed during all four cycles. Then, much less than a 10-degree temperature driving force will be needed, and the hydride bed performance will be better than predicted by our modeling results.

Figure 13 shows the volume-averaged concentration of hydrogen in the bed for four complete load/unload cycles. After the first couple of cycles, the high and low bed concentrations stabilize for all subsequent cycles at 41.02 and 29.71 moles H_2 per liter of bed volume. To store 3 kWh of heat during every 24-h cycle, about 72 moles of H_2 must be loaded and unloaded. Each mole of H_2 represents roughly 150 kJ (41.7 Wh) of heat, because the heat of reaction is 150 kJ/mole. The total required bed volume can be estimated by dividing the required 72 moles of H_2 by the difference between high and low bed H_2 concentrations ($41.02 - 29.71 = 11.31$ moles per liter). The result is an estimated required hydride volume of 6.36 liters. If each hydride bed is 46 cm (18 in) long and 1.9 cm (0.75 in) in diameter with a 0.63 cm (0.25 in)-diameter porous tube along the axis, then a total of 55 such tubes will be required to achieve the targeted 3 kWh of heat storage per cycle.

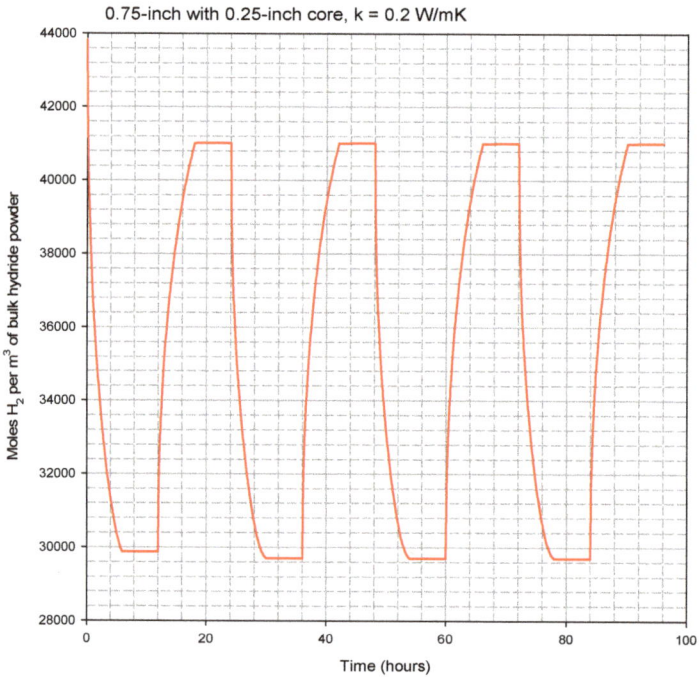

Figure 13. Modeled four complete cycles for the 1.9 cm (0.75 in)-diameter hydride bed.

The 55 tubes can be arranged in an array roughly 19 cm (7.5 in) wide, 16.5 cm (6.5 in) high and 46 cm (18 in) long, as shown in Figure 14. Each tube has an inside diameter of 1.9 cm (0.75 in) and an outside diameter of 2.22 cm (0.875 in). The total volume occupied by the array of tubes is 14.4 liters.

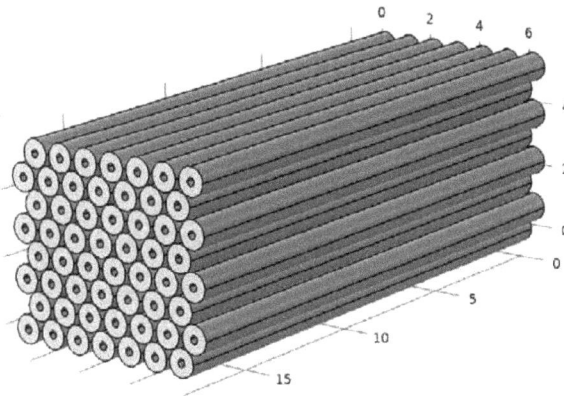

Figure 14. Close-packed array of 56 tubes, each 2.2 cm (0.875 in) in diameter and 46 cm (18 in) long. The dimensions shown on the background grid are inches.

To reduce the volume occupied by the bed and allow for better heat transfer and more design flexibility, we considered a second geometry in which the hydride powder is loaded into a single, larger-diameter cylinder that includes 8% dense (92% void space) reticulated copper foam with an estimated thermal conductivity of 9.3 W/mk. Simulations were performed for multiple cylinder diameters to determine the diameter that yields the desired hydrogen release time of approximately 6 h; when the cycles stabilize, the hydrogen concentration shifts between a high of 37.35 moles per liter and a low of 27.49 moles per liter. The modeled hydrogen release from cylinders with various inside diameters containing copper foam is shown in Figure 15. Using the same calculation method described earlier for the 1.9-cm tubes, we determine that a total of 7.3 liters of bed volume is required for 3 kWh of heat storage per cycle. For the 12.7-cm (5 in) cylinder diameter, a total cylinder length of 58 cm (23 in) is required for 7.3 liters of bed volume, and the external volume of the device is just under 7.7 liters. The cylinder containing copper foam, then, has a volume just over half that of the array of 55 tubes (14.4 liters).

2.4.3. First Generation Metal Hydride Thermal Energy Storage Prototype Size

Based on the modeled bed performance, we recommended the demonstration system to use a 12.7 cm (5 in)-diameter bed containing copper foam for enhanced heat transfer. Use of a single cylinder rather than an array of 55 cylinders will facilitate testing of the demonstration system by simplifying the test arrangement and reducing assembly costs. Further, the performance of the demonstration system with the copper foam will not be susceptible to changes in the thermal conductivity of the hydride powder that might occur with successive cycles.

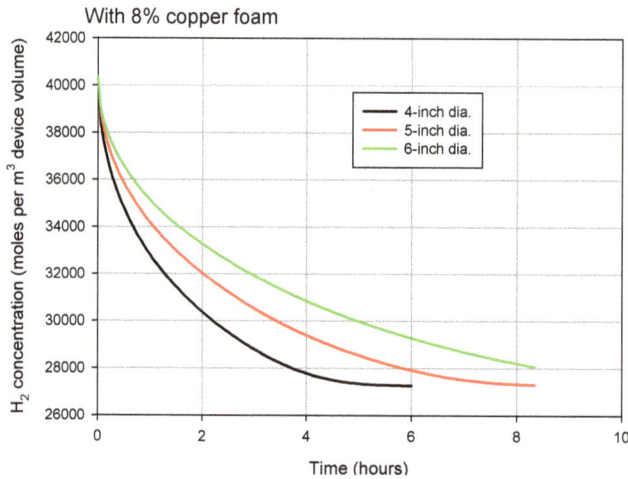

Figure 15. Modeled hydrogen release from cylinders with various inside diameters containing copper foam.

To allow for the use of an existing high-temperature furnace, the demonstration system comprised a single, 29 cm-long cylinder with a 12.7-cm diameter to show high-efficiency heat storage of at least 1.5 kWh per 24 h cycle. The cylinder has a 1.3-cm (0.5 in) porous metal tube down the axis.

2.4.4. Copper Foam Fill for Hydride Cylinder

Since the thermal conductivity of Ti/TiH is anticipated to not be sufficient for efficient thermal heat management, our bench-scale prototype was constructed with an internal copper foam to show the proof of concept. We chose to use an open cell 10 ppi Duocel copper foam with interstitial spaces that were filled with hydride powder. Using a copper foam allows for a simple construction of the first generation test cylinder, and it is easy to fill with hydride powder. The foam enhances conductivity in both radial and axial directions.

2.4.5. Hydride Test Bed Design Details

For the design of the hydride test bed, we used a column made from a 5" S40 pipe, 24-inch long cylindrical section. The pipes dimensions are pipe with an inner diameter ID = 5.047" and pipe with outer diameter OD = 5.563", with pipe caps welded onto each end. Screens at the ends of the straight cylinder maintain powder in copper foam, with the space in the caps unutilized. The bed consists of 8% dense,

10 ppi Duocel copper foam, filled with titanium powder. The cylindrical section is wrapped with a Ni-80 heater insulated with ceramic "Salamander" beads. The heater is covered by controlled thickness insulation layer with the ends well insulated. A porous metal, $\frac{1}{2}''$ OD tube at the bed centerline will add/remove H_2 from the bed. The structure of the container is shown in detail in Figure 16.

2.5. Prototype Installation

The prototype was fabricated in 306 L stainless steel, and the copper foam was filled with titanium powder. Portions of about 200 grams of the titanium powder were inserted into the 8% dense copper foam of 10 ppi by vibrating the test tube to facilitate homogenous compaction until the foam was filled up to the brim. Argon was flowed through the test tube during filling, and the setup was contained in a hood with protective Plexiglas. Due to the powder having a lower density by a factor of ~2 compared to a compacted pellet, we were not able to load as much as we initially had planned. We loaded 9.44 kilograms in total.

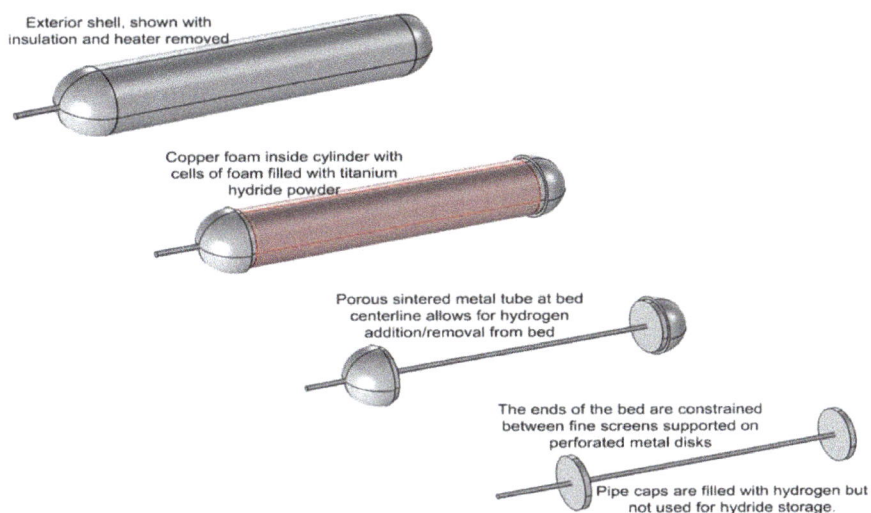

Figure 16. Internal structure of the hydride test bed.

The stainless steel tube was wrapped with a Ni-80 heater and insulated with ceramic "Salamander" beads for heating. The prototype was installed on our custom-built hydrogenation system and equipped with a mass flow controller (MFC) to regulate hydrogen flow in and a mass flow meter (MFM) to measure the hydrogen flow out. The system was also equipped with a back pressure regulator (BPR) to prevent back flow. The temperature controller was set up to control the center thermocouple near the wall (TC2). A LabVIEW program was programmed to operate the system and to collect and save data, including TC readings, pressures and mass flow rate. The test tube was leak tested with helium, and the MFC was calibrated with helium. Before starting the experiment, the test tube was purged with hydrogen gas, using the vacuum pump to evacuate.

To monitor the temperature throughout the bed, we equipped the test tube with three thermocouples: three near the wall at the ends and in the center and one thermocouple throughout the core of the bed, as can be seen in Figure 17. We installed a mass flow controller (MFC) to enable the control of hydrogen flow into the bed. It was calibrated with hydrogen gas. We also installed a wet test meter to monitor the flow of hydrogen out of the bed. By monitoring the temperature profile and measuring the hydrogen content in and out of the test bed to learn the heat diffusion and energy density capacity, which is in the hydrogen content, we can calculate efficiencies.

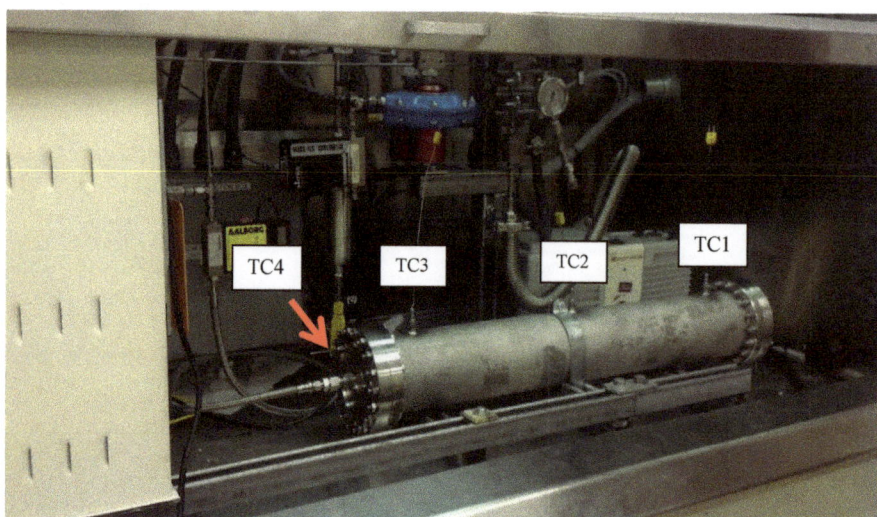

Figure 17. Stainless steel test tube filled with titanium powder and attached to the hydrogenation system. The tube was equipped with thermocouples (TC).

Since we have a cooler side of the prototype where room temperature hydrogen gas enters the hot bed, we anticipated the need for insulation to reduce the temperature gradient throughout the bed. We placed the test tube within a box of insulation boards, wrapped the top with a layer of fiberfrax and covered the ends with aluminum tape. This significantly improved the temperature differences. The hydrogen inlet end (TC3) was about 40–60 °C cooler than the center during operation, and the opposite end (TC1) was about 15–25 °C cooler than the center during operation. The core temperature (TC4) was about 10–25 °C cooler than the center wall temperature (TC2).

2.6. Showing the Proof of Concept of the Bench-Scale Prototype

The first time that the titanium bed was charged with hydrogen gas, the temperature was slowly increased on the temperature controller to avoid a potential thermal runaway reaction. The pressure of the hydrogen gas was set to 16.1 psia and the temperature controller to 350 °C. Hydrogen started to be absorbed by titanium at about 320–330 °C. The exotherm was significant and allowed the reaction to occur slowly under monitoring. Thereafter, the bed was discharged followed by performing five full cycles. The temperature was set to 645 °C for discharging and to 635 °C for charging, while maintaining pressure as close to 16.1 psia (one bar) as possible.

Since we do not have an LT hydride bed, the mass flow controller (MFC) was continuously adjusted to maintain the hydrogen pressure as close to 16.1 psia as possible to be in accordance with the model. With an LT hydride desorbing at the same rate as the HT hydride is absorbing, the pressure will be able to remain at a constant pressure, regulated by the hydrogen diffusion rate for absorption and desorption. Therefore, it is important to identify a LT-hydride with the same hydrogen diffusion rate for charging and discharging.

2.7. Model Predictions of Charge and Discharge Cycles

The models had predicted that the cycles would occur upon a forced 10 °C shift in wall temperature to keep pressure constant. To operate at 16.1 psia, a temperature swing between 635 and 645 °C was predicted as described above. The experiments were performance at constant pressure in accordance with the model predictions, i.e., repeating cycles of 6 h heat, 6 h rest, 6 h return heat and 6 h rest.

The COMSOL model predicts that the H_2 storage (in standard liters) for a 10-degree swing of the exterior wall varies, as shown below in Figure 18. Once the temperature drops below about 638 °C, the amount of H_2 stored drops quickly.

2.8. Experimental Validation of Models

By performing cycles with a 10 °C shift in wall temperature, we were indeed able to operate the test bed at 635–645 °C at a constant pressure close to 16.1 psia (up to 17 psia). As mentioned above, since we are using a hydrogen gas bottle instead of an LT hydride without an automated pressure control, the flow rate of hydrogen into the bed had to be manually controlled and adjusted to keep a constant pressure. During discharging, we did not have any means of controlling the flow rate out during operation. Similar to when charging, this will be self-regulated when the HT bed is paired up with an LT bed that has the same hydrogen diffusion rates.

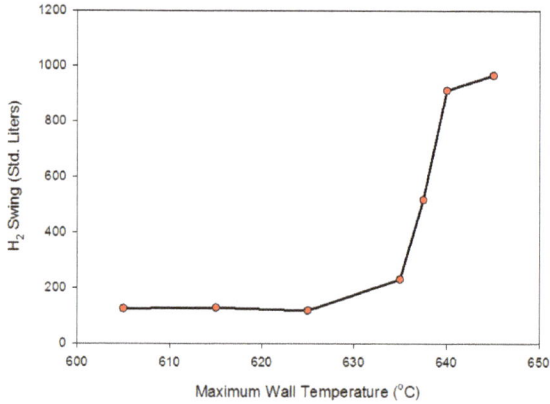

Figure 18. COMSOL model predicts the H_2 storage (in liters) for a 10-degree swing of the exterior wall.

The experimental test data from Cycle 5 are shown in Figures 19–21 below for charging and discharging of hydrogen.

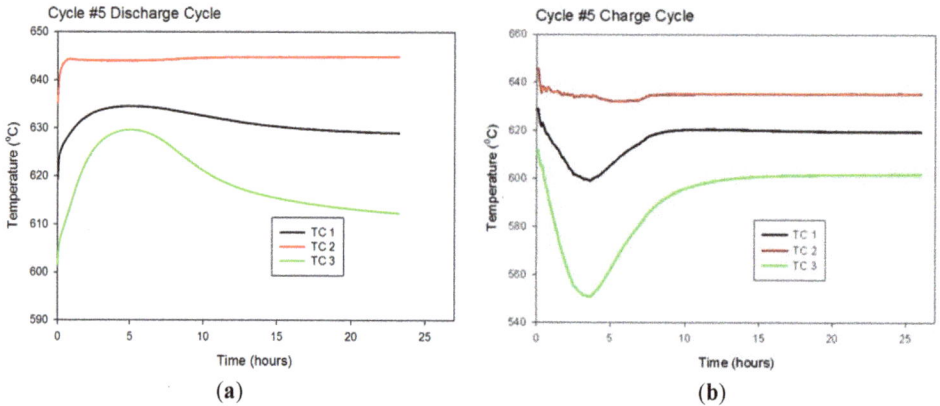

Figure 19. Wall temperatures for three thermocouples TC1, 2 and 3 with time for (a) discharging and (b) charging for the fifth cycle (#5).

Figure 19 shows the three wall thermocouples (TC1, 2 and 3) with time. The temperature is fairly constant for TC2 (center wall TC) at 635 °C for charging and 645 °C for discharging, but it is obvious that the test bed has a temperature gradient. As mentioned, hydrogen comes in at room temperature to the end with TC1, and therefore, the bed is cooler at that end. For future large-scale prototypes, a heat transfer fluid will pre-heat the hydrogen gas.

Figure 20 shows the hydrogen content during charging. As can be seen in the plot for the charging cycle, the absorbed amount of hydrogen was about 700 L. The predicted amount was 400 L, so we are absorbing more than expected by a factor of 1.75, indicating that the prototype performed better than predicted. The reason could however be that the temperature profile is different than in the models. The wet test meter recorded about 640 L desorbed, which indicates that our flow rate recording needs to be improved for our second generation prototype that will be tested with an LT hydride bed for follow-on funding. Figure 20 also tells us that the pressure was fairly constant, but swinging between 16 and 17 psia during charging. Since the diffusion rate during charging is controlled manually by regulating the flow rate on the MFC, it is difficult to keep it constant, but it is possible to improve on it by adjusting the flow rate more often. During discharging, the pressure was constant at 17 psia, which is a little higher than in the models.

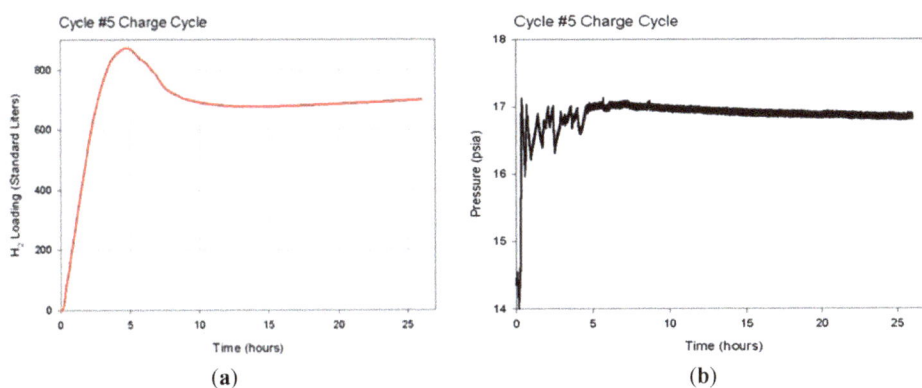

Figure 20. (a) Hydrogen loading with time and (b) hydrogen pressure with time for the fifth charge cycle (cycle #5).

Studying Figure 21, it can be seen that the hydrogen flow rate upon charging is considerably faster than discharging. The flow rate was regulated to keep the pressure as close to the model value as possible. The test bed absorbed most of the hydrogen within 5 h; thereafter, the bed discharged an amount before reaching equilibrium. The test bed desorbed within 10 h.

2.9. Summary of Experimental Results from Prototype Testing

Table 2 below summarizes the obtained preliminary data relative to the targets. It is important to take into account that due to the lower density of the titanium powder by a factor of 2.22 compared to the assumed density, we cannot meet the predicted energy density target at this point; however, as mentioned, in the future, we will use compacted disks of titanium powder in order to increase the density by at least two times. The model predicts that we should store/release about 400 liters of H_2 for each 10-degree swing, assuming that the internal temperatures vary linearly between the center temperature (TC2 = 645 °C) and the temperatures near the ends (TC1 = 630 °C, TC3 = 610 °C). For the next prototype, the temperature profile needs to be improved. The actual quantity of H_2 stored appears to be about 600–700 liters, so the prototype is performing better than expected based on the model with the linear temperature assumption; it is likely that the temperature assumption is overly conservative.

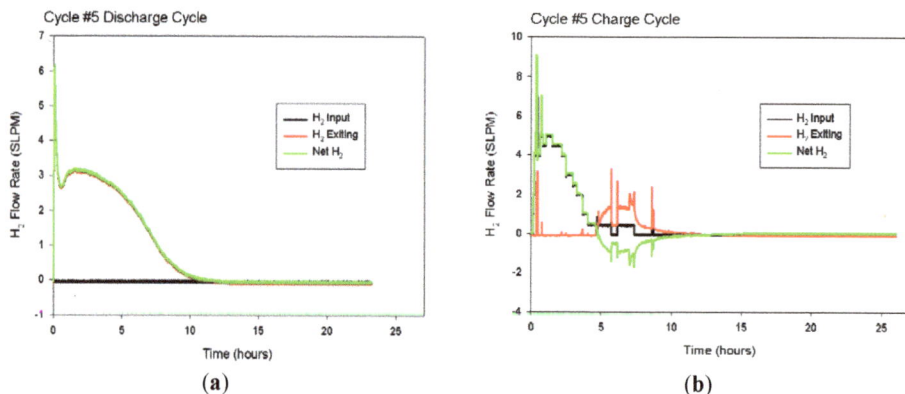

Figure 21. (a) Hydrogen flow rate during discharging and (b) charging towards time for cycle #5.

Table 2. Summary of measured parameters compared to metrics. Note that the assumed density changed from 3.068 g/mL to 1.384 g/mL achieved, which impacts the volumetric storage by a factor of two.

Metric Goal	Charge	Discharge
>600 °C	635 °C	645 °C
<5 bar/72.5 psi H_2	~16.5 psia	~17 psia
Charging time 6 h	5 h	10 h
700 kJ/kg	305	300
400 kWh/m³ on system	178	200

For the exergy calculation, we need to accurately know how much was adsorbed and desorbed and at what wall temperature. Both the flowmeter discrepancy and the uneven temperature profile make this difficult. At this point, we have not seen anything that leads us to believe that the original exergy predictions are not achievable, but we cannot really validate the predictions until we have accurately measured data, which will be collected during the testing of a future prototype.

The planned operating line was to go from $TiH_{1.0}$–$TiH_{1.35}$ (see Figure 22 below). If all of the material were cycling on this plateau, we would obtain 0.35/2 = 0.175 mol H_2 adsorbed per mol $TiH_{1.0}$, which works out to ~0.175 × 150 kJ/((47.90 + 1.01)/

1000) = (26.25)/(0.0489 kg) = 536.7 kJ/kg. Hence, our theoretical limit if we hit our plateau is 536.7 kJ/kg.

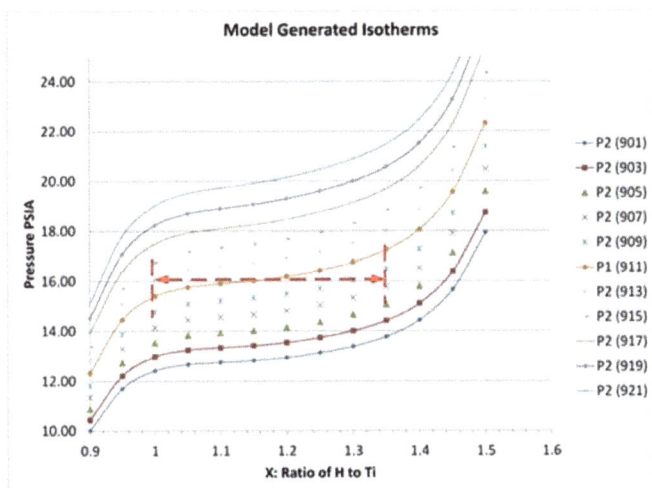

Figure 22. Model isotherms for a swing from $TiH_{1.0}$–$TiH_{1.35}$ at 650 °C and 16.1 psia (1.1 atm).

3. Experimental Section

For the evaluation of the performance of the titanium powders, we utilized a custom-built hydrogenation system to perform high-temperature cycling experiments at 600–800 °C of selected metal hydrides. We designed and fabricated a reactor for gram-sized samples and purchased an autoclave/stainless steel tube rated for 800 °C with the necessary high-temperature fittings. The reactor was installed on PNNL's custom built Sievert's system, shown in Figure 23. Thereafter, we proceeded with volume calibration with helium to be able to accurately calculate hydrogen pressure differences for determining reversible hydrogen content.

We established a program in LabVIEW to be able to easily operate and monitor experimental sequences, including measuring kinetics, life cycle, hydrogen storage content and pressure-temperature-concentration (PCT) experiments automatically.

Figure 23. High-temperature reactor installed on PNNL's Sievert's system for measuring isotherms and for life cycle testing of metal hydride thermal energy storage materials.

To assess the need for thermal enhancement, we explored the thermal diffusivity of a series of Ti-based pellets. We utilized an (Anter FlashLine™) [XP][S2] Thermal Properties Analyzer, xenon flash pulse source. We have an add-on furnace module for operation from room temperature (RT) to 500 °C that allows for three 0.5"-diameter specimens. The detector is an liquid nitrogen (LN2)-cooled IR detector. The Windows™ operating and data analysis software also include specific heat capacity testing software and thermal conductivity calculation software. The thermophysical property of the rate of conductive heat propagation is measured at various temperatures over time. Thermal diffusivity is given as: L $\alpha = \lambda/\rho Cp$, where α is thermal diffusivity, λ is conductivity, ρ density and Cp is specific heat.

We calibrated our instrument with standards provided by Anter. Powders of titanium and titanium hydride were compressed into pellets in our argon-filled glove box to protect samples from oxidation. The density of the pellets was 3.2–3.4 kg/m^3. The samples were heated up to 500 °C, and the thermal diffusivity was measured in intervals of every 50 or 100 °C under the flow through of argon.

4. Conclusions

The goal was to show the proof of concept of a thermal energy storage prototype to operate at >600 °C, <5 bar H$_2$-pressure, with feasibility for 95% efficiency. We demonstrated the ability to meet or exceed the DOE targets by demonstrating the prototype to reversibly operate at 635–645 °C under one bar H$_2$-pressure for at least 60 cycles with a practical gravimetrical energy density of about 800 kJ/kg and a volumetric energy density of ~200 kWh/m^3 on the HT bed (neglecting the 2.2 factor lower density than assumed). Our metal hydride thermal energy storage exceeds the DOE energy density target by eight times. A future bench-scale prototype will include a low-temperature bed for hydrogen storage.

Acknowledgments: We acknowledge the U.S. Department of Energy ARPA-E HEATS program for financially supporting this research under Award 0471-1554. We acknowledge Heavystone Lab LLC and Ronald White, CEO, for support. Kevin Simmons (formerly of PNNL) is appreciated for valuable discussions. Gary Maupin, PNNL, provided valuable help with the Sievert's system. Ben Roberts, PNNL, provided valuable help with installing the prototype and setting up the LabVIEW-based program to operate it.

Author Contributions: Ewa Rönnebro conceived of, designed and performed all experiments, performed all materials characterization on the Sievert's system and tested the prototype. Greg Whyatt and Mike Powell performed the modeling, designed the prototype and evaluated the experimental data compared to the models. Matthew Westman performed the measurements on the thermal diffusivity instrument. Feng (Richard) Zheng made the LabVIEW software program for the Sievert's system. Rönnebro wrote most of the paper with input from Whyatt and Powell. Zhigang Zak Fang provided materials for characterization and performed the materials scale up.

Conflicts of Interest: The authors declare no conflict of interest.

References

1. International Energy Agency (IEA). *Technology Roadmap—Energy Storage*; Report; IEA: Paris, France; 19; March; 2014.
2. International Energy Agency (IEA). *Technology Roadmap—Solar Thermal Electricity*; Report; IEA: Paris, France; September; 2014.
3. Hauer, A.; Simbolotti, G.; Tosato, G.; Gielen, D. *Thermal Energy Storage. IEA-ETSAP and IRENA© Technology Policy Brief E17*; International Renewable Energy Agency: Abu Dhabi, UAE; January; 2012.
4. Chu, S.; Majumdar, A. Opportunities and challenges for a sustainable energy future. *Nature* **2012**, *488*, 294–303.
5. Rönnebro, E.; Majzoub, E. Recent advances in metal hydrides for clean energy applications. *MRS Bull.* **2013**, *38*, 452–461.
6. Meng, X.; Yang, F.; Bao, Z.; Deng, J.; Serge, N.N.; Zhang, Z. Theoretical study of a novel solar trigeneration system based on metal hydrides. *Appl. Energy* **2010**, *87*, 2050–2061.
7. Muthukumar, P.; Groll, M. Metal hydride based heating and cooling systems: A review. *Int. J. Hydrog. Energy* **2010**, *35*, 3817–3831.
8. Satesh, A.; Muthukumar, P. Performance investigations of a single-stage metal hydride heat pump. *Int. J. Hydrog. Energy* **2010**, *35*, 6950–6958.
9. Felderhoff, M.; Bogdanović, B. High Temperature Metal Hydrides as Heat Storage Materials for Solar and Related Applications. *Int. J. Mol. Sci.* **2009**, *10*, 325–344.
10. Wang, W.-E. Thermodynamic evaluation of the titanium-hydrogen system. *J. Alloys Compd.* **1996**, *238*, 6–12.
11. Borchers, C.; Khomenko, T.I.; Leonov, A.V.; Morozova, O.S. Interrupted thermal desorption of TiH2. *Thermochim. Acta* **2009**, *493*, 80.

12. Ershova, O.G.; Dobrovolsky, V.D.; Solonin, Yu.M.; Khyzhun, O.Yu. Hydrogen-sorption and thermodynamic characteristics of mechanically grinded TiH1.9 as studied using thermal desorption spectroscopy. *J. Alloys Compd.* **2011**, *509*, 128–133.

13. San-Martin, A.; Manchester, F.D. The H-Ti (hydrogen-titanium) system. *Bull Alloy Phase Diagr.* **1987**, *8*, 863–873.

14. Dantzer, P. High temperature thermodynamics of H2 and D2 in titanium, and in dilute titanium oxygen solid solutions. *Phys. Chem. Solids* **1983**, *44*, 913–923.

15. Ito, M.; Setoyama, D.; Matsunaga, J.; Muta, H.; Kurosaki, K.; Uno, M.; Yamanaka, S. Electrical and Thermal properties of Titanium Hydrides. *J. Alloys Compd.* **2006**, *420*, 25–28.

16. Chaise, A.; de Rango, P.; Marty, Ph.; Fruchart, D.; Miraglia, S.; Olives, R.; Garrier, S. Enhancement of hydrogen sorption in magnesium hydride using expanded natural graphite. *Int. J. Hydrog. Energy* **2009**, *34*, 8589–8596.

Increasing Hydrogen Density with the Cation-Anion Pair $BH_4{}^-$-$NH_4{}^+$ in Perovskite-Type $NH_4Ca(BH_4)_3$

Pascal Schouwink, Fabrice Morelle, Yolanda Sadikin, Yaroslav Filinchuk and Radovan Černý

Abstract: A novel metal borohydride ammonia-borane complex $Ca(BH_4)_2 \cdot NH_3BH_3$ is characterized as the decomposition product of the recently reported perovskite-type metal borohydride $NH_4Ca(BH_4)_3$, suggesting that ammonium-based metal borohydrides release hydrogen gas via ammonia-borane-complexes. For the first time the concept of proton-hydride interactions to promote hydrogen release is applied to a cation-anion pair in a complex metal hydride. $NH_4Ca(BH_4)_3$ is prepared mechanochemically from $Ca(BH_4)_2$ and NH_4Cl as well as NH_4BH_4 following two different protocols, where the synthesis procedures are modified in the latter to solvent-based ball-milling using diethyl ether to maximize the phase yield in chlorine-free samples. During decomposition of $NH_4Ca(BH_4)_3$ pure H_2 is released, prior to the decomposition of the complex to its constituents. As opposed to a previously reported adduct between $Ca(BH_4)_2$ and NH_3BH_3, the present complex is described as NH_3BH_3-stuffed α-$Ca(BH_4)_2$.

Reprinted from *Energies*. Cite as: Schouwink, P.; Morelle, F.; Sadikin, Y.; Filinchuk, Y.; Černý, R. Increasing Hydrogen Density with the Cation-Anion Pair $BH_4{}^-$-$NH_4{}^+$ in Perovskite-Type $NH_4Ca(BH_4)_3$. *Energies* **2015**, *8*, 8286–8299.

1. Introduction

Complex hydrides material design has undergone major progress since the first reports suggesting $LiBH_4$ as a solid state hydrogen storage material [1]. Semi-empirical chemical concepts combining the stabilities of the less stable and more stable borohydrides [2], for instance, have given way to more profound concepts that exploit different kinds of weak interaction [3]. Recently, we reported on a large family of mixed-metal borohydrides crystallizing in the perovskite structure type, the latter being a prime example of functional materials design [4]. As a very stable structure type, the perovskite form allows one to study the behavior of a large range of different metals in a borohydride environment, but also to apply an immense toolbox developed throughout decades of metal-oxide perovskite design. The most simple concept comprises cation-substitution on the cuboctahedral *A*- and octahedral *B*-sites of a material generalized as $AB(BH_4)_3$. A logical further step is to explore anion-mixing on the ligand site. It was shown that the borohydride

29

perovskite host readily takes up halide anions on the borohydride site [4]. In this context, it comes to mind to test the nitrogen-based hydrogen molecules, which are currently being vigorously investigated as means to tailor hydrogen release temperatures [3]. Recently we described interesting lattice instabilities in the metal boroyhdride perovskites where supposedly close homopolar dihydrogen contacts between adjacent borohydride anions stabilize lattice distortions in the high-temperature phase of $AB(BH_4)_3$, while these are most commonly destabilized by temperature in metal oxide or halide perovskites [4].

Aside from ammonium borohydride NH_4BH_4 itself [5–7], the ammonium cation has been rather neglected by the hydrogen storage community, and so far considered only in conjunction with higher boranes such as $NH_4B_3H_8$ [8], $(NH_4)_2B_{10}H_{10}$ or $(NH_4)_2B_{12}H_{12}$ [9] which impede the borazine evolution during thermolysis thanks to their stable borane cages, but never as a building block in a metal-based material. Attempts to tailor the thermal stability of NH_4BH_4, which decomposes slowly at room temperature (RT), have also been carried out to develop it as solid state hydrogen storage material [7,10].

The underlying concept of incorporating NH_4^+ was to increase the gravimetric hydrogen capacity ρ_m of perovskite-type $KCa(BH_4)_3$ by monovalent cation substitution on the K-site or divalent cation substitution on the Ca-site. Attempts to achieve this by the substitution K^+-Na^+ or Ca^{2+}-Mg^{2+} fail, as the hypothetical compounds $NaCa(BH_4)_3$, $KMg(BH_4)_3$ or $NaMg(BH_4)_3$ all come to lie outside the experimentally established stability field of the structure type within borohydrides [4]. The ionic radius of monovalent NH_4^+ (1.43 Å), on the other hand, is close to that of K^+ (1.33 Å) and ammonium easily substitutes for potassium in many metal halides. The calculated tolerance and octahedral factors suggest $NH_4Ca(BH_4)_3$ to be a stable compound. Its thermal stability and structure were recently reported in [4], accomplishing the long-term RT-stabilization of ammonium borohydride NH_4BH_4, with the extreme hydrogen density of 24.5 wt % H_2 in a metal ammonium borohydride, $NH_4Ca(BH_4)_3$, with a high density of ρ_m = 15.7 wt % H_2. Herein, we modify the reported synthesis based on ammonium chloride NH_4Cl to produce chlorine-free samples, taking up this study and focus on the actual decomposition mechanism of $NH_4Ca(BH_4)_3$ *via* an unknown ammonia-borane complex, which is obtained as a single phase sample thanks to the modified procedure and whose structure is solved on the basis of synchrotron X-ray powder diffraction. The vividly applied concept to tailor hydrogen release properties, the hetero-polar contact between protic and hydridic hydrogen, is thus extended for the first time to a cation-anion pair, NH_4^+-BH_4^-, in a metal borohydride. The presented concept of cation-substitution by ammonium is easily extended to a large number of previously reported mixed-metal borohydrides [11].

2. Results and Discussion

The substitution of K^+ for NH_4^+ results in a stable compound $NH_4Ca(BH_4)_3$, which is presently investigated by means of two different synthetic mechanochemistry routes using different sources of the ammonium molecule. While the first employs a metathesis reaction based on ammonium chloride, in the second, chlorine-free, approach, NH_4BH_4 is loaded at low temperatures in the milling jar and reacted directly with $Ca(BH_4)_2$ in an addition reaction. The Cl-approach results in better crystallized compounds and a less complex phase composition (of novel phases) in as-milled samples, which is why the decomposition is discussed for these samples, despite the chlorine-content. On the other hand, the main decomposition product occurring during thermolysis of both samples is obtained as a well crystallized single phase composition in the NH_4BH_4-based one, when milled in the presence of a solvent (ether). The crystal structure of the main decomposition product is hence solved and discussed on the basis of these data.

2.1. Formation of $NH_4Ca(BH_4)_3$

In samples synthesized from NH_4Cl, perovskite-type $NH_4Ca(BH_4)_3$ is formed according to a general metathesis reaction used for borohydride-perovskites, developed for cases when one of the precursors cannot be provided as a borohydride:

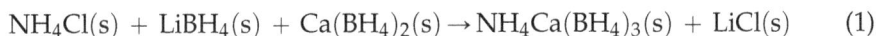

$$NH_4Cl(s) + LiBH_4(s) + Ca(BH_4)_2(s) \rightarrow NH_4Ca(BH_4)_3(s) + LiCl(s) \qquad (1)$$

A number of different nominal NH_4Cl:$Ca(BH_4)_2$:$LiBH_4$ compositions were investigated, $NH_4Ca(BH_4)_3$ forming in the less NH_4Cl-rich mixtures 1:1:1, 1:2:1 and 1:2:2, but neither in 2:2:1 nor 3:1:3. At least three other phases with very low decomposition temperatures were observed as impurities or the main phase in 2:2:1 and 3:1:3 mixtures in as-milled samples prepared from the chloride, presumably containing NH_4^+ as indicated by their low decomposition temperatures, but these could not be identified and will not be further discussed presently.

The formation of $NH_4Ca(BH_4)_3$ from the borohydrides, *i.e.*, $Ca(BH_4)_2$ and NH_4BH_4, was attempted in various trials. To this end milling jars were preliminarily cooled to 243 K, in order to stabilize NH_4BH_4 (unstable at RT) in a ratio 1:1 next to solid $Ca(BH_4)_2$. Soft milling (30 min at 500 rpm) of this mixture resulted in a multiphase powder with an unsatisfactory phase yield of $NH_4Ca(BH_4)$ (black curve in Figure 1). The low yield may be owed to partial decomposition of NH_4BH_4, since the jars were not cooled during milling and the inside-temperature reached approximately 310 K.

The procedure was subsequently modified to a solvent-assisted protocol where diethyl ether was added to the mixture and the inside-temperature was kept below 270 K, resulting in a significantly improved phase yield of $NH_4Ca(BH_4)_3$ (blue curve in Figure 1). At temperature below RT, and neglecting impurities, the formation from the borohydrides can hence ideally be approximated as:

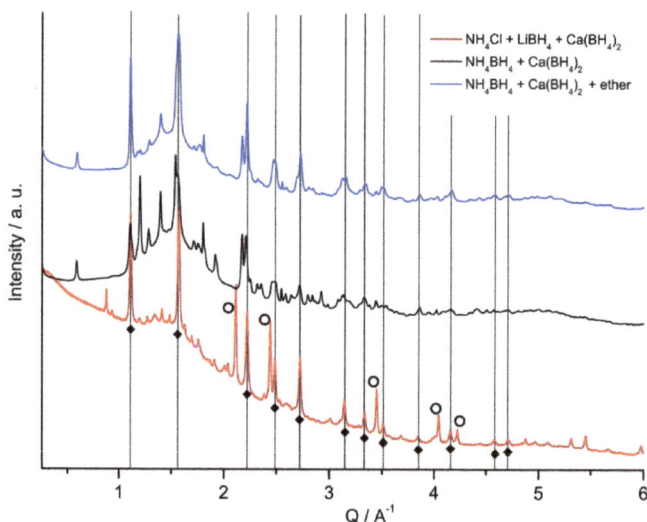

$$NH_4BH_4(s) + Ca(BH_4)_2(s) \xrightarrow{ether} NH_4Ca(BH_4)_3(s) \qquad (2)$$

Figure 1. Room temperature synchrotron X-ray powder diffraction patterns shown for chloride-based (bottom curve, mixture 1:1:1), dry NH_4BH_4-ballmilled (middle) and ether-NH_4BH_4-ballmilled samples (top). Squares and vertical lines mark peak position of $NH_4Ca(BH_4)_3$, open circles correspond to LiCl.

Despite differences in the phase composition with respect to impurities the perovskite-type borohydride $NH_4Ca(BH_4)_3$ is the main phase in both the presently discussed chloride (1:1:1) and ether-based approaches. Furthermore, the fact that the lattice parameter *a* is identical within error (5.655(3) and 5.651(5) Å at 313 K for $NH_4Ca(BH_4)_3$ produced by chloride- and ether-based ball milling) as indicated by identical Bragg peak position in Figure 1 demonstrates the efficiency of Reaction (1) in terms of its capability to produce a borohydride-pure product without chloride on the anion-site, showing that the metathesis is completed in such an approach.

2.2. Thermolysis of $NH_4Ca(BH_4)_3$

The ammonium metal-borohydride perovskite $NH_4Ca(BH_4)_3$ starts decomposing at approximately 380 K in both, chloride and chloride-free samples. The respective temperature-dependent *in-situ* synchrotron diffraction data are shown for the chloride sample in Figure 2. The suggested decomposition mechanism for the chloride synthesis involves the formation of β-Ca(BH$_4$)$_2$ and LiBH$_4$ *via* an intermediate decomposition product, which occurs as an impurity already in some as milled mixtures in both approaches (Supplementary Figure S1 and Figure S2). This decomposition intermediate is an unreported AB-complex of Ca(BH$_4$)$_2$ (AB = NH$_3$BH$_3$) and is discussed in detail below.

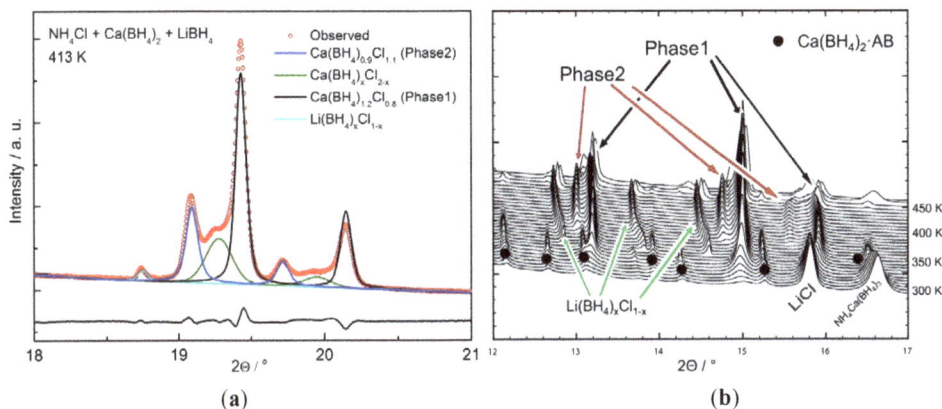

Figure 2. (a) Selected part of Rietveld plot at 413 K for the chloride-based mixture 1:1:1. Three different compositions of Ca(BH$_4$)$_x$Cl$_{2-x}$ were modeled in addition to Li(BH$_4$)$_x$Cl$_{1-x}$. The difference curve to the fit is shown on the bottom; (b) Close-up on *in-situ* diffraction data showing crystallization and chlorine uptake by Ca(BH$_4$)$_x$Cl$_{2-x}$. The models of Ca(BH$_4$)$_{0.9}$Cl$_{1.1}$ and Ca(BH$_4$)$_{1.2}$Cl$_{0.8}$ at 413 K (left) correspond to Phase 1 and Phase 2 in the temperature ramp.

As temperature is increased, calcium borohydride then proceeds to take up chlorine according to Ca(BH$_4$)$_{2-x}$Cl$_x$ in Cl-mixtures, as discussed below. Interestingly, chlorine seems to stabilize the β-phase of Ca(BH$_4$)$_2$, which has previously been observed [12], whereas in ether-milled samples the phase composition *post-decomposition* is a mixture of β- and γ-phases. Ideally, the decomposition mechanism for the chloride-free protocol consists in inverting Reaction (2), *i.e.*, the stored NH$_4$BH$_4$ would be released. In reality, however, the decomposition involves the intermediate 1, *i.e.*, the novel AB-complex Ca(BH$_4$)$_2$· AB. Hence, while NH$_4$Ca(BH$_4$)$_3$ may be considered a means of storing (irreversibly) ammonium borohydride in the solid state, it cannot be released. The idealized decomposition mechanism of NH$_4$Ca(BH$_4$)$_3$ can therefore be written as:

$$NH_4Ca(BH_4)_3(s) \rightarrow Ca(BH_4)_2 \cdot NH_3BH_3(s) + H_2(g) \qquad (3)$$

While this reaction is stoichiometric and observed in the chloride free samples, the situation is more complex when the presence of chlorine is taken into account, leading to multiple processes releasing and absorbing chlorine which are outlined below.

Figure 3 shows the previously published [4] analysis of volatile decomposition products with the corresponding thermal and gravimetric signals, measured on the chloride-sample. Notably, there is no diborane released during the whole thermal decomposition of $NH_4Ca(BH_4)_3$, which in turn does release one molecule of hydrogen around 380 K (observed on Cl sample) and transforms to $Ca(BH_4)_2 \cdot$ AB, in both chloride and ether milled samples, according to Reaction (3).

In chloride-based samples the decomposition proceeds directly to the AB-complex. This gas release is readily identified as the first mass loss in Figure 3, where mass spectrometry (top panel) shows that the released product's composition is nearly pure hydrogen (very small signal of NH_3 at 380 K). $Ca(BH_4)_2 \cdot$ AB then decomposes to release additional hydrogen above 400 K, accompanied by a minor amount of ammonia NH_3 (Figure 3), similar to previously reported reaction mechanisms for AB complexes involving recovery of β-$Ca(BH_4)_2$ [13,14]. This corresponds to the second mass loss and is visible as a continuous evolution in the spectrometric detection of H_2. The decomposition temperature of the AB complex compares quite well with those of the first two reported members of the metal borohydride AB family [15].

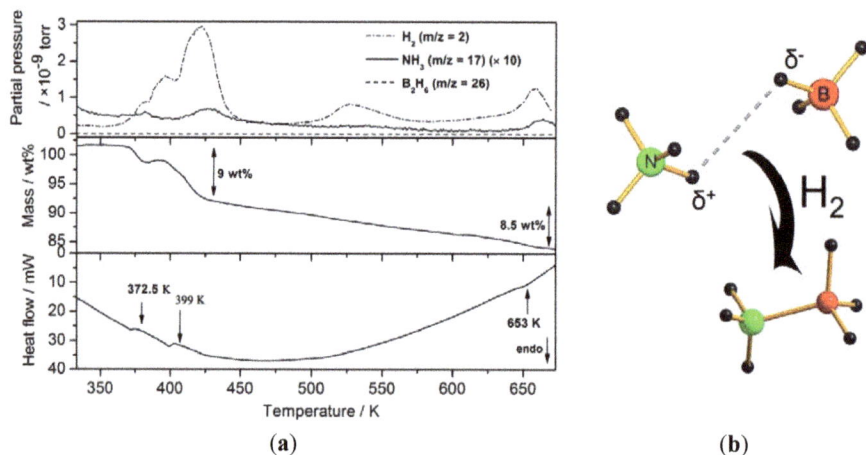

Figure 3. Simultaneous gas release and thermal analysis of the decomposition of $NH_4Ca(BH_4)_3$ under nitrogen flow (heating rate 5 K/min, chloride sample, reproduced from [4]). (a) Results from mass-spectrometry (MS), thermal gravimetric analysis (TGA) and differential scanning calorimetry (DSC), the signal corresponding to NH_3 has been multiplied $\times 10$; (b) Close heteronuclear dihydrogen contact between protic hydrogen pertaining to the ammonium cation and hydridic hydrogen from the borohydride anion facilitates hydrogen releases via AB.

34

2.2.1. Chloride Substitution in β-Ca(BH$_4$)$_2$

In the following we briefly discuss the release and uptake of chlorine by different phases during thermal decomposition of the mixture NH$_4$Cl:Ca(BH$_4$)$_2$:LiBH$_4$ 1:1:1 and referring to the *in-situ* diffraction data shown in Figure 2. Chloride substitution has recently been investigated by ball-milling CaCl$_2$ + Ca(BH$_4$)$_2$ [16]. Therein, anion-substitution is achieved by heating as-milled mixtures of unreacted precursors, which results in halide-substituted β-Ca(BH$_4$)$_{2-x}$Cl$_x$. The authors report different compositions with discrete lattice parameters depending on the nominal composition of precursors in the respective sample. The chlorine uptake by β-Ca(BH$_4$)$_2$ during the decomposition of NH$_4$Ca(BH$_4$)$_3$ is fundamentally different, since it is not CaCl$_2$ acting as a chlorine-source. In this case, the chlorine is provided by LiCl, which is formed during the metathesis Reaction (1). Concurring with the decomposition of the ammonium perovskite LiCl continuously transforms into hexagonal HT-phase Li(BH$_4$)$_x$Cl$_{1-x}$, which crystallizes together with halide-substituted β-Ca(BH$_4$)$_2$ (Figure 2) according to Reaction (4):

$$\beta - Ca(BH_4)_2 + nLiCl \rightarrow Ca(BH_4)_xCl_{2-x} + nLi(BH_4)_{n-x}Cl_{1-n-x} \qquad (4)$$

While the dissolution of chloride is known to stabilize the hexagonal HT-phase of LiBH4, we observe a hitherto unreported behavior of the HT-phase (β) of Ca(BH$_4$)$_2$. A detailed examination of the *in-situ* diffraction data provided in Figure 2 is essential to the following discussion. At approximately 350 K the remnant Ca(BH$_4$)$_2$ not reacted but amorphized during ball-milling starts recrystallizing, accompanying the formation and crystallization of the AB-complex Ca(BH$_4$)$_2$· AB. Despite the temperature increase the lattice parameters of β-Ca(BH$_4$)$_2$ decrease as crystallinity increases (see Bragg signal at 15° shifting to higher angles with temperature, in Figure 2), due to chlorine-uptake originating from LiCl, while the Li is used to form Li(BH$_4$)$_x$Cl$_{1-x}$. At temperatures of approximately 400 K, coinciding with the decomposition of the AB-complex in the chloride-based sample, halide-substituted β-Ca(BH$_4$)$_2$ abruptly forms two phases of well-defined chemical composition and discrete lattice parameters (labeled Phase1 and Phase2 in Figure 2), accompanied by a third phase of intermediate composition and much broader diffraction peaks. Rietveld refinement was performed on data collected at 413 K and including 3 different β-Ca(BH$_4$)$_x$Cl$_{2-x}$ compositions. The result of the refinement is shown in Figure 2. Two compositions Ca(BH$_4$)$_{0.9}$Cl$_{1.1}$ and Ca(BH$_4$)$_{1.2}$Cl$_{0.8}$ could be refined reliably corresponding to unit cell volumes of 194.8(9) and 185.3(7) Å^3, respectively. the evolution of the cell volumes at higher temperature is shown in Supplementary Figure S3. These values are far below that of the reported unit cell volume of pure homoleptic β-Ca(BH$_4$)$_2$ at 305 K, 208.1 Å^3 [17], in agreement with the smaller anion radius of Cl$^-$. The anion-composition of the third component corresponding to the broad signals in Figure 2 could not be determined reproducibly but is thought to lie between the two determined stoichiometries. The line width is approximately a factor 2.5 larger than in both other compositions. This may be due to a continuous distribution of chlorine dissolution on the BH_4-site. But also a shorter coherence length could be at the origin of the signal-width, suggesting intermediate chemical compositions on very short length scales, for instance at the grain-boundaries between compositions $Ca(BH_4)_{0.9}Cl_{1.1}$ and $Ca(BH_4)_{1.2}Cl_{0.8}$. To our knowledge the coexistence of various discrete anion-compositions has not been reported for β-$Ca(BH_4)_2$, despite anion-substitution having been quite well studied in this compound [12,16].

2.2.2. Ether-Samples

In the samples which were prepared by ball milling solidified NH_4BH_4 in ether at low temperatures the decomposition mechanism is not influenced by the kinetics of chlorine-exchange between the involved pseudobinary compounds β-$Ca(BH_4)_2$ and $LiBH_4$.

While the transformation of $NH_4Ca(BH_4)_3$ to the novel AB-complex is observed as in the chloride-sample, the decomposition now involves a further intermediate **2** (Supplementary Figure S2) that has not been characterized (T-interval 370–385 K, given the vicinity to the solved structure of $Ca(BH_4)_2 \cdot AB$ we assume **2** to be a further AB-complex, possibly a phase transformation then takes place at 380 K), but indexed to an orthorhombic cell of $a = 11.602$, $b = 14.339$, $c = 14.296$ Å, $V = 2378$ Å^3. This phase does not occur in chloride-syntheses. It is unlikely that there is a hydrogen loss occurring between intermediate 2 and $Ca(BH_4)_2 \cdot AB$, since the reported product of the first two decomposition steps of NH_4BH_4 is the diammoniate of diborane $[(NH_3)_2BH_2](BH_4)$, which itself is an isomer in equilibrium with AB [18], and no intermediate deprotonated molecule exists. Given the small stability field of approximately 5 K and the vicinity to the stability field of $Ca(BH_4)_2 \cdot AB$, intermediate 2 may be a further AB-derivative of different stoichiometry, or merely a superstructure to $Ca(BH_4)_2 \cdot AB$, related by a polymorphic transformation.

In ether milled mixtures $Ca(BH_4)_2 \cdot AB$ begins to decompose to β- and γ-$Ca(BH_4)_2$ at 400 K, proceeding until 430 K (Supplementary Figure S2), which is in surprising agreement with the decomposition-onset of the previously reported AB-complex of $Ca(BH_4)_2$, determined at 398 K, and ending at 428 K [14]. Hence, while the second decomposition step follows the established mechanism of AB-complexes, the perovskite adds an additional hydrogen release prior to this.

We have not identified further decomposition products. The previously reported AB-complexed $Ca(BH_4)_2$ decomposes to its constituents. AB melts at 377 K therefore its possible existence cannot be determined from the diffraction patterns. A difference in structure between the two (discussed below) could suggest a slightly modified decomposition pathway, the reported AB-complex [14] possibly showing a greater predisposition to dissociate to $Ca(BH_4)_2$ and AB due to its structural topology, which is built of defined layers of both precursors, while $Ca(BH_4)_2 \cdot$ AB reported herein contains AB molecules whose nearest neighbor interactions are considered to be weaker with respect to those taking place in AB itself.

2.3. Crystal Structure of $NH_4Ca(BH_4)_3$

In the ammonium calcium borohydride perovskite the NH_4^+ cation occupies the *A*-site of the perovskite structure, where Ca is the B-cation and the BH_4^- anion is the ligand. We have chosen to model the structure in an ordered manner in a subgroup symmetry which allows for an identical unit cell metric to the primitive perovskite unit. A fully disordered structure can be modelled in *Pm-3m*, however the fit is not better. The ordered description is supported, albeit not fully justified, by X-ray diffraction data collected down to temperatures of 100 K which did not show a change in diffraction pattern that would suggest a phase transition due to ordering. All reported phase transitions in borohydride perovskites occur above RT without any further events down to 100 K [4,19].

Furthermore, the dominant weak interaction in $NH_4Ca(BH_4)_3$ is of attractive nature and structurally points towards an ordered model. Future neutron studies, as planned also for other metal borohydride perovskites may however result in a correction of this space group symmetry. Our structural description currently is in *P-43m*, which is a subgroup to *Pm-3m*, and allows maximizing homopolar dihydrogen contacts BH$\cdots$HB between hydride anions, which are constrained to lower values by symmetry in other subgroups to *Pm-3m*, as for instance *P432*. One boron position and one nitrogen position provide orientation between the molecular species that are rotated by 90° around the principal cubic direction with respect to each other (Figure 4).

2.4. Crystal Structure of $Ca(BH_4)_2 \cdot NH_3BH_3$

The dehydrogenation product of the ammonium borohydride perovskite is the calcium borohydride ammonia-borane complex $Ca(BH_4)_2 \cdot$ AB, which contains one AB molecule per $Ca(BH_4)_2$ unit and whose structure was solved *ab-initio* on a single phase sample from diffraction data collected at 414 K (Figure 5) during the decomposition of ether-milled $NH_4Ca(BH_4)_3$.

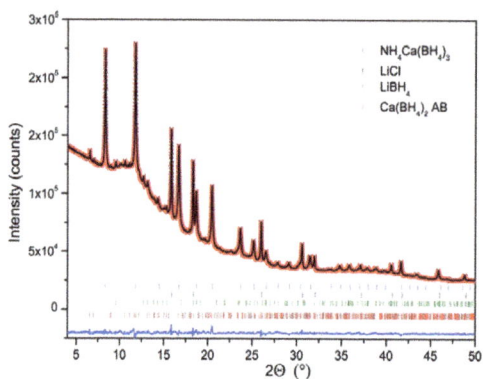

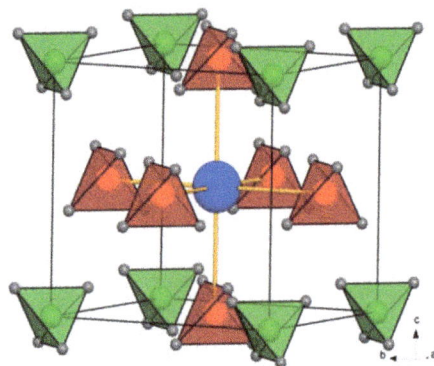

(a)

(b)

Figure 4. (a) Rietveld plot for the chloride-based synthesis of $NH_4Ca(BH_4)_3$, refined at 298 K; and (b) structural model in space group $P\text{-}43m$. NH_4^+ green, BH_4^- red, Ca blue.

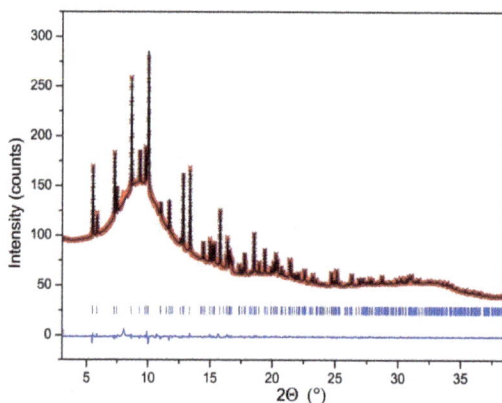

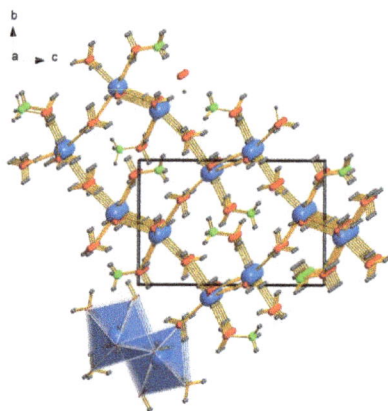

(a)

(b)

Figure 5. (a) Rietveld plot for the ether-milled sample, refined at 414 K, showing a pure phase composition of $Ca(BH_4)_2 \cdot$ AB; (b) Corresponding structural model. N green, B red, H grey, Ca blue.

Interestingly, an AB complex of $Ca(BH_4)_2$ has previously been reported, which contains two AB molecules per unit cell [14] and was observed not during thermal composition, but obtained as an adduct by ball-milling NH_3BH_3 with $Ca(BH_4)_2$. The structure therein was described as containing layers of AB alternating with layers (square lattice) of borohydride stacked along the long axis, b.

A close inspection of the structural model of $Ca(BH_4)_2 \cdot$ AB (Figure 6) reveals that the borohydride layers formed by a square lattice are conserved, however buckled by

$90°$. A further relationship is found in α-Ca(BH$_4$)$_2$-like ribbons, running parallel to the a-axis in Ca(BH$_4$)$_2\cdot$ AB. These Ca(BH$_4$)$_2$-ribbons are highlighted in the grey box in Figure 6. In this description the NH$_3$BH$_3$ molecules form a further one-dimensional structural element (zig-zag chain), which alternates with Ca(BH$_4$)$_2$ ribbons (see dark blue box in Figure 6). In this context, the relationship between the AB-complex Ca(BH$_4$)$_2\cdot$ AB and the reported Ca(BH$_4$)$_2\cdot$ 2AB may therefore be described best as a reduction in dimensionality of the structural elements of the precursors in the former.

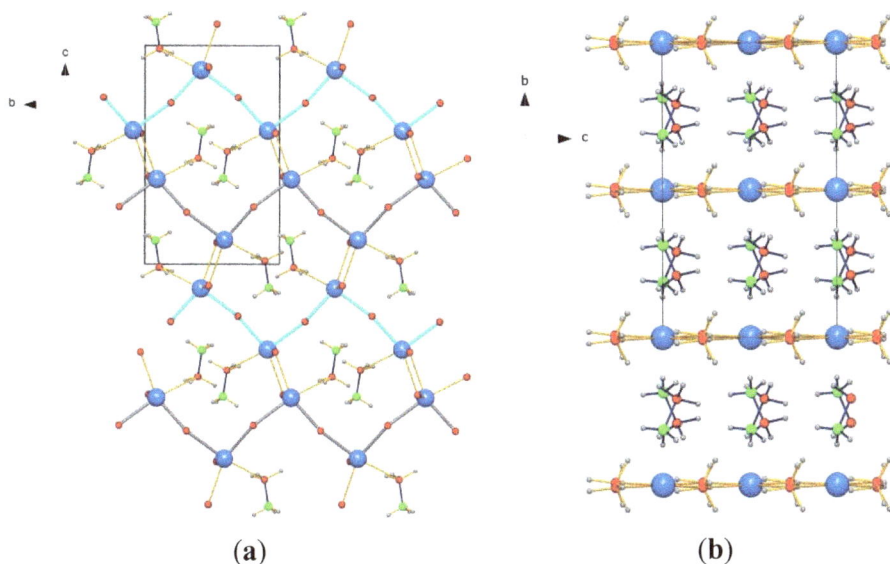

(a) (b)

Figure 6. (a) Comparison of the crystal structure of Ca(BH$_4$)$_2\cdot$ AB; and (b) the reported Ca(BH$_4$)$_2\cdot$ 2AB-complex [14]. On the left, the buckled square lattice of Ca(BH$_4$)$_2$ layers is drawn in light-blue and grey. Ca(BH$_4$)$_2$ ribbons are shown in a dark blue box; N green, B red, H grey, Ca blue. H atoms have been omitted on BH$_4$ groups for sake of clarity on the left.

In α-Ca(BH$_4$)$_2$ the square lattice (layers) of Ca-BH$_4$ conserved in both AB-complexes is oriented perpendicular to the b-axis, while the ribbons conserved only in Ca(BH$_4$)$_2\cdot$ AB connect these layers and run along the a-axis. If the buckled Ca-BH$_4$ square lattice (blue and grey sheets in Figure 6) is flattened out the α-Ca(BH$_4$)$_2$ structure is recovered in Ca(BH$_4$)$_2\cdot$ AB, however with 50% of the ribbons replaced by zig-zag chains of NH$_3$BH$_3$. There is therefore no major reconstruction of the α-Ca(BH$_4$)$_2$ structure in this AB-complex.

The B-N lengths in Ca(BH$_4$)$_2\cdot$ AB at 414 K amount to 1.613(5) Å, significantly longer than those in the reported Ca(BH$_4$)$_2\cdot$ 2AB (1.546 Å at 298 K) and very close to those in solid NH$_3$BH$_3$ (1.592 Å) [13,14]. The intermolecular distances between hydride and proton ions BH···HN pertaining to adjacent AB molecules amount to 2.05 Å, as opposed to 1.736 Å in the reported $Ca(BH_4)_2 \cdot 2AB$, and are slightly larger to those in AB (2.02 Å). This attests to a larger degree of interaction to BH_4 groups as opposed to $Ca(BH_4)_2 \cdot 2AB$, which could lead to a slightly modified decomposition mechanism. The closest refined BH···HN distance between NH_3BH_3 and adjacent hydridic H^- of BH_4 groups is 2.02 Å, which is in quite good agreement with the values of 1.986 and 2.037 Å reported for $Ca(BH_4)_2 \cdot 2AB$. The main difference hence lies in intermolecular AB contacts, which are significantly larger in $Ca(BH_4)_2 \cdot AB$. Nevertheless, we need to recall that the determination of interatomic hydrogen distances from X-ray powder diffraction data are severely biased by the application of restraints during refinement, and the determined values need not be very accurate.

3. Experimental Section

3.1. Synthesis

$Ca(BH_4)_2$, $LiBH_4$ and NH_4Cl were used as purchased from Sigma-Aldrich (Buchs, Switzerland). Chloride based samples were prepared by high-energy ball milling (mechanochemistry) in a Pulverisette P7 planetary ball mill (Fritsch, Idar-Oberstein, Germany). The ball-to-powder mass ratio was approximately 50, 60 milling cycles at 600 rpm of 2 min each were interrupted by 5 min cooling breaks to prevent heating and powder agglomeration on the walls of the milling jar. For the second approach based on NH_4BH_4, the latter was prepared according to ref. [20] with some minor modifications. Milling was performed in the Fritsch Pulverisette P7 planetary ball mill in a stainless steel vial cooled down to approximatively 240 K prior to milling. The milling speed was set to 500 rpm and the ball to powder mass ratio was approximately 50 for both dry and liquid assisted (500 μL (350 mg) of Et_2O at 500 mg of sample mass) milling. The temperature of the milling jar was not kept low during milling. In the case of dry milling, the milling time was 30 min and the temperature of the jar was around 35 °C (308 K) at the end of the milling. In the case of the ether assisted milling, the milling time was of 5 min, resulting in a much smaller temperature increase during milling. The ether was removed by evacuating the sample directly after the milling.

3.2. Synchrotron Radiation X-ray Powder Diffraction (SR-XPD)

Synchrotron radiation powder X-ray diffraction (SR-PXD) data were collected at the Swiss Norwegian Beamlines of the European Synchrotron Radiation Facility (ESRF, Grenoble, France) and at the Materials Science Beamline of the Swiss Light Source (Paul Scherrer Institute, Villigen, Switzerland) 0.6888 and 0.8271 Å, respectively, as calibrated by a Si standard. Samples were loaded into 0.5 mm borosilicate capillaries. Structures were solved in direct space using the software

FOX [21] and treating BH_4 as a rigid body. After structure solution the respective models were refined with Fullprof [22] or TOPAS [23].

4. Conclusions

$NH_4Ca(BH_4)_3$ is to our knowledge the first perovskite-type inorganic compound that contains two inherently dynamic hydrogen-rich molecules, both on the anion and cation sites. The usage of ammonium NH_4^+ as A-site cation allows stabilizing ammonium borohydride for the first time at room temperature and pressure. This describes an approach that is not restricted to perovskite-type borohydrides and the incorporation of NH_4^+ may be generalized to search for new hydrogen-rich compounds. Besides the chemical benefit to hydrogen storage applications, the ammonium cation also enriches the spectrum of structural tools by a further parameter, providing, next to the non-spherical borohydride anion, a cation with tetrahedral charge distribution. The combination of non-spherical ions with chemically distinct hydrogen species may be exploited in complex hydrides to design polar structures.

$NH_4Ca(BH_4)_3$ decomposes at 380 K to new ammonia-borane complex $Ca(BH_4)_2 \cdot NH_3BH_3$ and releases one hydrogen molecule in the process. A major drawback of boron and nitrogen-based hydrogen storage systems is the possible evolution of ammonia NH_3 and diborane B_2H_6, which both lead to the failure of fuel cells. Neither of them are observed during the hydrogen release of $NH_4Ca(BH_4)_3$, while minor NH_3 evolution occurs during decomposition of $Ca(BH_4)_2 \cdot NH_3BH_3$, similar to previously reported AB-complexes of metal-borohydrides. $NH_4Ca(BH_4)_3$ is the first compound of a series of novel metal borohydrides which is expected to grow quickly thanks to a concept that is easily applied to a large and growing number of mixed-metal borohydride salts that contain cations similar in size and thus readily replaced by NH_4^+. For instance, replacing NH_4^+ for K in $K_3Mg(BH_4)_5$ [24] or $KAl(BH_4)_4$ [25,26] would result in hydrogen densities of 21.2 and 19 wt %, which is already very close to NH_4BH_4 itself. Generally, the decomposition of such proposed compounds is assumed to be dominated by the transformation of NH_4^+ and BH_4^- to AB-complexes. While this mechanism is bound to release pure hydrogen, the subsequent decomposition of AB-complexes poses the known problems of rehydrogenation and fuel cell contamination.

Supplementary Materials: Supplementary materials can be accessed at: http://www.mdpi.com/1996-1073/8/8/8286/s1.

Acknowledgments: This work was supported by the Swiss National Science Foundation under project 200020_149218, by the Fonds National de la Recherche Scientifique (FNRS, Belgium) via instruments CC, PDR, EQP and by FRIA (Fonds pour la formation à la recherche dans l'Industrie et l'Agriculture, Belgium) funding FM's PhD works. The authors thank SNBL at the ESRF and SLS at the PSI for the beamtime allocation and Dmitry Chernyshov (SNBL), Vadim Dyadkin (SNBL) and Nicola Casati (SLS) for their assistance.

Author Contributions: P.S. and Y.S. performed the chloride based synthesis, F.M. performed the ammonium borohydride based synthesis. The synchrotron X-ray powder diffraction data were measured by P.S., Y.S., F.M., Y.F. and R.C. The diffraction data were analyzed by P.S. and R.C., crystal structures were solved by P.S. The manuscript has been written by P.S., and was commented by all authors.

Conflicts of Interest: The authors declare no conflict of interest.

References

1. Schlapbach, L.; Züttel, A. Hydrogen-storage materials for mobile applications. *Nature* **2001**, *414*, 353–358.

2. Schrauzer, G.N. Ueber ein periodensystem der metallboranate. *Naturwissenschaften* **1955**, *42*, 438.

3. Jepsen, L.H.; Ley, M.B.; Su-Lee, Y.; Cho, Y.W.; Dornheim, M.; Jensen, J.O.; Filinchuk, Y.; Jorgensen, J.E.; Besenbacher, F.; Jensen, T.R. Boron-nitrogen based hydrides and reactive composites for hydrogen storage. *Mater. Today* **2014**, *17*, 129–135.

4. Schouwink, P.; Ley, M.B.; Tissot, A.; Hagemann, H.; Jensen, T.R.; Smrčok, L.; Černý, R. Structure and properties of a new class of complex hydride perovskite materials. *Nat. Commun.* **2014**, *5*, 5706.

5. Dixon, D.A.; Gutowski, M. Thermodynamic Properties of Molecular Borane Amines and the $[BH_4^-][NH_4^+]$ Salt for Chemical Hydrogen Storage Systems from *ab Initio* Electronic Structure Theory. *J. Phys. Chem. A* **2005**, *109*, 5129–5139.

6. Karkamkar, A.; Kathmann, S.M.; Schenter, G.K.; Heldebrant, D.J.; Hess, N.; Gutowski, M.; Autrey, T. Thermodynamic and Structural Investigations of Ammonium Borohydride, a Solid with a Highest Content of Thermodynamically and Kinetically Accessible Hydrogen. *Chem. Mater.* **2009**, *21*, 4356–4358.

7. Flacau, R.; Ratcliffe, C.I.; Desgreniers, S.; Yao, Y.; Klug, D.D.; Pallister, P.; Moudrakowski, I.L.; Ripmeester, J.A. Structure and dynamics of ammonium borohydride. *Chem. Commun.* **2010**, *46*, 9164–9166.

8. Huang, Z.; Chen, X.; Yisgedu, T.; Meyers, E.A.; Shore, S.G.; Zhao, J.-C. Ammonium Octahydrotriborate ($NH_4B_3H_8$): New Synthesis, Structure, and Hydrolytic Hydrogen Release. *Inorg. Chem.* **2011**, *50*, 3738–3742.

9. Yisgedu, T.B.; Huang, Z.; Chen, X.; Lingam, H.K.; King, G.; Highley, A.; Maharrey, S.; Woodward, P.M.; Behrens, R.; Shore, S.G.; *et al.* The structural characterization of $(NH_4)_2B_{10}H_{10}$ and thermal decomposition studies of $(NH_4)_2B_{10}H_{10}$ and $(NH_4)_2B_{12}H_{12}$. *Int. J. Hydrog. Energy* **2012**, *37*, 4267–4273.

10. Nielsen, T.K.; Karkamkar, A.; Bowden, M.; Besenbacher, F.; Jensen, T.R.; Autrey, T. Methods to stabilize and destabilize ammonium borohydride. *Dalton Trans.* **2013**, *42*, 680–687.

11. Rude, L.H.; Nielsen, T.K.; Ravnsbæk, D.B.; Bösenberg, U.; Ley, M.B.; Richter, B.; Arnbjerg, L.M.; Dornheim, M.; Filinchuk, Y.; Besenbacher, F.; *et al.* Tailoring properties of borohydrides for hydrogen storage: A review. *Phys. Status Solidi* **2011**, *208*, 1754–1773.

12. Rude, L.H.; Filinchuk, Y.; Sørby, M.H.; Hauback, B.C.; Besenbacher, F.; Jensen, T.R. Anion Substitution in $Ca(BH_4)_2$-CaI_2: Synthesis, Structure and Stability of Three New Compounds. *J. Phys. Chem. C* **2011**, *115*, 7768–7777.

13. Xiong, Z.; Keong Yong, C.; Wu, G.; Chen, P.; Shaw, W.; Karkamkar, A.; Autrey, T.; Owen Jones, M.; Johnson, S.R.; Edwards, P.P.; *et al.* High-capacity hydrogen storage in lithium and sodium amidoboranes. *Nat. Mater.* **2007**, *7*, 138–141.

14. Wu, H.; Zhou, W.; Yildirim, H. Alkali and Alkaline-Earth Metal Amidoboranes: Structure, Crystal Chemistry, and Hydrogen Storage Properties. *J. Am. Chem. Soc.* **2008**, *130*, 14834–14839.

15. Wu, H.; Zhou, W.; Pinkerton, F.E.; Meyer, M.S.; Srinivas, G.; Yildirim, T.; Udovic, T.J.; Rush, J.J. A new family of metal borohydride ammonia borane complexes: Synthesis, structures, and hydrogen storage properties. *J. Mater. Chem.* **2010**, *20*, 6550–6556.

16. Grove, H.; Rude, L.H.; Jensen, T.R.; Corno, M.; Ugliengo, P.; Baricco, M.; Sørby, M.H.; Hauback, B.C. Halide substitution in $Ca(BH_4)_2$. *RSC Adv.* **2014**, *4*, 4736–4742.

17. Finlinchuk, Y.; Rönnebro, E.; Chandra, D. Crystal structures and phase transformations in $Ca(BH_4)_2$. *Acta Mater.* **2009**, *57*, 732–738.

18. Bowden, M.; Autrey, T. Characterization and mechanistic studies of the dehydrogenation of NH_xBH_x materials. *Curr. Opin. Solid State Mater. Sci.* **2011**, *15*, 73–79.

19. Schouwink, P.; Hagemann, H.; Embs, J.P.; D'Anna, V.; Černý, R. Di-hydrogen contact induced lattice instabilities and structural dynamics in complex hydride perovskites. *J. Phys. Condens. Matt.* **2015**, *27*, 265403.

20. Parry, R.W.; Schultz, D.R.; Girardot, P.R. The Preparation and Properties of Hexamminecobalt (III) Borohydride, Hexamminechromium (III) borohydride and Ammonium Borohydride. *J. Am. Chem. Soc.* **1958**, *1*, 1–3.

21. Favre-Nicolin, V.; Černý, R. FOX, "free objects for crystallography": A modular approach to *ab initio* structure determination from powder diffraction. *J. Appl. Cryst.* **2002**, *35*, 734–743.

22. Rodriguez-Carvajal, J. *FULLPROF SUITE*; LLB Sacley & LCSIM: Rennes, France, 2003.

23. Coelho, A.A. TOPAS-Academic V5. 2012. Available online: http://www.topas-academic.net/ (accessed on 15 February 2013).

24. Schouwink, P.; D'Anna, V.; Ley, M.B.; Lawson Daku, L.M.; Richter, B.; Jensen, T.R.; Hagemann, H.; Černý, R. Bimetallic Borohydrides in the System $M(BH_4)_2$-KBH_4 (M = Mg, Mn): On the Structural Diversity. *J. Phys. Chem. C* **2012**, *116*, 10829–10840.

25. Dovgaliuk, I.; Ban, V.; Sadikin, Y.; Černý, R.; Aranda, L.; Casati, N.; Devillers, M.; Filinchuk, Y. The first halide-free bimetallic aluminum borohydride: Synthesis, structure, stability, and decomposition pathway. *J. Phys. Chem. C* **2014**, *118*, 145–153.

26. Knight, D.A.; Zidan, R.; Lascola, R.; Mohtadi, R.; Ling, C.; Sivasubramanian, P.; Kaduk, J.A.; Hwang, S.-J.; Samanta, D.; Jena, P. Synthesis, Characterization, and Atomistic Modeling of Stabilized Highly Pyrophoric $Al(BH_4)_3$ via the Formation of the Hypersalt $K[Al(BH_4)_4]$. *J. Phys. Chem. C* **2013**, *117*, 19905–19915.

Combined X-ray and Raman Studies on the Effect of Cobalt Additives on the Decomposition of Magnesium Borohydride

Olena Zavorotynska, Stefano Deledda, Jenny G. Vitillo, Ivan Saldan, Matylda N. Guzik, Marcello Baricco, John C. Walmsley, Jiri Muller and Bjørn C. Hauback

Abstract: Magnesium borohydride ($Mg(BH_4)_2$) is one of the most promising hydrogen storage materials. Its kinetics of hydrogen desorption, reversibility, and complex reaction pathways during decomposition and rehydrogenation, however, present a challenge, which has been often addressed by using transition metal compounds as additives. In this work the decomposition of $Mg(BH_4)_2$ ball-milled with $CoCl_2$ and CoF_2 additives, was studied by means of a combination of several *in-situ* techniques. Synchrotron X-ray diffraction and Raman spectroscopy were used to follow the phase transitions and decomposition of $Mg(BH_4)_2$. By comparison with pure milled $Mg(BH_4)_2$, the temperature for the $\gamma \rightarrow \varepsilon$ phase transition in the samples with CoF_2 or $CoCl_2$ additives was reduced by 10–45 °C. *In-situ* Raman measurements showed the formation of a decomposition phase with vibrations at 2513, 2411 and 766 cm^{-1} in the sample with CoF_2. Simultaneous X-ray absorption measurements at the Co K-edge revealed that the additives chemically transformed to other species. CoF_2 slowly reacted upon heating till ~290 °C, whereas $CoCl_2$ transformed drastically at ~180 °C.

Reprinted from *Energies*. Cite as: Zavorotynska, O.; Deledda, S.; Vitillo, J.G.; Saldan, I.; Guzik, M.N.; Baricco, M.; Walmsley, J.C.; Muller, J.; Hauback, B.C. Combined X-ray and Raman Studies on the Effect of Cobalt Additives on the Decomposition of Magnesium Borohydride. *Energies* **2015**, *8*, 9173–9190.

1. Introduction

Metal borohydrides (or tetrahydroborates) are complex hydrides containing BH_4^- anions counterbalanced by metal cations. Group I and II borohydrides (except Be) have pure ionic interactions between the cations and BH_4^- and thus strong B-H bonding, which renders the compounds particularly stable, decomposing only above 200 °C with typically pure H_2 release [1,2]. In transition metal (TM) or mixed cation borohydrides, on the contrary, the interaction between the cation(s) and H^- of the BH_4^- group can be partially covalent, which significantly destabilizes the B-H bonding and promotes decomposition even below room temperature (RT), but with release of mostly B_2H_6 [3–8]. TM borohydrides can, in many cases, be obtained by

44

reaction of the corresponding chloride with an alkali borohydride [5,8]. Due to their exceptionally high gravimetric hydrogen content (up to 18 wt% in LiBH$_4$), metal borohydrides, both stable and unstable, have been extensively studied for hydrogen storage applications [2,9,10].

Magnesium borohydride (Mg(BH$_4$)$_2$) has one of the highest hydrogen densities (14.5 wt.% and >100 kg·m^{-3}) [11–14]. DFT calculations have predicted its decomposition to a mixture of MgB$_x$ (or MgB$_{12}$H$_{12}$) with MgH$_2$ below 100 °C, accompanied by H$_2$ release [15–18]. Experimentally Mg(BH$_4$)$_2$ decomposes only at ~200 °C, which is still a low temperature compared to other stable borohydrides. The decomposition is a multi-step reaction, the path depending on the experimental conditions, and involves the formation of amorphous boron hydride compounds at 285 °C–320 °C [19–27]. Nevertheless, it has been shown that H→D substitution at the surface occurs at ~100 °C in γ-Mg(BH$_4$)$_2$ [28], which is a much lower temperature than seen in other ionic borohydrides [29–31], confirming a comparatively low stability of the B-H bonds in the compound. These inconsistencies between the theoretical stability of Mg(BH$_4$)$_2$ and experimental T_{dec} can be explained by kinetic barriers, such as H diffusion in the bulk [28], and/or the formation of intermediate phases disregarded in the calculations [32].

Addition of TM-based additives, e.g., oxides, halides, metal nanoparticles, has been one of the strategies used to enhance the hydrogen storage performance of borohydrides. After the report of Bogdanovic *et al.* [33] that Ti-based additives resulted in significantly improved hydrogen cycling properties of NaAlH$_4$, this approach has been applied to a wide range of similar compounds and composites, including metal borohydrides [2,26,34–42]. The addition of several TM-compounds has indeed resulted in improved hydrogen storage properties of stable borohydrides and related composites, including decrease in T_{dec} [35,41], in some cases remarkably by as much as 200 °C [36]. Enhancement of the desorption and absorption reactions kinetics [26,42], improved hydrogen purity, reversibility [37] and borohydride synthesis from the decomposition products [39] has also been documented. A few studies have addressed the mechanisms of the additives' "catalytic" activity [26,34,42,43]. It has been shown that the additives undergo chemical transformations and form alkali- or alkaline-earth metal halides [34], finely dispersed metal clusters or/and boride-like structures [26,42,43]. The latter can be stable upon cycling [26,42] or change the metal-B coordination numbers reversibly [43].

In our previous works [26,42] we have shown that Ni- and Co-based additives have an effect on the kinetics of hydrogen desorption and absorption in γ-Mg(BH$_4$)$_2$ [13]. *Ex-situ* X-ray absorption spectroscopy (XAS) studies have shown that the additives were chemically modified to compounds similar to Ni- and Co-borides after first decomposition of Mg(BH$_4$)$_2$. In this work we present a series of *in-situ* measurements involving quasi simultaneous monitoring of the changes in both the Co-additives and in γ-Mg(BH$_4$)$_2$. XAS spectroscopy was applied to follow the modifications in the additives, and synchrotron radiation powder X-ray diffraction (SR-PXD) with Raman spectroscopy were used to contemporaneously characterize the borohydride matrix. Additional characterization of the samples was performed by transmission electron microscopy (TEM) *ex situ*.

2. Results

2.1. Synchrotron Radiation Powder X-ray Diffraction (SR-PXD) Study of Mg(BH$_4$)$_2$ Decomposition at 2.5 Bar H$_2$

Figure 1 shows *in-situ* SR-PXD data upon heating the γ-Mg(BH$_4$)$_2$ at 2.5 bar H$_2$ backpressure until the formation of amorphous phase(s) and/or melting at ~285 °C. The sample was ball-milled before the thermal treatment, following the same procedure adopted for the samples with the additives. The RT pattern of γ-Mg(BH$_4$)$_2$ (Figure 1) has a high background at small 2θ, which can be assigned to amorphous Mg(BH$_4$)$_2$ formed upon material storage [44] and milling [25]. In the 150–200 °C range, pure γ-Mg(BH$_4$)$_2$ underwent two phase transitions.

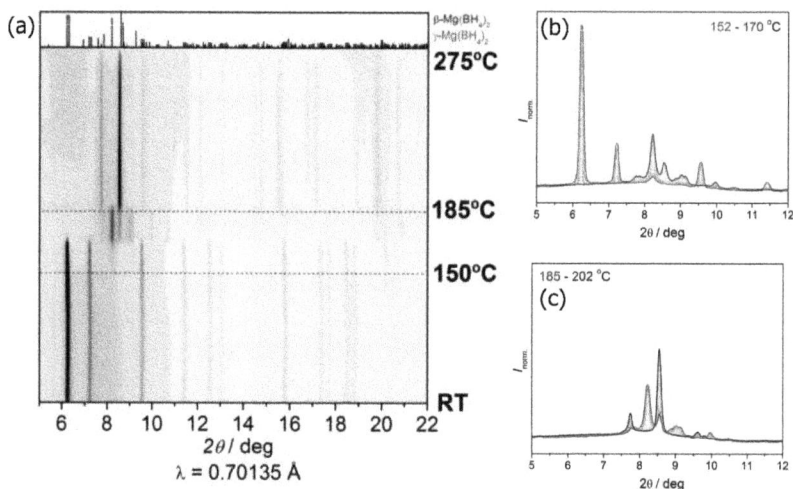

Figure 1. (a) *In-situ* synchrotron radiation powder X-ray diffraction (SR-PXD) of milled γ-Mg(BH$_4$)$_2$ obtained by heating the sample from room temperature (RT) to 285 °C at 5 °C/min and 2.5 bar H$_2$. Detailed view: (b) γ → ε; and (c) ε → β' phase transitions in γ-Mg(BH$_4$)$_2$.

The first phase transition (Figure 1b) to the commonly named ε-phase occurred in the 150–170 °C temperature range. The PXD pattern of this phase is similar to the one observed earlier [24]. The second phase transition (Figure 1c), to the disordered β-Mg(BH$_4$)$_2$ or β'-Mg(BH$_4$)$_2$ [45], was observed at 185–202 °C. At about 200 °C the peak intensities of β'-Mg(BH$_4$)$_2$ started decreasing, together with the simultaneous increase in the amorphous background, which indicated the onset of sample decomposition. At ~285 °C all crystalline reflections disappeared, in agreement with sample decomposition to the amorphous phases and/or melting [24].

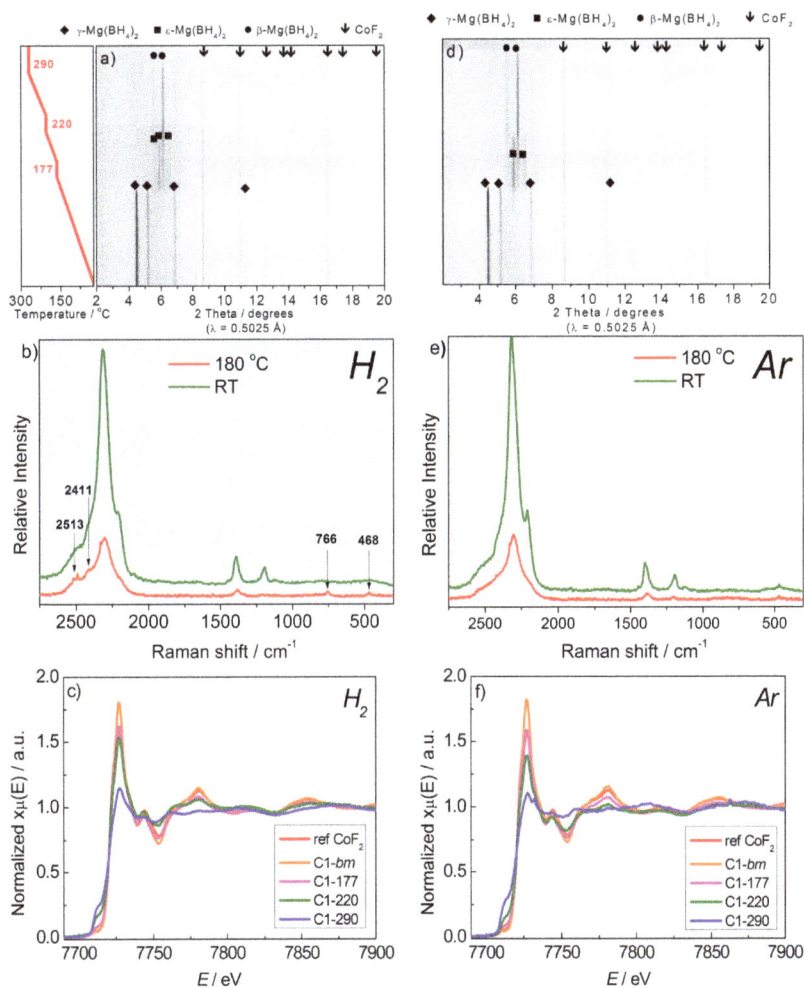

Figure 2. *In-situ* characterization using (from top to bottom) SR-PXD, Raman and X-ray absorption (XAS) spectroscopies of the thermal decomposition of Mg(BH$_4$)$_2$ + 2 mol% CoF$_2$ (C1 sample) in the RT-300 °C range in two different annealing environments: (**a–c**) 2.5 bar H$_2$; and (**d–f**) 1 bar Ar.

2.2. In-Situ X-ray Diffraction (XRD)/X-ray Absorption Spectroscopy (XAS)/Raman Study of Mg(BH₄)₂+Co_add Decomposition

2.2.1. $Mg(BH_4)_2$ + CoF_2

Figure 2a,d show the series of *in-situ* SR-PXD data obtained while decomposing the sample $Mg(BH_4)_2$ + 2 mol% CoF_2 (named C1) in 2.5 bar H_2 (C1-H_2) and 1 bar Ar (C1-Ar). The $\gamma \to \varepsilon$ phase transition in Ar was observed at a lower temperature than that in H_2, in the ~105–160 °C range *vs.* ~130–175 °C in H_2. The $\varepsilon \to \beta'$ phase transition occurred at ~175–190 °C in the C1-H_2 sample and at ~175–200 °C in the C1-Ar sample. Starting from about 220 °C, a change was observed in the amorphous background profile and in the β'-$Mg(BH_4)_2$ peaks intensities in the C1-H_2 sample. Such changes can be related to the decomposition of $Mg(BH_4)_2$ to amorphous compounds.

Raman spectroscopy can be used to characterize both amorphous and crystalline phases. Raman spectra of the C1 sample in H_2 and Ar obtained at RT before the measurements (Figure 2b,e) in the 2900–600 cm^{-1} region show peaks due to the vibrations of BH_4^- molecular ions [46]. The B–H stretching modes are centered at 2316 cm^{-1} (symmetric stretching, ν_{sym}), and the bending modes are located at ca. 1400 (symmetric bending, δ_{sym}), 1197 and 1120 cm^{-1} (asymmetric bending, δ_{asym}), respectively [46]. Scattering due to the overtones and combinations of BH_4^- bending were observed at ~2520 and 2208 cm^{-1}. The weak broad peaks >500 cm^{-1} can be tentatively assigned to the librations of $(BH_4)^-$. The small peak at 468 cm^{-1} might be due to CoF_2 [47]. Heating to 180 °C caused a gradual decrease of the intensities of $(BH_4)^-$ stretching and bending bands, due mainly to thermal effects since no decomposition is expected below 200 °C. No evident new peaks were observed during heating. At 180 °C, new vibrations appeared at 2513, 2411, and 766 cm^{-1} in samples heated in both Ar and in H_2, although the intensities of these peaks were more pronounced in the spectrum of C1-H_2. Unfortunately, further decomposition at >180 °C could not be followed with Raman measurements, the spectrum being obscured by a strong fluorescence background. This background may be related to the decomposition phase fluorescent with the used laser.

The crystalline additive CoF_2 was observed in the SR-PXD patterns after ball-milling and throughout the *in-situ* measurement until 290 °C, when the $Mg(BH_4)_2$ either melted or decomposed to amorphous phases. The intensity of CoF_2 peaks started to decrease only at ~220 °C, indicating a reaction of the fluoride during heating both in Ar and in H_2.

Co K-edge XAS spectra of sample C1, obtained at the isothermal steps during the *in-situ* measurement in H_2 and Ar are shown in Figure 2 (c and f, respectively). The spectrum obtained at RT (orange curve) is very similar to the reference CoF_2, indicating that no chemical reactions occurred during ball-milling. No significant modification of the spectra was observed during heating up to 220 °C. However, the spectra obtained at 290 °C were considerably different, reflecting changes in the oxidation and coordination state of cobalt atoms. The high intensity of the white line in the spectrum at RT is characteristic for cobalt atoms in high oxidation states (CoF_2, CoF_3) and its decrease is an indication of reduction of cobalt atoms [26]. The observed modifications in the post-edge region (*ca.* above 7730 eV) suggest also changes in the structural environment around Co atoms. The final XAS spectrum was not similar to either Co_2B, Co or CoF_2, indicating that the additive had reacted to an intermediate phase or mixture of phases.

In order to clarify the nature of the intermediate phases, reference spectra of pure Co_2B, Co or CoF_2 were collected. The linear combination fitting (LCF) of the spectra of C1 decomposed in H_2 and in Ar are reported in Figure 3a–f, respectively, for three temperatures: 177 (a,d), 220 (b,e) and 290 °C (c,f).

LCFs of C1-H_2 show that the additive was still primarily composed of CoF_2 at 177 °C with a residual part converted in a new phase, which was present in too small quantity to be properly fitted. At 220 °C, the best fit was obtained with the CoF_2 and Co_2B references, whereas at 290 °C CoF_2 and Co had to be used. Notably, the amounts of CoF_2 regularly decreased in the spectra collected at increasing temperatures. However, the attempts to obtain better fits with consistently Co_2B or Co for the 220 °C and 290 °C spectra were unsuccessful, resulting in unacceptably large shifts in E_0 (>9 eV) and visibly bad fits. The poor fit at the two highest temperatures can be an indication of other than the references phase(s). On the basis of the LCF analysis, it can be suggested that at ~200 °C CoF_2 formed borides which were transformed to metallic nano-clusters (giving no diffraction peaks) at higher temperature. The spectrum of C1-Ar sample (Figure 3d–f) obtained at 177 °C was best fitted with CoF_2 and Co foil references, although the fit with CoF_2 and Co_2B (not shown) was only slightly worse. Notable is the large shift in the edge energy, ΔE_0, (both in the fit with Co foil and Co_2B) and small amount of the new phase. The large ΔE_0 can indicate that the reference set was not strictly adequate to the actual composition of the additive and/or that the amount of the new cobalt phase was too low to obtain meaningful fit. The satisfactory fit with CoF_2 and Co references of the C1-Ar spectra obtained at 220 °C and 290 °C indicates a gradual transformation of CoF_2 to metallic cobalt with increasing temperature.

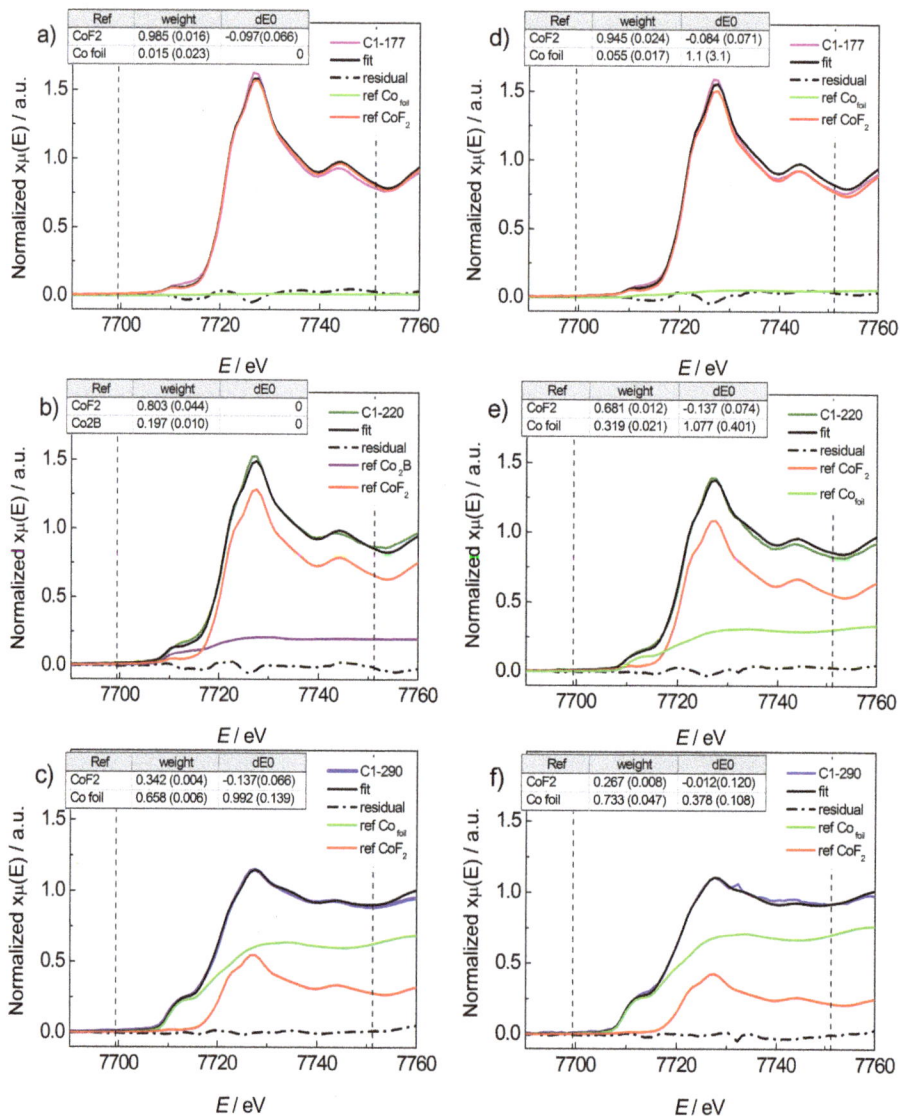

Ref	weight	dE0
CoF2	0.985 (0.016)	-0.097(0.066)
Co foil	0.015 (0.023)	0

a) — C1-177, — fit, –··– residual, — ref Co_foil, — ref CoF_2

Ref	weight	dE0
CoF2	0.945 (0.024)	-0.084 (0.071)
Co foil	0.055 (0.017)	1.1 (3.1)

d) — C1-177, — fit, –··– residual, — ref Co_foil, — ref CoF_2

Ref	weight	dE0
CoF2	0.803 (0.044)	0
Co2B	0.197 (0.010)	0

b) — C1-220, — fit, –··– residual, — ref Co_2B, — ref CoF_2

Ref	weight	dE0
CoF2	0.681 (0.012)	-0.137 (0.074)
Co foil	0.319 (0.021)	1.077 (0.401)

e) — C1-220, — fit, –··– residual, — ref CoF_2, — ref Co_foil

Ref	weight	dE0
CoF2	0.342 (0.004)	-0.137(0.066)
Co foil	0.658 (0.006)	0.992 (0.139)

c) — C1-290, — fit, –··– residual, — ref Co_foil, — ref CoF_2

Ref	weight	dE0
CoF2	0.267 (0.008)	-0.012(0.120)
Co foil	0.733 (0.047)	0.378 (0.108)

f) — C1-290, — fit, –··– residual, — ref Co_foil, — ref CoF_2

Figure 3. Linear combination fitting (LCF) of C1-H_2 (**a–c**) and C1-Ar (**d–f**) XAS spectra recorded (from top to bottom) at 177 °C, 220 °C and 290 °C using CoF_2, Co, and Co_2B as references. All combinations of the three references were used in the fit, with the weight sum set to 1; the edge energy parameter, E_0, was freed and ΔE_0 denotes its shift after fitting. The fit range is shown by vertical dashed lines. The errors are indicated in parenthesis.

2.2.2. Mg(BH$_4$)$_2$ + CoCl$_2$

Mg(BH$_4$)$_2$ + 2 mol% CoCl$_2$ (named C2 in the following) was annealed only in Ar. The $\gamma \to \varepsilon$ and $\varepsilon \to \beta'$ phase transitions in Mg(BH$_4$)$_2$ were observed at ~107–150 °C and ~177–190 °C, respectively (Figure 4a). The onset temperature for the decrease in the intensity of diffraction peaks due to β'-Mg(BH$_4$)$_2$ was ~220 °C. The Raman spectrum of C2 at RT was similar to that of samples C1 (Figure 4b, green line) but, at about 180 °C, the sample was already fluorescent and all vibrational features were obscured (red curve in Figure 4b).The diffraction peaks of the crystalline CoCl$_2$ additive were observed in the PXD pattern of the sample after milling indicating that the additive had not reacted with Mg(BH$_4$)$_2$.

In the *in-situ* SR-PXD measurements (Figure 4a), the intensities of the peaks for CoCl$_2$ remained unaltered up to 200 °C. At this temperature they started to decrease, and disappeared completely by 290 °C. On the contrary, *in-situ* XAS spectra (Figure 4c) showed a larger temperature dependence of Co coordination and oxidation state. A strong change in the XAS spectra was observed during the 177 °C isotherm. In particular, after 5 min at 177 °C the spectrum (Figure 4c, brown curve) was very different from the one recorded at RT. It suggests that the local environment around cobalt in the CoCl$_2$ additive was significantly altered already at this temperature. LCF of the first spectrum in the sequence at 177 °C (Figure 4d) indicated that Co was present mostly as CoCl$_2$ (~88%). However, the spectrum recorded only after ~5 min shows that the amount of CoCl$_2$ was reduced by half (Figure 4e). A nano-scale dimension of the Co particles could explain the absence of the peaks due to metallic Co in the SR-PXD patterns. The residuals in the fitting of the XAS-data indicate that, as for the C1-Ar and C1-H$_2$ sample, a considerable amount of Co atoms were in the oxidation/coordination states different from those of the references. Further changes were observed in the spectrum obtained at 290 °C. Fitting this spectrum with the available references was unsuccessful, indicating that the Co atoms were mainly bound in unidentified compounds. The XAS spectrum obtained by quenching C2-Ar at RT after the experiment (light green curve in Figure 4c) showed further chemical transformations of the Co-additive. Also in this case, LCF did not give any meaningful results with the available references.

The as-milled sample was additionally characterized by TEM (Figure 5). A TEM image (Figure 5a) illustrates variable size of the sample particles ranging from some tens of nanometers to over a hundred nanometers. Figure 5b shows a dark-field scanning (S)TEM image of several particles in the sample, along with the corresponding energy dispersive spectroscopy (EDS) maps. The Mg map identifies the main Mg/B phase particles. Figure 5c shows electron energy loss spectroscopy (EELS) spectra obtained from three different representative positions in the sample. The positions of the B-K, Cl-L$_{2,3}$, C-K, O-K and Co-L$_{2,3}$ peaks are indicated. Curve 1 from a discrete particle of the additive phase exhibits strong peaks due to Cl and Co.

Carbon in the spectrum is due to the slight contamination of the sample and the fact that the electron beam was passing through the carbon support film. Curves 2 and 3 were obtained from the positions indicated in the STEM image of an aggregate of particles shown in the dark-field STEM image of Figure 5d.

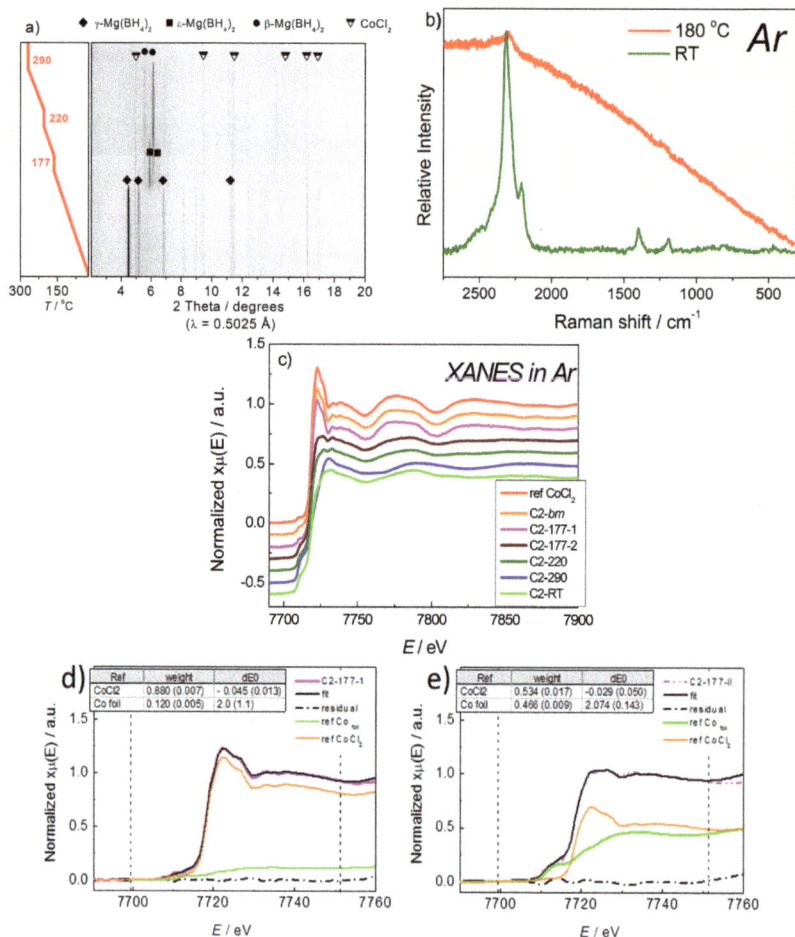

Figure 4. *In-situ* characterization of the thermal decomposition of $Mg(BH_4)_2$ + 2 mol% $CoCl_2$ in Ar in the RT-290 °C range. (a) SR-PXD at increasing T; (b) Raman spectra at RT (green) and at 180 °C (red); (c) Co K-edge XAS spectra of $CoCl_2$ reference (at RT) and C2 recorded at different temperatures and at RT after the measurements (light green). XAS spectra are y-offset for clarity. LCF of C2-Ar with CoF_2, Co, and Co_2B references: at 177 °C (d) and after 5 min at 177 °C (e). All combinations of the three references were used in the fit with the weight sum set to 1; the E_0 parameter was freed. The fit range is shown by dashed lines. The errors are indicated in parenthesis.

Curve 2 is from the main part of the particle. The prominent oxygen peaks indicate some oxidation of the sample which could have happened during transportation and handling. A strong B peak is also present and the EELS map for this element is also shown on the Figure 5d. There is no Mg peak in the energy range of the EELS spectra, but the presence of this element was confirmed by the Mg map of Figure 5b, which includes the aggregate of Figure 5d.

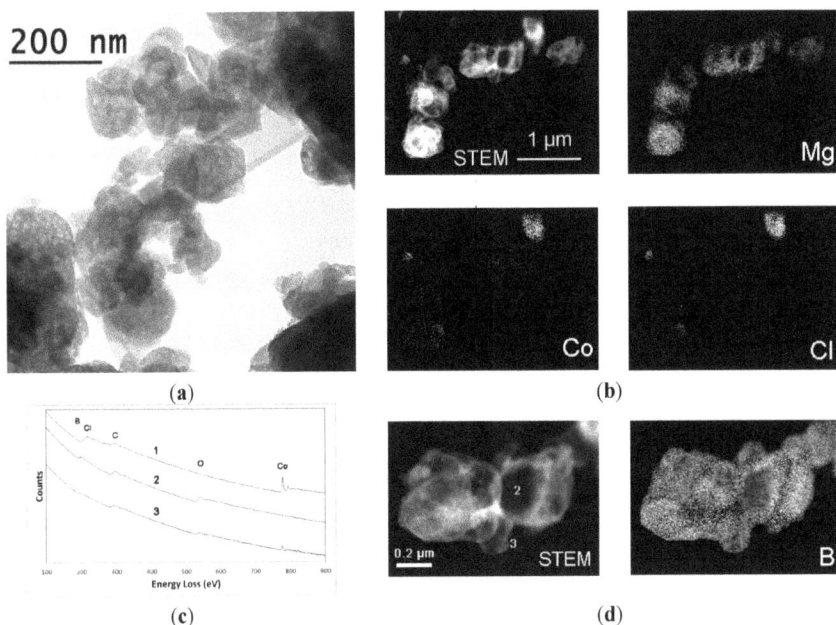

(a)

(b)

(c)

(d)

Figure 5. Transmission electron microscopy (TEM) observations of C2-*bm* sample. (a) TEM image showing $Mg(BH_4)_2$ particles dispersed on C support film; (b) STEM image and EDS maps; (c) EELS spectra from different positions in the specimen; (d) STEM image and Boron EELS map from region included in (b).

The Co and Cl maps indicate similar distributions of these elements confirming that most of the additive remained as $CoCl_2$ phase after milling and confirming the PXD results. However, the maps show also that a small amount of the elements is associated with the Mg/B phase, which suggests breaking up of $CoCl_2$ and its incorporation into the borohydride matrix. It should be mentioned that the EDS maps are susceptible to variations in background intensity when the level of elements is low.

Spectrum 3 of Figure 5c was obtained from a small area at the particle surface. It shows the presence of Co, in addition to B and O, and relatively smaller amount of Cl, which suggests that a small amount of a chlorine-less phase of cobalt had formed at the particle surfaces during ball milling. In this way, TEM and SR-PXD analyses (Figure 4) show that most of $CoCl_2$ additive did not react with the sample during ball-milling.

A comparison between the XAS spectra of the C1 and C2 samples obtained at 290°C is shown in Figure 6a. It is clear that the CoF_2 and $CoCl_2$ additives did not form identical compounds at this stage (violet and pink curves). In particular, cobalt in CoF_2 at 290 °C both in H_2 and Ar preserves higher oxidation state than the cobalt in $CoCl_2$. This is evident both from the slightly higher edge energy (7721 eV *vs.* 7718 eV in CoF_2 and $CoCl_2$, respectively) and more intense white line. Meanwhile the XAS spectra of $CoCl_2$ obtained at 290 °C and at RT after the measurements also show some differences which can probably be explained by temperature-related changes in the cobalt coordination state.

Fourier transform of the extended X-ray absorption fine structure (FT-EXAFS) spectra are presented in Figure 6b, c for samples C1 and C2, respectively. These graphs show the moduli part of the FT of the k^2-weighted EXAFS signal. The latter represents the radial distribution function centered on Co atom. The position of each peak in the FT corresponds to an average length (after phase correction, $\Delta\varphi$) of a single scattering path or to the convolution of more than one path, whereas the intensity is related to the number of the nearest neighbors and structural disorder in the system.

In the spectrum of C1-bm (CoF_2 additive, Figure 6b), the first (and most intense) peak at 1.56 Å corresponds to the backscattering from the six fluorine atoms in the first coordination shell of cobalt [26]. These atoms are located at 2.01 and 2.06 Å, thus giving the phase difference, $\Delta\varphi$, of 0.45–0.5 Å. This first peak shifts to 1.63 and 1.96 Å upon heating to 177 °C and 290 °C, respectively (Figure 6b). The reduction in the peak intensity can be explained by structural disorder around cobalt and/or decrease in the coordination number. The peaks at 2.5–4.2 Å are composed by several contributions from different scatterers with predominant contributions from cobalt atoms at 3.2 Å and 3.7 Å and from F atoms at 3.6–3.8 Å [26]. The intensity of these peaks decreased significantly with temperature (Figure 6b) reflecting more disorder in the distant coordination shells. At 220 °C, a new peak at 2.2 Å appeared in the spectra, which shifted to 2.0 Å at 290 °C and became the only feature in the spectra. These changes in cobalt environment upon heating can be explained by atomic rearrangements around cobalt. Firstly, the Co-F distances increase by ~0.1 Å due to the thermal lattice expansion and a simultaneous decrease in the number of the first shell neighbors. Above 200 °C new neighboring atoms appeared around cobalt with the new coordination shell formed at ~2.45 Å ($2.0 + \Delta\varphi$). This value is very close to the interatomic distances in cobalt metal (2.42 Å). Furthermore it can be noted that the EXAFS spectra (Figure 6b) obtained at 290 Å do not exhibit the shoulder at low radial distance which would correspond to the first-shell B neighbors [26]. Thus, it can be suggested that cobalt metal clusters, rather than cobalt boride species, were the predominant phases formed by the CoF_2 additive in this study. The changes in cobalt environment observed upon C1 decomposition in Ar and H_2 were very similar, suggesting a small effect of the atmosphere on the chemical transformations in the additive in the present temperature range. Therefore, the slight difference in the γ-ε phase-transition and decomposition temperature of $Mg(BH_4)_2$, observed in the *in-situ* SR-PXD (Section 2.2.1) should be attributed to the different atmosphere.

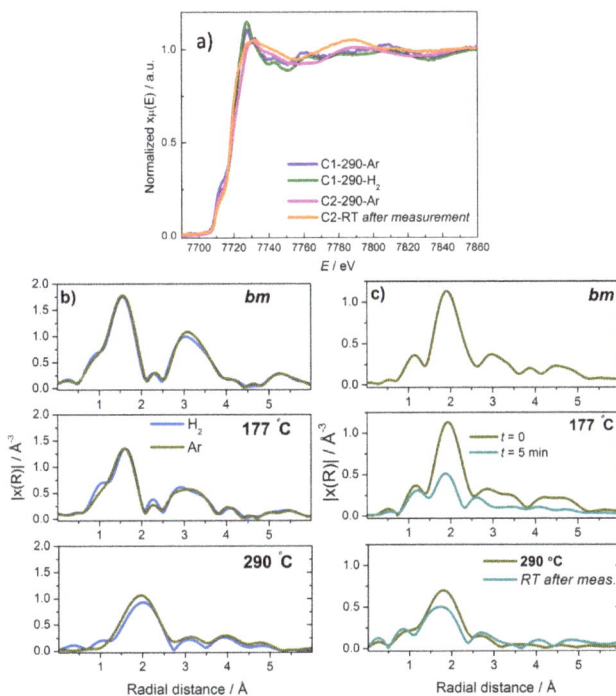

Figure 6. (a) Co K-edge XAS spectra of the C1 sample at 290 °C obtained in Ar (violet line) and H_2 (green line) and XAS spectrum of C2 sample at 290 °C in Ar pink curve). The spectrum recorded after quenching C2 at RT is also reported for comparison (yellow line). Co K-edge k^2-weighted FT-transform moduli of the EXAFS signals (phase-uncorrected) at different temperatures of (b) C1-H2 and C1-Ar and (c) C2-Ar.

The analysis of the EXAFS spectra conducted for the C2 sample ($CoCl_2$ additive) is shown in Figure 6c. Also in this case the radial distribution function for the as milled sample indicates that the additive was essentially unreacted during the milling. In fact, the most intense peaks at 1.9 Å and 3.0 Å are due to the scattering from Cl and Co neighbors located in $CoCl_2$ at 2.5 and 3.6 Å, respectively ($\Delta\varphi = 0.6$ Å) [26]. Up to 177 °C (Figure 6c), the FT-EXAFS spectra indicates the presence of only one cobalt species, $CoCl_2$. In the spectra collected after 5 min at 177 °C, a significant change in the local cobalt environment is evident (Figure 6c, 177 °C cyan curve). A strong reduction in the intensity indicates also a highly disordered coordination environment which was preserved upon heating up to 290 °C and after cooling the sample down to room temperature. In the new cobalt phase the nearest neighbors were located at the distances 2.3–2.4 Å, if the $\Delta\varphi$ obtained from fitting the $CoCl_2$ reference is used.

2.3. Discussion

It has been repeatedly shown that TM-additives, particularly halides, react with the host complex hydride matrix [26,34,42,43]. Even the borides lose long-range order, although preserving the local environment around the metal [26,42]. Therefore, these types of compounds cannot be called catalysts in the classical meaning. Our *in-situ* study demonstrates that this is also the case for Co-based additives, although some differences can be noticed between CoF_2 and $CoCl_2$. $CoCl_2$ already reacted with $Mg(BH_4)_2$ at 177 °C, whereas CoF_2 was slowly reduced upon heating above this temperature. At 290 °C both the additives produce highly disordered phases that, however, differ slightly from each other. The EXAFS extraction allowed some information about the cobalt environment in these compounds to be obtained, but their precise nature remains somewhat unclear. In particular, CoF_2 forms the species with the first coordination shell at ~2.45 Å whereas for $CoCl_2$ a 2.3–2.4 Å distance was suggested for the first neighbors. Linear combination fits of the Co K-edge XANES spectra indicate that cobalt metal rather than cobalt boride is formed upon reduction, with the concomitant presence of other, unidentified, phases. In our previous *ex-situ* study [26], we have found that cobalt additives transform into boride-like species upon cycling with $Mg(BH_4)_2$. Nevertheless, in that study the *ex-situ* XAS measurements were performed at least one month after cycling. It may indicate that borides could be formed upon further cycling in H_2 and/or ageing of the samples. The *in-situ* XAS spectra have shown also a gradual change in the cobalt environment in the additives which rules out the low-temperature formation and decomposition of cobalt borohydride. TEM observations suggest that some of the Co was incorporated into the hydride surfaces.

Table 1 summarizes the phase-transitions and decomposition temperatures of the samples, as observed by *in-situ* SR-PXD in this study. It is notable that the $\gamma \rightarrow \varepsilon$ phase transition occurred about 45 °C lower in the samples with CoF_2 and $CoCl_2$ decomposed in Ar than in pure milled $Mg(BH_4)_2$. In the C1-H_2 sample this temperature was lowered by only 20 °C. A destabilization of the γ phase (and then a decrease in the temperature of $\gamma \rightarrow \varepsilon$ transition) can be explained on the basis of the effect of the additive on the metastable nature of the porous γ-$Mg(BH_4)_2$. A smaller effect was observed also in the onset temperature of the $\varepsilon \rightarrow \beta'$ phase transition, which was lowered by 10 °C in all cases. The T_{dec}, which was defined as the onset of the decrease in the diffraction peaks intensities of the β' phase and increase in the amorphous halo intensity, was similar for all the samples. C1-H_2 appeared to be stabilized by the CoF_2 additive. It is noteworthy as only in this case the LCF of the XANES spectra did not evidence the partial transformation of the Co-additive in Co metal particles for T lower than 220°C.

Table 1. Summary of phase transition and decomposition temperatures of the $Mg(BH_4)_2$ samples obtained in this study.

Event	$Mg(BH_4)_2$ Milled-H_2	C1 ($Mg(BH_4)_2$ + 0.2 mol% CoF_2)/H_2	C1 ($Mg(BH_4)_2$ + 0.2 mol% CoF_2)/Ar	C2 ($Mg(BH_4)_2$ + 0.2 mol% $CoCl_2$)/Ar
γ → ε phase transition	150–170 °C	130–175 °C	105–160 °C	107–150 °C
ε → β' phase transition	185–202 °C	175–190 °C	175–200 °C	175–190 °C
Decomposition	>200 °C	>220 °C	>200 °C	>200 °C

Raman spectra of C2-Ar have shown that the sample became strongly fluorescent at 180 °C, whereas C1-H_2 and C1-Ar still had well defined Raman scattering. This could indicate that the $CoCl_2$ lowers T_{dec} of $Mg(BH_4)_2$. At the same time, since the PXD data have shown similar decomposition temperatures for both C1 and C2, the fluorescence could also originate from the reacted cobalt additive. Raman spectra of the C1-H_2 gave also the peaks at 2513, 2411, and 766 cm^{-1}, belonging to an unidentified boron hydride phase. The values at 2513 cm^{-1} and 2411 cm^{-1} are similar to the BH stretching in higher boron hydrides whereas the peak at 766 cm^{-1} is in the range of the B–B stretching of these compounds. It is notable that no peaks were observed in the 1800–2200 cm^{-1} region where the B–H–B stretching of bridged hydrogen atoms should appear [48].

The question on how Co-additives affect the decomposition process of $Mg(BH_4)_2$ is intriguing, especially taking into account the literature results on Ni and Co-additives effect on the decomposition of borohydrides and related compounds. It should be noted though that the most prominent results, such as dramatic decrease in T_{dec}, was observed for the samples where additives were combined with other destabilization approaches, for example, dispersion in the highly porous media [36,43] or mixing with electron-rich compounds [41]. The effects of pure additives on pure borohydrides could thus be attributed simply to chemical reactions between the two and the ball-milling process usually used to prepare the composites. The large effect of the ball milling alone on the partial decomposition and rehydrogenation of pure $Mg(BH_4)_2$ was recently demonstrated [26].

3. Experimental Section

All procedures were carried out in a glove box under a continuously purified Ar atmosphere (O_2, H_2O < 1 ppm) if not stated otherwise. Commercial γ-$Mg(BH_4)_2$ (95%, Sigma-Aldrich, St. Louis, MO, USA) was used. XRD analysis of this sample showed that the crystalline fraction of this batch was constituted by γ-$Mg(BH_4)_2$. $CoCl_2$ (99.999%), and CoF_2 were purchased from Sigma-Aldrich. The starting powder mixtures consisted of $Mg(BH_4)_2$ and 2 mol% of an additive (Co_{add}). The milling was carried out in stainless steel vials and balls with a 40:1 ball-to-powder weight ratio. Pure $Mg(BH_4)_2$ and the additives were milled for 1 h at 280 rpm under 1 bar Ar pressure.

In-situ SR-PXD of the ball-milled γ-$Mg(BH_4)_2$ decomposition was performed at the Swiss-Norwegian Beam Line (station BM01A) at the European Synchrotron Radiation Facility (ESRF) in Grenoble, France. The X-ray wavelength λ = 0.70153 Å was used. The combined *in-situ* SR-PXD, XAS, and Raman spectroscopies data were collected at the station BM01B at ESRF [49]. For the measurements in Ar, the samples were placed in borosilicate capillaries (1.0 mm in diameter) and sealed in the glove-box with the air-tight glue. For the measurements in H_2, the capillaries were attached to a sample holder enabling connection to a gas line. Before the measurements in H_2, the samples were degassed in vacuum at RT, and 2.5 bar H_2 pressure was set in the capillaries. The capillaries were heated by a calibrated hot blower. Temperature calibration was carried out with In and Sn standards (experimental melting points were obtained at the nominal temperatures 145 (In) (theoretical. 156.61 °C) and 219 (Sn), theoretical 231.9 °C). The samples were heated up to 290 °C with a 5 °C /min heating rate with two isothermal pauses at 177 °C and 220 °C of 60 min. After reaching 290 °C, the samples were kept at that temperature for several hours. SR-PXD data were collected every 0.5 min during the heating ramps and every 4 min at the isothermal steps. XAS scans were obtained during the isothermal steps every 5 min. Raman spectra were collected during the experiments until the samples became fluorescent at about 180 °C.

SR-PXD data were collected with a wavelength of 0.5025 Å. 2D images were obtained with an exposure time of 30 s using an image plate detector. A number of dark-current images (17–90 for different ramps and isothermal steps) were obtained before the sample scans, averaged, and subsequently subtracted from the sample images.

In-situ XAS data were collected in the XANES and EXAFS regions in the energy range 7.6–8.6 keV comprising Co K-edge at 7.7089 keV in transmission mode using a double crystal Si (111) monochromator. The spectrum of the Co foil (hcp) was used as a reference for data calibration and alignment. The spectra of the reference powders—CoF_2, CoF_3, $CoCl_2$, used in the preparation of the composites, and additionally Co powder (99.5%, Alfa Aesar, Ward Hill, MA, USA), Co_2B (American

Elements, Los Angeles, CA, USA), and CoO (>99.99%, Aldrich) were measured diluted with boron nitride to 2 mol% of Co_{add}. XAS data analysis was carried out with DEMETER software pack (ATHENA and ARTEMIS, B. Ravel and M. Newville, Brookhaven National Laboratory, Upton NY, USA) [50]. The spectra were pre-processed in ATHENA (background subtracted, aligned, the step scans were averaged and used for final plots and fitting). Liner combination analyses (fits, LCF) in the range from -30 eV to $+20$ eV were performed with the ATHENA software. Calculation of the theoretical scattering paths in the EXAFS region and fitting were performed with the FEFF6 code [51] using ARTEMIS. Due to the low concentration of the additive and the small sample quantities required by the capillaries in the experimental set-up, the energy step was low and the spectra obtained were rather noisy, especially in the EXAFS region. Therefore, several consequent spectra, obtained at the higher temperatures were merged in order to increase the signal to noise ratio.

Raman spectra were collected on an RA 100 Raman analyzer (Renishaw, New Mills, UK) using a 532 nm (green) excitation wavelength in backscattering mode with exposure times of 200 s and step of 1.2 cm^{-1} over a 3200–200 cm^{-1} range.

TEM was performed in the NORTEM TEM Gemini Centre, Norwegian University of Science and Technology (NTNU), Norway, using an ARM200F instrument (JEOL, Tokyo, Japan) operating at 200 kV. A JEOL Centurio detector was used for EDS analysis and a Gatan and EELS using a Gatan Quantum imaging filter. As-milled sample was studied. Sample preparation was performed in a glove box. A small quantity of the powder was crushed in a mortar and dispersed onto a standard holey carbon film on a Cu support grid. The sample was rather stable under the electron beam.

4. Conclusions

The *in-situ* study combining X-ray diffraction and absorption, and Raman spectroscopy allowed us to follow several aspects of the decomposition of $Mg(BH_4)_2$-Co_{add} system simultaneously, *i.e.*, changes in the crystalline and amorphous $Mg(BH_4)_2$ matrix and in the cobalt-based additives. This approach can provide valuable knowledge on the decomposition process of complex (molecular) hydrides and effect of the metal-based additives on this process. It allowed us to follow the reaction during the thermal treatment, pointing out that the cobalt was reduced by $Mg(BH_4)_2$. CoF_2 was found to be more stable at a higher temperature than $CoCl_2$. *In-situ* XAS measurements suggested that Co-additives form metal clusters rather than CoB_x species upon reduction. In case of $CoCl_2$ additive, TEM observations suggest that small amount of the Co was incorporated into the hydride surfaces already after ball-milling although the main additive phase remained as $CoCl_2$. The identification of the amorphous fraction of the $Mg(BH_4)_2$ decomposition products by means of Raman scattering was hindered by fluorescence with the

532 nm laser, and the data obtained at higher temperature were not very informative. However, we were still able to observe that one of the decomposition products had the vibrational modes at 2513, 2411, and 766 cm^{-1}, belonging to an unidentified boron hydride phase. These vibrations are very similar to the ones measured by means of infrared spectroscopy in a previous study on $Mg(BH_4)_2$ decomposition [52].

Acknowledgments: This work was financed by the European Fuel Cells and Hydrogen Joint Undertaking under collaborative project "BOR4STORE" (Grant Agreement No. 303428). Partial financial support from SYNKNØYT program in the Research Council of Norway is greatly acknowledged. TEM analysis was supported by the NORTEM (Grant 197405) within the program INFRASTRUCTURE of the Research Council of Norway (RCN). NORTEM was co-funded by the RCN and the project partners NTNU, UiO and SINTEF. We thank the beamline scientists Paula Abdala, Herman Emerich, and Wouter van Beek at the BM01B SNBL for their help with the measurements.

Author Contributions: Stefano Deledda, Bjørn C. Hauback, Jiri Muller, Ivan Saldan, and Olena Zavorotynska have initiated the research and planned the experiments. Olena Zavorotynska, Ivan Saldan, Jenny G. Vitillo, and Matylda N. Guzik have carried out most of the experimental work and analyzed the results. Olena Zavorotynska wrote the manuscript with substantial revisions from Jenny G. Vitillo and Stefano Deledda; John C. Walmsley carried out TEM measurements and data analysis. All the authors significantly contributed to the editing and improvement of the manuscript.

Conflicts of Interest: The authors declare no conflict of interest.

References

1. George, L.; Saxena, S.K. Structural stability of metal hydrides, alanates and borohydrides of alkali and alkali-earth elements: A review. *Int. J. Hydrog. Energy* **2010**, *35*, 5454–5470.

2. Li, H.-W.; Yan, Y.; Orimo, S.; Zuttel, A.; Jensen, C.M. Recent Progress in Metal Borohydrides for Hydrogen Storage. *Energies* **2011**, *4*, 185–214.

3. Hummelshoj, J.S.; Landis, D.D.; Voss, J.; Jiang, T.; Tekin, A.; Bork, N.; Dulak, M.; Mortensen, J.J.; Adamska, L.; Andersin, J.; *et al.* Density functional theory based screening of ternary alkali-transition metal borohydrides: A computational material design project. *J. Chem. Phys.* **2009**, *131*.

4. Besora, M.; Lledos, A. Coordination modes and hydride exchange dynamics in transition metal tetrahydroborate complexes. In *Contemporary Metal Boron Chemistry I: Borylenes, Boryls, Borane*; Marder, T.B., Lin, Z., Eds.; Springer: Berlin, Germany, 2008; pp. 149–202.

5. Marks, T.J.; Kolb, J.R. Covalent transition metal, lanthanide, and actinide tetrahydroborate complexes. *Chem. Rev.* **1977**, *77*, 263–293.

6. Callini, E.; Borgschulte, A.; Ramirez-Cuesta, A.J.; Zuettela, A. Diborane release and structure distortion in borohydrides. *Dalton Trans.* **2013**, *42*, 719–725.

7. Chong, M.; Callini, E.; Borgschulte, A.; Zuettel, A.; Jensen, C.M. Dehydrogenation studies of the bimetallic borohydrides. *RSC Adv.* **2014**, *4*, 63933–63940.

8. Albanese, E.; Civalleri, B.; Casassa, S.; Baricco, M. Investigation on the Decomposition Enthalpy of Novel Mixed $Mg_{(1-x)}Zn_x(BH_4)_2$ Borohydrides by Means of Periodic DFT Calculations. *J. Phys. Chem. C* **2014**, *118*, 23468–23475.

9. Schouwink, P.; Ley, M.B.; Tissot, A.; Hagemann, H.; Jensen, T.R.; Smrčok, L.; Černý, R. Structure and properties of complex hydride perovskite materials. *Nat. Commun.* **2014**, *5*.

10. Wang, J.; Li, H.W.; Chen, P. Amides and borohydrides for high capacity solid-state hydrogen storage—Materials design and kinetics improvement. *MRS Bull.* **2013**, *38*, 480–487.

11. Li, H.-W.; Kikuchi, K.; Sato, T.; Nakamori, Y.; Ohba, N.; Aoki, M.; Miwa, K.; Towata, S.; Orimo, S. Synthesis and Hydrogen Storage Properties of a Single-Phase Magnesium Borohydride $Mg(BH_4)_2$. *Mater. Trans.* **2008**, *49*, 2224–2228.

12. Matsurtaga, T.; Buchter, F.; Miwa, K.; Towata, S.; Orimo, S.; Zuttel, A. Magnesium borohydride: A new hydrogen storage material. *Renew. Energy* **2008**, *33*, 193–196.

13. Filinchuk, Y.; Richter, B.; Jensen, T.R.; Dmitriev, V.; Chernyshov, D.; Hagemann, H. Porous and Dense Magnesium Borohydride Frameworks: Synthesis, Stability, and Reversible Absorption of Guest Species. *Angew. Chem. Int. Ed.* **2011**, *50*, 11162–11166.

14. Roennebro, E. Development of group II borohydrides as hydrogen storage materials. *Curr. Opin. Solid State Mater. Sci.* **2011**, *15*, 44–51.

15. Ozolins, V.; Akbarzadeh, A.R.; Gunaydin, H.; Michel, K.; Wolverton, C.; Majzoub, E.H. First-principles computational discovery of materials for hydrogen storage. *J. Phys. Conf. Ser.* **2009**, *180*.

16. Ozolins, V.; Majzoub, E.H.; Wolverton, C. First-principles prediction of a ground state crystal structure of magnesium borohydride. *Phys. Rev. Lett.* **2008**, *100*.

17. Van Setten, M.J.; de Wijs, G.A.; Fichtner, M.; Brocks, G. A density functional study of α-$Mg(BH_4)_2$. *Chem. Mater.* **2008**, *20*, 4952–4956.

18. Pinatel, E.R.; Albanese, E.; Civalleri, B.; Baricco, M. Thermodynamic modelling of $Mg(BH_4)_2$. *J. Alloys Compd.* **2015**, *645*, S64–S68.

19. Riktor, M.D.; Sorby, M.H.; Chlopek, K.; Fichtner, M.; Buchter, F.; Zuttel, A.; Hauback, B.C. *In situ* synchrotron diffraction studies of phase transitions and thermal decomposition of $Mg(BH_4)_2$ and $Ca(BH_4)_2$. *J. Mater. Chem.* **2007**, *17*, 4939–4942.

20. Hanada, N.; Chlopek, K.; Frommen, C.; Lohstroh, W.; Fichtner, M. Thermal decomposition of $Mg(BH_4)_2$ under He flow and H_2 pressure. *J. Mater. Chem.* **2008**, *18*, 2611–2614.

21. Soloveichik, G.L.; Gao, Y.; Rijssenbeek, J.; Andrus, M.; Kniajanski, S.; Bowman, R.C., Jr.; Hwan, S.-J.; Zhao, J.-C. Magnesium borohydride as a hydrogen storage material: Properties and dehydrogenation pathway of unsolvated $Mg(BH_4)_2$. *Int. J. Hydrog. Energy* **2009**, *34*, 916–928.

22. Yan, Y.; Li, H.-W.; Maekawa, H.; Aoki, M.; Noritake, T.; Matsumoto, M.; Miwa, K.; Towata, S.; Orimo, S. Formation Process of $[B_{12}H_{12}]^{2-}$ from $[BH_4]^-$ during the Dehydrogenation Reaction of $Mg(BH_4)_2$. *Mater. Trans.* **2011**, *52*, 1443–1446.

23. Yang, J.; Zhang, X.; Zheng, J.; Song, P.; Li, X. Decomposition pathway of $Mg(BH_4)_2$ under pressure: Metastable phases and thermodynamic parameters. *Scr. Mater.* **2011**, *64*, 225–228.

24. Paskevicius, M.; Pitt, M.P.; Webb, C.J.; Sheppard, D.A.; Filso, U.; Gray, E.M.; Buckley, C.E. *In-Situ* X-ray Diffraction Study of γ-Mg(BH$_4$)$_2$ Decomposition. *J Phys. Chem. C* **2012**, *116*, 15231–15240.

25. Guo, S.; Chan, H.Y.L.; Reed, D.; Book, D. Investigation of dehydrogenation processes in disordered γ-Mg(BH$_4$)$_2$. *J. Alloys Compd.* **2013**, *580*, S296–S300.

26. Zavorotynska, O.; Saldan, I.; Hino, S.; Humphries, T.D.; Deledda, S.; Hauback, B.C. Hydrogen cycling in γ-Mg(BH$_4$)$_2$ with cobalt-based additives. *J. Mater. Chem. A* **2015**, *3*, 6592–6602.

27. Chong, M.; Karkamkar, A.; Autrey, T.; Orimo, S.-I.; Jalisatgi, S.; Jensen, C.M. Reversible dehydrogenation of magnesium borohydride to magnesium triborane in the solid state under moderate conditions. *Chem. Commun.* **2011**, *47*, 1330–1332.

28. Zavorotynska, O.; Deledda, S.; Li, G.; Matsuo, M.; Orimo, S.-I.; Hauback, B.C. Isotopic Exchange in Porous and Dense Magnesium Borohydride. *Angew. Chem. Int. Ed.* **2015**.

29. Brown, W.G.; Kaplan, L.; Wilzbach, K.E. The exchange of hydrogen gas with lithium and sodium borohydrides. *J. Am. Chem. Soc.* **1952**, *74*, 1343–1344.

30. Gremaud, R.; Lodziana, Z.; Hug, P.; Willenberg, B.; Racu, A.-M.; Schoenes, J.; Ramirez-Cuesta, A.J.; Clark, S.J.; Refson, K.; Züttel, A.; *et al.* Evidence for hydrogen transport in deuterated LiBH$_4$ from low-temperature Raman-scattering measurements and first-principle calculations. *Phys. Rev. B* **2009**, *80*, 1–4.

31. Mesmer, R.E.; Jolly, W.L. The Exchange of Deuterium with Solid Potassium Hydroborate. *J. Am. Chem. Soc.* **1962**, *84*, 2039–2042.

32. Van Setten, M.J.; Lohstroh, W.; Fichtner, M. A new phase in the decomposition of Mg(BH$_4$)$_2$: First-principles simulated annealing. *J. Mater. Chem.* **2009**, *19*, 7081–7087.

33. Bogdanovic, B.; Schwickardi, M. Ti-doped alkali metal aluminium hydrides as potential novel reversible hydrogen storage materials. *J. Alloys Compd.* **1997**, *253–254*, 1–9.

34. Mao, J.F.; Yu, X.B.; Guo, Z.P.; Liu, H.K.; Wu, Z.; Ni, J. Enhanced hydrogen storage performances of NaBH$_4$-MgH$_2$ system. *J. Alloys Compd.* **2009**, *479*, 619–623.

35. Zhang, B.J.; Liu, B.H. Hydrogen desorption from LiBH$_4$ destabilized by chlorides of transition metal Fe, Co, and Ni. *Int. J. Hydrog. Energy* **2010**, *35*, 7288–7294.

36. Wahab, M.A.; Jia, Y.; Yang, D.; Zhao, H.; Yao, X. Enhanced hydrogen desorption from Mg(BH$_4$)$_2$ by combining nanoconfinement and a Ni catalyst. *J. Mater. Chem. A* **2013**, *1*, 3471–3478.

37. Xu, J.; Qi, Z.; Cao, J.; Meng, R.; Gu, X.; Wang, W.; Chen, Z. Reversible hydrogen desorption from LiBH$_4$ catalyzed by graphene supported Pt nanoparticles. *Dalton Trans.* **2013**, *42*, 12926–12933.

38. Humphries, T.D.; Kalantzopoulos, G.N.; Llamas-Jansa, I.; Olsen, J.E.; Hauback, B.C. Reversible Hydrogenation Studies of NaBH$_4$ Milled with Ni-Containing Additives. *J. Phys. Chem. C* **2013**, *117*, 6060–6065.

39. Au, Y.S.; Yan, Y.; de Jong, K.P.; Remhof, A.; de Jongh, P.E. Pore Confined Synthesis of Magnesium Boron Hydride Nanoparticles. *J. Phys. Chem. C* **2014**, *118*, 20832–20839.

40. Zhang, Z.G.; Wang, H.; Liu, J.W.; Zhu, M. Thermal decomposition behaviors of magnesium borohydride doped with metal fluoride additives. *Thermochim. Acta* **2013**, *560*, 82–88.

41. Zhang, Y.; Liu, Y.; Pang, Y.; Gao, M.; Pan, H. Role of Co_3O_4 in improving the hydrogen storage properties of a $LiBH_4$-$2LiNH_2$ composite. *J. Mater. Chem. A* **2014**, *2*, 11155–11161.

42. Saldan, I.; Hino, S.; Humphries, T.D.; Zavorotynska, O.; Chong, M.; Jensen, C.M.; Deledda, S.; Hauback, B.C. Structural changes observed during the reversible hydrogenation of $Mg(BH_4)_2$ with Ni-based additives. *J. Phys. Chem. C* **2014**, *118*, 23376–23384.

43. Ngene, P.; Verkuijlen, M.H.W.; Zheng, Q.; Kragten, J.; van Bentum, P.J.M.; Bitter, J.H.; de Jongh, P.E. The role of Ni in increasing the reversibility of the hydrogen release from nanoconfined $LiBH_4$. *Faraday Discuss.* **2011**, *151*, 47–58.

44. Ban, V.; Soloninin, A.V.; Skripov, A.V.; Hadermann, J.; Abakumov, A.; Filinchuk, Y. Pressure-Collapsed Amorphous $Mg(BH_4)_2$: An Ultradense Complex Hydride Showing a Reversible Transition to the Porous Framework. *J. Phys. Chem. C* **2014**, *118*, 23402–23408.

45. David, W.I.F.; Callear, S.K.; Jones, M.O.; Aeberhard, P.C.; Culligan, S.D.; Pohl, A.H.; Johnson, S.R.; Ryan, K.R.; Parker, J.E.; Edwards, P.P.; *et al.* The structure, thermal properties and phase transformations of the cubic polymorph of magnesium tetrahydroborate. *Phys. Chem. Chem. Phys.* **2012**, *14*, 11800–11807.

46. Giannasi, A.; Colognesi, D.; Ulivi, L.; Zoppi, M.; Ramirez-Cuesta, A.J.; Bardaji, E.G.; Roehm, E.; Fichtner, M. High Resolution Raman and Neutron Investigation of $Mg(BH_4)_2$ in an Extensive Temperature Range. *J. Phys. Chem. A* **2010**, *114*, 2788–2793.

47. Shimanouchi, T.; Nakagawa, I. Infrared spectroscopic study on the co-ordination bond-I: Infrared spectra of cobalt hexammine, pentammine and trans-tetraammine complexes. *Spectrochim. Acta* **1962**, *18*, 89–100.

48. Tomkinson, J.; Ludman, C.J.; Waddington, T.C. IR Raman and Inelastic Neutron-Scattering Spectra of Cesium Octahydroborate, $C_sB_3H_8$. *Spectrochim. Acta Part A Mol. Biomol. Spectrosc.* **1979**, *35*, 117–122.

49. Newton, M.A.; van Beek, W. Combining synchrotron-based X-ray techniques with vibrational spectroscopies for the *in situ* study of heterogeneous catalysts: A view from a bridge. *Chem. Soc. Rev.* **2010**, *39*, 4845–4863.

50. Ravel, B.; Newville, M. ATHENA, ARTEMIS, HEPHAESTUS: Data analysis for X-ray absorption spectroscopy using IFEFFIT. *J. Synchrotron Radiat.* **2005**, *12*, 537–541.

51. Rehr, J.J.; Albers, R.C. Theoretical approaches to x-ray absorption fine structure. *Rev. Mod. Phys.* **2000**, *72*, 621–654.

52. Vitillo, J.G.; Bordiga, S.; Baricco, M. Spectroscopic and structural characterization of thermal decomposition of γ-$Mg(BH_4)_2$: Dynamic vacuum *vs.* H_2 atmosphere. *J. Phys. Chem. C* **2015**. in press.

Thermal Decomposition of Anhydrous Alkali Metal Dodecaborates $M_2B_{12}H_{12}$ (M = Li, Na, K)

Liqing He, Hai-Wen Li and Etsuo Akiba

Abstract: Metal dodecaborates $M_{2/n}B_{12}H_{12}$ are regarded as the dehydrogenation intermediates of metal borohydrides $M(BH_4)_n$ that are expected to be high density hydrogen storage materials. In this work, thermal decomposition processes of anhydrous alkali metal dodecaborates $M_2B_{12}H_{12}$ (M = Li, Na, K) synthesized by sintering of MBH_4 (M = Li, Na, K) and $B_{10}H_{14}$ have been systematically investigated in order to understand its role in the dehydrogenation of $M(BH_4)_n$. Thermal decomposition of $M_2B_{12}H_{12}$ indicates multistep pathways accompanying the formation of H-deficient monomers $M_2B_{12}H_{12-x}$ containing the icosahedral B_{12} skeletons and is followed by the formation of $(M_2B_{12}H_z)_n$ polymers. The decomposition behaviors are different with the *in situ* formed $M_2B_{12}H_{12}$ during the dehydrogenation of metal borohydrides.

Reprinted from *Energies*. Cite as: He, L.; Li, H.-W.; Akiba, E. Thermal Decomposition of Anhydrous Alkali Metal Dodecaborates $M_2B_{12}H_{12}$ (M = Li, Na, K). *Energies* **2015**, *8*, 12429–12438.

1. Introduction

Metal dodecaborates $M_{2/n}B_{12}H_{12}$ (n is the valence of M) have been widely regarded as a dehydrogenation intermediate of metal borohydrides $M(BH_4)_n$ with a high gravimetric hydrogen density of 10 mass% [1–16]. The formation of $M_{2/n}B_{12}H_{12}$, despite is still controversial, largely depends on the dehydrogenation temperature, hydrogen backpressure, particle size and sample pretreatment [11–21]. Due to the strong B-B bonds in an icosahedral boron cage B_{12}, the intermediate comprising of polyatomic anion $[B_{12}H_{12}]^{2-}$ has been widely regarded at the main obstacle for the rehydrogenation of $M(BH_4)_n$ [5,22–24]. Systematic investigation on the thermal decomposition of $M_{2/n}B_{12}H_{12}$ is therefore of great importance to understand their role in the dehydrogenation of $M(BH_4)_n$.

$M_{2/n}B_{12}H_{12}$ is generally synthesized using liquid phase reactions, followed by careful dehydration processes [25]. However, in some $M_{2/n}B_{12}H_{12}$ such as $MgB_{12}H_{12}$, the coordination water tends to form a hydrogen bond with the polyatomic anion $[B_{12}H_{12}]^{2-}$, resulting in the failure of dehydration [26]. To solve such problems, we have recently successfully developed a novel solvent-free synthesis process, *i.e.*, sintering of $M(BH_4)_n$ and $B_{10}H_{14}$ with stoichiometric molar ratio. Several anhydrous metal dodecaborates $M_{2/n}B_{12}H_{12}$ (M = Li, Na, K, Mg, Ca, LiNa), so far, have been successfully synthesized via the newly developed method [21,27,28].

In this work, we carefully investigate the thermal decomposition behaviors of anhydrous alkali metal dodecaborates $M_2B_{12}H_{12}$ (M = Li, Na, K) that are synthesized using the reported method [27]. Furthermore, the roles of $M_2B_{12}H_{12}$ (M = Li, Na, K) in the dehydrogenation of corresponding borohydrides MBH_4 are discussed, based on the comparison of the decomposition pathways of $M_2B_{12}H_{12}$ and those *in situ* formed during the decomposition of MBH_4.

2. Results and Discussion

2.1. Decomposition of Anhydrous $Li_2B_{12}H_{12}$

Figure 1 shows the thermogravimetry (TG) and mass spectrometry (MS) measurement results of anhydrous $Li_2B_{12}H_{12}$. Only hydrogen is detected in MS, indicating that the weight loss upon heating results from the dehydrogenation. The weight loss starts at approximately 200 °C and the dehydrogenation amount reaches 5.1 mass% (approximately 66% of the theoretical hydrogen content in $Li_2B_{12}H_{12}$) when heated up to 700 °C. The decomposition proceeds with multistep reactions, as shown in TG and MS results.

In order to investigate the decomposition process of anhydrous $Li_2B_{12}H_{12}$, the sample was heated to respective temperatures and subsequently cooled down to room temperature. The changes examined by X-ray diffraction (XRD), Raman and nuclear magnetic resonance (NMR) are shown in Figures 2 and 3 respectively. When the temperature is increased to 250 °C, no obvious changes of XRD patterns, Raman spectra and the main resonance at −15.3 ppm for $Li_2B_{12}H_{12}$ are observed, whereas the resonance at −41.3 ppm originated from residual $LiBH_4$ decreases significantly and those at 11.2 ppm and −36.0 ppm for the unknown side product disappears (Table 1). This indicates that anhydrous $Li_2B_{12}H_{12}$ is stable up to 250 °C and the weight loss of 0.3 mass% up to 250 °C is originated from decomposition of the residual $LiBH_4$ and side product. When the temperature is increased to 300 °C, intensities for the diffraction peaks (2θ = 15.8° and 18.4°) and Raman spectra (between 500–1000 cm^{-1} and around 2500 cm^{-1}) attributed to $Li_2B_{12}H_{12}$ decrease. This indicates that anhydrous $Li_2B_{12}H_{12}$ starts to decompose above 250 °C, similar to the reported temperature [16]. The resonance at −29.8 ppm originated from $Li_2B_{10}H_{10}$ and that at −17.5 ppm for the unknown side product become significantly weaker when heated up to 300 °C and completely disappear when heated up to 500 °C. When the temperature is increased to 600 °C, diffraction peaks and Raman spectra attributed to $Li_2B_{12}H_{12}$ nearly disappear, and the main resonances at −15.3 ppm for ^{11}B and at 1.4 ppm for ^{1}H originated from $Li_2B_{12}H_{12}$

decrease significantly without any change in the chemical shift. This indicates that a major part of B–H bond in the icosahedral polyatomic anion $[B_{12}H_{12}]^{2-}$ has been broken to release hydrogen [21]. The dehydrogenation amount reaches 3.7 mass% including the contribution from the residual LiBH$_4$ and side products, suggesting that the decomposition product is probably H-deficient $Li_2B_{12}H_{12-x}$ ($x < 5.3$) that remains the icosahedral B_{12} skeletons [16,21]. No signals in the solution-state ^{11}B NMR were detected, implying that the formed $Li_2B_{12}H_{12-x}$ is DMSO insoluble. When the temperature is further increased to 700 °C, the dehydrogenation amount reaches 5.1 mass%, the major resonance of $Li_2B_{12}H_{12}$ at −15.3 ppm in ^{11}B MAS NMR shifts to −11.9 ppm, and that at 1.4 ppm in ^{1}H MAS NMR changes to several weak resonance peaks from −10 ppm to 10 ppm. This suggests that $Li_2B_{12}H_{12-x}$ continuously releases hydrogen accompanied by the polymerization of the icosahedral B_{12} skeletons and the formation of $(Li_2B_{12}H_z)_n$ polymers [15,21,29], insoluble in water and DMSO.

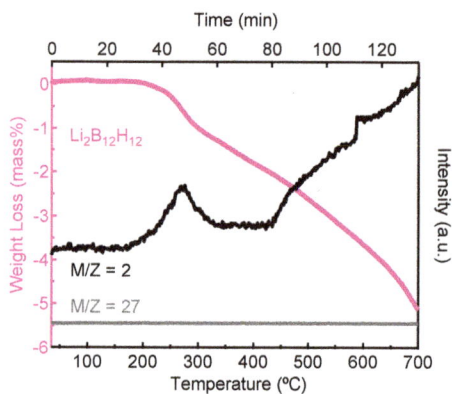

Figure 1. Thermogravimetry (TG) curve and mass spectrometry (MS) signals of anhydrous $Li_2B_{12}H_{12}$ (mass numbers 2 and 27 represent H_2 and B_2H_6).

Table 1. Relative amount of the B-H species in synthesized $Li_2B_{12}H_{12}$ when heated up to respective temperatures ($\leqslant$500 °C), estimated from the peak fitting of ^{11}B MAS NMR spectra shown in Figure 3.

Temperature, °C	$[B_{12}H_{12}]^{2-}$, %	$[B_{11}H_{11}]^{2-}$, %	$[B_{10}H_{10}]^{2-}$, %	$[BH_4]^-$, %	Unknown, %
200	60.93	9.03	18.76	5.14	6.14
225	65.82	9.50	12.16	3.76	8.76
250	70.17	10.00	17.21	2.62	0
300	71.62	11.70	16.20	0.48	0
400	86.52	0	13.48	0	0
500	100	0	0	0	0

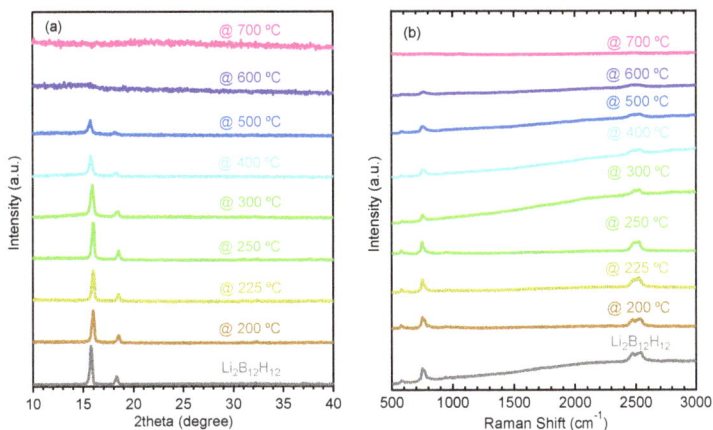

Figure 2. *Ex-situ* (**a**) X-ray diffraction (XRD) patterns and (**b**) Raman spectra of anhydrous $Li_2B_{12}H_{12}$ and heated up to respective temperatures.

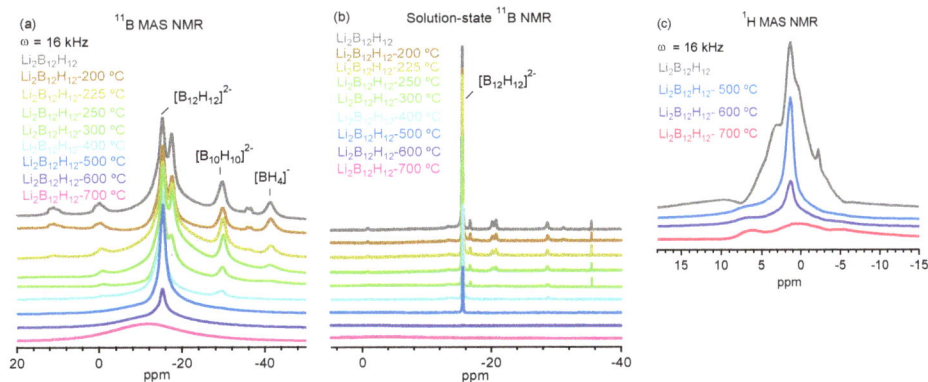

Figure 3. *Ex-situ* ^{11}B and ^1H NMR spectra of anhydrous $Li_2B_{12}H_{12}$ and heated up to respective temperatures: (**a**) solid-state ^{11}B MAS NMR spectra; (**b**) solution-state ^{11}B NMR spectra measured in DMSO-d_6 and (**c**) solid-state ^1H MAS NMR spectra. Resonance assignments of ^{11}B spectra: -15.6 ppm $[B_{12}H_{12}]^{2-}$, -35.6 ppm $[BH_4]^-$, -0.9 & -28.8 ppm $[B_{10}H_{10}]^{2-}$, -16.8 ppm $[B_{11}H_{11}]^{2-}$, -20.3 (-20.8) ppm $[B_9H_9]^{2-}$ [30]. Resonance assignments of ^1H spectra: 1.2 ppm $[B_{12}H_{12}]^{2-}$ [10,16].

The thermal decomposition pathway of anhydrous $Li_2B_{12}H_{12}$ up to 700 °C is, therefore, summarized based on the abovementioned experimental results:

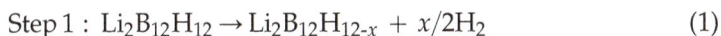

$$\text{Step 1}: \quad Li_2B_{12}H_{12} \rightarrow Li_2B_{12}H_{12-x} + x/2H_2 \tag{1}$$

$$\text{Step 2}: \quad nLi_2B_{12}H_{12-x} \rightarrow (Li_2B_{12}H_z)_n + z'H_2 \tag{2}$$

67

The decomposition pathway is similar to those of anhydrous $MgB_{12}H_{12}$ and $CaB_{12}H_{12}$ [21]. It is worth noting that the thermal decomposition behaviors of anhydrous $Li_2B_{12}H_{12}$ are different from that *in situ* formed during the dehydrogenation of $LiBH_4$, like those of anhydrous $MgB_{12}H_{12}$ and $CaB_{12}H_{12}$ [21]. Anhydrous $Li_2B_{12}H_{12}$ shows a lower initial decomposition temperature and a wider decomposition temperature range of 250 ~>700 °C than those *in situ* formed during dehydrogenation of $LiBH_4$. The formation of $Li_2B_{12}H_{12}$ during the dehydrogenation of $LiBH_4$ generally experiences complicated condensation process with sluggish kinetics [11], attributing to the higher initial decomposition temperature than that of anhydrous $Li_2B_{12}H_{12}$. On the other hand, the high activity of the *in situ* formed $Li_2B_{12}H_{12}$ together with the concurrent formation of LiH facilitate the decomposition of $Li_2B_{12}H_{12}$ [8], resulting in the lower temperature of complete decomposition than that of anhydrous $Li_2B_{12}H_{12}$.

2.2. Decomposition of Anhydrous $Na_2B_{12}H_{12}$

Figure 4 shows the TG and MS measurement results of anhydrous $Na_2B_{12}H_{12}$. Only hydrogen is detected in MS, indicating that the weight loss is originated from the dehydrogenation. The weight loss starts at approximately 580 °C and the dehydrogenation amount reaches 1.9 mass%, which is approximately 29% of the theoretical hydrogen capacity in $Na_2B_{12}H_{12}$. The value is comparable to the reported one (~1.5 mass% at 697 °C) [31].

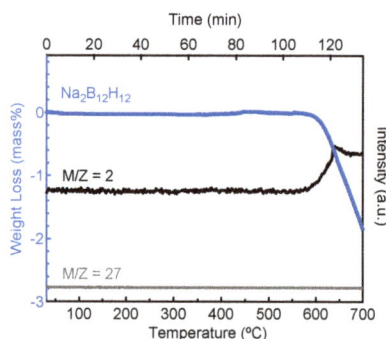

Figure 4. TG curve and MS signals of anhydrous $Na_2B_{12}H_{12}$ (mass numbers 2 and 27 represent H_2 and B_2H_6).

The changes of anhydrous $Na_2B_{12}H_{12}$ heated to respective temperatures and subsequently cooled down to room temperature examined by XRD, Raman and ^{11}B NMR are shown in Figures 5 and 6 respectively. When the temperature is increased to 500 °C, no obvious changes of diffraction peaks, Raman spectra and the major resonance at −15.7 ppm for $Na_2B_{12}H_{12}$ are observed, whereas the resonances originated from $NaBH_4$ and $Na_2B_{10}H_{10}$ nearly disappear. This suggests that the small amount of side product $Na_2B_{10}H_{10}$ and the residual $NaBH_4$ (<8 mass%) start to decompose below 500 °C without detected weight loss. When the temperature is increased to 600 °C, diffraction peaks and Raman spectra attributed to $Na_2B_{12}H_{12}$ becomes weak, indicating that $Na_2B_{12}H_{12}$ starts to decompose at 600 °C. When the temperature is further increased to 700 °C, diffraction peaks and Raman spectra from $Na_2B_{12}H_{12}$ are hardly observed, the main resonance at -15.7 ppm significantly weakens and a broad resonance between -12.4 ppm and -14.8 ppm appears. This suggests that the major dehydrogenation of $Na_2B_{12}H_{12}$ to H-deficient $Na_2B_{12}H_{12-x}$ and the polymerization of $Na_2B_{12}H_{12-x}$ to water and DMSO insoluble $(Na_2B_{12}H_z)_n$ polymers start to take place at 700 °C [15,21,29].

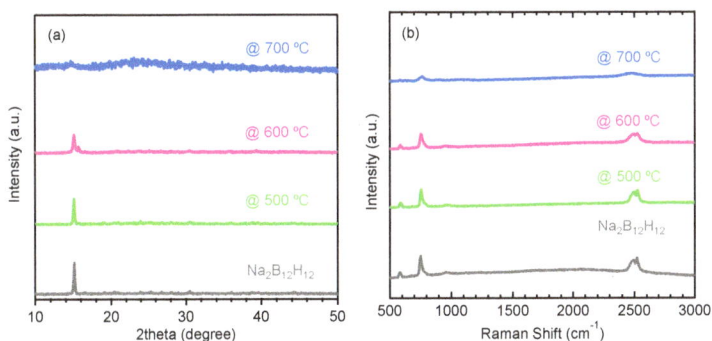

Figure 5. *Ex-situ* (**a**) XRD patterns and (**b**) Raman spectra of anhydrous $Na_2B_{12}H_{12}$ as synthesized and heated up to respective temperatures.

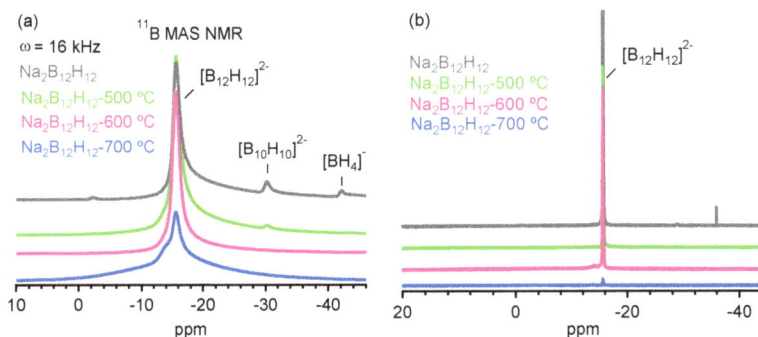

Figure 6. *Ex-situ* ^{11}B NMR spectra of anhydrous $Na_2B_{12}H_{12}$ as synthesized and heated up to respective temperatures: (**a**) solid-state ^{11}B MAS NMR spectra and (**b**) solution-state ^{11}B NMR spectra measured in DMSO-d_6. Resonance assignments: -15.6 ppm $[B_{12}H_{12}]^{2-}$, -35.9 ppm $[BH_4]^-$, -28.8 ppm $[B_{10}H_{10}]^{2-}$ [32].

2.3. Decomposition of Anhydrous $K_2B_{12}H_{12}$

Figure 7 shows the TG and MS results of anhydrous $K_2B_{12}H_{12}$. Only hydrogen is detected in MS, indicating that the weight loss upon heating results from the dehydrogenation. The weight loss starts at approximately 480 °C and the dehydrogenation amount reaches 4.4 mass% (approximately 80% of theoretical hydrogen capacity in $K_2B_{12}H_{12}$) when heated up to 700 °C. The dehydrogenation proceeds with multistep reactions, as shown in TG and MS results.

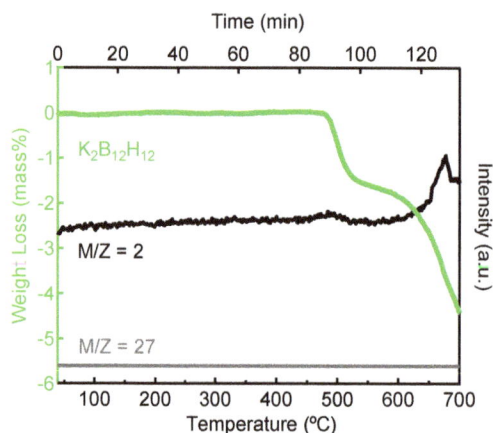

Figure 7. TG curve and MS signals of anhydrous $K_2B_{12}H_{12}$ (mass numbers 2 and 27 represent H_2 and B_2H_6).

The changes of anhydrous $K_2B_{12}H_{12}$ heated up to respective temperatures and subsequently cooled down to room temperature examined by XRD, Raman and ^{11}B NMR are shown in Figures 8 and 9 respectively. When the temperature is increased to 475 °C, no obvious changes of diffraction peaks, Raman spectra and the major resonance at −15.4 ppm attributed to $K_2B_{12}H_{12}$ are seen. The resonance originated from residual KBH_4 (−38.2 ppm in solid-state and −35.6 ppm in solution-state ^{11}B NMR) disappears when the temperature is increased to 550 °C, whereas no obvious changes of diffraction peaks, Raman spectra and the major resonance attributed to $K_2B_{12}H_{12}$ are observed. This suggests that the weight loss bellow 550 °C is responsible for the dehydrogenation of residual KBH_4. When the temperature is increased to 625 °C, diffraction peaks, Raman spectra and the main resonance attributed to $K_2B_{12}H_{12}$ become weak slightly, indicating the partial decomposition of $K_2B_{12}H_{12}$ at 625 °C. It is worth noting that the initial thermal decomposition temperature increases with the order of $Li_2B_{12}H_{12} < Na_2B_{12}H_{12} < K_2B_{12}H_{12}$, which shows the same trend to the dehydrogenation temperature of corresponding metal borohydrides [33].

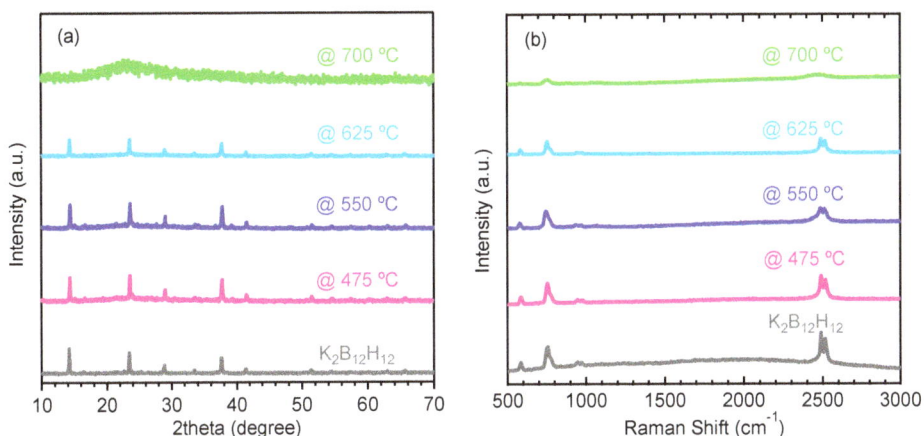

Figure 8. *Ex-situ* (**a**) XRD patterns and (**b**) Raman spectra of anhydrous $K_2B_{12}H_{12}$ as synthesized and heated up to respective temperatures.

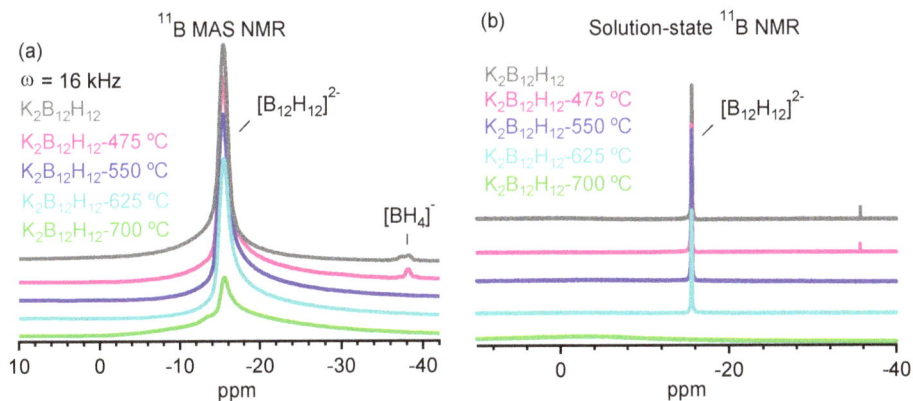

Figure 9. *Ex-situ* ^{11}B NMR spectra of anhydrous $K_2B_{12}H_{12}$ as synthesized and heated up to respective temperatures: (**a**) solid-state ^{11}B MAS NMR spectra and (**b**) solution-state ^{11}B NMR spectra measured in DMSO-d_6. Resonance assignments: −15.6 ppm $[B_{12}H_{12}]^{2-}$, −35.6 ppm $[BH_4]^-$ [32].

When the temperature is increased to 700 °C, the diffraction peaks and Raman spectra from $K_2B_{12}H_{12}$ are hardly observed, the main resonance at −15.4 ppm in ^{11}B MAS NMR decreases significantly without any change in the chemical shift and a broad resonance between −12.2 ppm and −14.2 ppm appears. It suggests that the major dehydrogenation of $K_2B_{12}H_{12}$ to $K_2B_{12}H_{12-x}$ and the polymerization of $K_2B_{12}H_{12-x}$ to $(K_2B_{12}H_z)_n$ polymers start to happen at 700 °C [21], similar to those of $Na_2B_{12}H_{12}$. Like $(Li_2B_{12}H_z)_n$ and $(Na_2B_{12}H_z)_n$ polymers, the produced $(K_2B_{12}H_z)_n$ polymers are also insoluble in water and DMSO. The decomposition behavior of anhydrous $K_2B_{12}H_{12}$ is different from that formed as a dehydrogenation intermediate of KBH_4 predicted by theoretical calculation [34], suggesting that the coexisting KH may facilitate the decomposition of $K_2B_{12}H_{12}$.

In summary, the thermal decomposition of anhydrous alkali metal dodecaborates $M_2B_{12}H_{12}$ (M = Li, Na, K) proceeds in two steps: (1) dehydrogenate to produce H-deficient $M_2B_{12}H_{12-x}$ containing the icosahedral B_{12} skeletons and (2) polymerization of $M_2B_{12}H_{12-x}$ to form $(M_2B_{12}H_z)_n$. Such behaviors are similar to those of anhydrous $MgB_{12}H_{12}$ and $CaB_{12}H_{12}$ [21], but fairly differ from those *in situ* formed during the dehydrogenation of $M(BH_4)_n$. These findings suggest that further investigations on the correlation between thermal decomposition behaviors of possible dehydrogenation intermediates and of the corresponding metal borohydrides are of great importance for the clarification of the dehydrogenation mechanism.

3. Experimental Section

Anhydrous $Li_2B_{12}H_{12}$, $Na_2B_{12}H_{12}$ and $K_2B_{12}H_{12}$ were synthesized by sintering of $B_{10}H_{14}$ with $LiBH_4$, $NaBH_4$ and KBH_4 (Sigma-Aldrich, Ichikawa, Japan), at 200–450 °C for 15–20 h [27]. All the synthesized anhydrous $Li_2B_{12}H_{12}$, $Na_2B_{12}H_{12}$ and $K_2B_{12}H_{12}$ were stored in glove box filled with purified Ar gas.

Powder XRD patterns were recorded by a Rigaku Ultima IV X-ray diffractometer with Cu-Kα radiation (Rigaku, Tokyo, Japan), and the accelerating voltage/tube current were set as 40 kV/40 mA. The sample powders were placed on a zero diffraction plate sealed by Scotch tape to prevent air exposure during the measurement. Raman spectra were obtained from a RAMAN-11 VIS-SS (Nanophoton, Osaka, Japan) using a green laser with 532 nm wavelength. Thermal decomposition was analyzed by TG (Rigaku), with a heating rate of 5 °C/min under a 200 mL/min flow of helium gas. The gas released during the TG measurement was analyzed by a quadrupole mass spectrometer coupled with TG. Solid-state MAS NMR spectra were recorded by a Bruker Ascend-600 spectrometer (Bruker, Yokohama, Japan), at room temperature. NMR sample preparations were always operated in a glove box filled with purified Ar gas and sample spinning was conducted using dry N_2 gas. ^{11}B MAS NMR spectra were obtained at excitation pulses of 6.5 μs ($\pi/2$ pulse) and with strong ^{1}H decoupling pulses. ^{11}B NMR chemical shifts were referenced to BF_3OEt_2 (δ = 0.00 ppm). ^{1}H MAS NMR spectra were obtained at excitation pulses of 6.5 μs ($\pi/2$ pulse) and the chemical shifts were referred to deuterated water (δ = 4.75 ppm). Solution-state ^{11}B NMR experiments were carried out using the same apparatus of Bruker Ascend-600 (Bruker), dimethyl sulfoxide (DMSO-d_6) was used as solvent and saturated $B(OH)_3$ aqueous solution at 19.4 ppm was used as external standard sample.

4. Conclusions

Systematic investigations of thermal decomposition indicate that anhydrous alkali metal dodecaborates $M_2B_{12}H_{12}$ (M = Li, Na, K) firstly dehydrogenate to produce the H-deficient $M_2B_{12}H_{12-x}$ containing the icosahedral B_{12} skeletons, followed by the polymerization of $M_2B_{12}H_{12-x}$ to form $(M_2B_{12}H_z)_n$ polymers, similar to those of anhydrous $MgB_{12}H_{12}$ and $CaB_{12}H_{12}$ [21]. No amorphous B was detected in all $M_2B_{12}H_{12}$ samples upon heating up to 700 °C, suggesting that higher temperature is needed for the complete decomposition of $(M_2B_{12}H_z)_n$. The initial thermal decomposition temperature increases with the order of $Li_2B_{12}H_{12}$ < $Na_2B_{12}H_{12}$ < $K_2B_{12}H_{12}$, which shows the same trend to the dehydrogenation temperature of corresponding borohydrides. The thermal decomposition behaviors of anhydrous $M_2B_{12}H_{12}$ are fairly different with those *in situ* formed during the dehydrogenation of corresponding metal borohydrides. Further investigations on the correlation between thermal decomposition of possible dehydrogenation intermediates and of the corresponding metal borohydrides are expected in order to clarify the exact dehydrogenation mechanism of metal borohydrides.

Acknowledgments: We would like to sincerely thank Miho Yamauchi and Motonori Watanabe in I2CNER for their great help on ^{11}B MAS NMR measurement. This study was partially supported by JSPS KAKENHI Grant No. 25709067 and the International Institute for Carbon Neutral Energy Research (WPI-I2CNER), sponsored by the Japanese Ministry of Education, Culture, Sports, Science and Technology.

Author Contributions: All of the authors contributed to this work. Liqing He and Hai-Wen Li designed and carried out the experiments. Liqing He, Hai-Wen Li and Etsuo Akiba analyzed the experimental results and wrote the manuscript.

Conflicts of Interest: The authors declare no conflict of interest.

References

1. Li, H.-W.; Yan, Y.; Orimo, S.-I.; Züttel, A.; Jensen, C.M. Recent progress in metal borohydrides for hydrogen storage. *Energies* **2011**, *4*, 185–214.

2. Rude, L.H.; Nielsen, T.K.; Ravnsbaek, D.B.; Bösenberg, U.; Ley, M.B.; Richter, B.; Arnbjerg, L.M.; Dornheim, M.; Filinchuk, Y.; Besenbacher, F.; *et al.* Tailoring properties of borohydrides for hydrogen storage: A review. *Phys. Status Solidi A* **2011**, *208*, 1754–1773.

3. Orimo, S.-I.; Nakamori, Y.; Ohba, N.; Miwa, K.; Aoki, M.; Towata, S.-I.; Zuttel, A. Experimental studies on intermediate compound of $LiBH_4$. *Appl. Phys. Lett.* **2006**, *89*.

4. Ohba, N.; Miwa, K.; Aoki, M.; Noritake, T.; Towata, S.-I.; Nakamori, Y.; Orimo, S.-I.; Züttel, A. First-principles study on the stability of intermediate compounds of $LiBH_4$. *Phys. Rev. B* **2006**, *74*.

5. Li, H.-W.; Kikuchi, K.; Nakamori, Y.; Ohba, N.; Miwa, K.; Towata, S.; Orimo, S. Dehydriding and rehydriding processes of well-crystallized $Mg(BH_4)_2$ accompanying with formation of intermediate compounds. *Acta Mater.* **2008**, *56*, 1342–1347.

6. Hwang, S.-J.; Bowman, R.C.; Reiter, J.W.; Rijssenbeek, J.; Soloveichik, G.L.; Zhao, J.-C.; Kabbour, H.; Ahn, C.C. NMR confirmation for formation of $[B_{12}H_{12}]^{2-}$ complexes during hydrogen desorption from metal borohydrides. *J. Phys. Chem. C* **2008**, *112*, 3164–3169.

7. Her, J.-H.; Yousufuddin, M.; Zhou, W.; Jalisatgi, S.S.; Kulleck, J.G.; Zan, J.A.; Hwang, S.-J.; Bowman, R.C., Jr.; Udovic, T.J. Crystal structure of $Li_2B_{12}H_{12}$: A possible intermediate species in the decomposition of $LiBH_4$. *Inorg. Chem.* **2008**, *47*, 9757–9759.

8. Ozolins, V.; Majzoub, E.; Wolverton, C. First-Principles Prediction of Thermodynamically reversible hydrogen storage reactions in the Li-Mg-Ca-B-H system. *J. Am. Chem. Soc.* **2009**, *131*, 230–237.

9. Li, H.-W.; Miwa, K.; Ohba, N.; Fujita, T.; Sato, T.; Yan, Y.; Towata, S.; Chen, M.; Orimo, S. Formation of an intermediate compound with a $B_{12}H_{12}$ cluster: Experimental and theoretical studies on magnesium borohydride $Mg(BH_4)_2$. *Nanotechnology* **2009**, *20*.

10. Stavila, V.; Her, J.-H.; Zhou, W.; Hwang, S.-J.; Kim, C.; Ottley, L.A.M.; Udovic, T.J. Probing the structure, stability and hydrogen storage properties of calcium dodecahydro-*closo*-dodecaborate. *J. Solid State Chem.* **2010**, *183*, 1133–1140.

11. Yan, Y.; Li, H.-W.; Maekawa, H.; Aoki, M.; Noritake, T.; Matsumoto, M.; Miwa, K.; Towata, S.-I.; Orimo, S.-I. Formation Process of $[B_{12}H_{12}]^{2-}$ from $[BH_4]^-$ during the Dehydrogenation Reaction of $Mg(BH_4)_2$. *Mater. Trans.* **2011**, *52*, 1443–1446.

12. Bonatto Minella, C.; Garroni, S.; Olid, D.; Teixidor, F.; Pistidda, C.; Lindemann, I.; Gutfleisch, O.; Baró, M.D.; Bormann, R.; Klassen, T.; Dornheim, M. Experimental evidence of $Ca[B_{12}H_{12}]$ formation during decomposition of a $Ca(BH_4)_2 + MgH_2$ based reactive hydride composite. *J. Phys. Chem. C* **2011**, *115*, 18010–18014.

13. Garroni, S.; Milanese, C.; Pottmaier, D.; Mulas, G.; Nolis, P.; Girella, A.; Caputo, R.; Olid, D.; Teixdor, F.; Baricco, M. Experimental evidence of $Na_2[B_{12}H_{12}]$ and Na formation in the desorption pathway of the $2NaBH_4 + MgH_2$ system. *J. Phys. Chem.C* **2011**, *115*, 16664–16671.

14. Yan, Y.; Li, H.-W.; Maekawa, H.; Miwa, K.; Towata, S.; Orimo, S. Formation of intermediate compound $Li_2B_{12}H_{12}$ during the dehydrogenation process of the $LiBH_4$-MgH_2 system. *J. Phys. Chem. C* **2011**, *115*, 19419–19423.

15. Yan, Y.; Remhof, A.; Hwang, S.-J.; Li, H.-W.; Mauron, P.; Orimo, S.-I.; Züttel, A. Pressure and temperature dependence of the decomposition pathway of $LiBH_4$. *Phys. Chem. Chem. Phys.* **2012**, *14*, 6514–6519.

16. Pitt, M.P.; Paskevicius, M.; Brown, D.H.; Sheppard, D.A.; Buckley, C.E. Thermal Stability of $Li_2B_{12}H_{12}$ and Its Role in the Decomposition of $LiBH_4$. *J. Am. Chem. Soc.* **2013**, *135*, 6930–6941.

17. Yan, Y.; Remhof, A.; Rentsch, D.; Züttel, A. The role of $MgB_{12}H_{12}$ in the hydrogen desorption process of $Mg(BH_4)_2$. *Chem. Commun.* **2015**, *51*, 700–702.

18. Xia, G.; Meng, Q.; Guo, Z.; Gu, Q.; Liu, H.; Liu, Z.; Yu, X. Nanoconfinement significantly improves the thermodynamics and kinetics of co-infiltrated 2LiBH$_4$-LiAlH$_4$ composites: Stable reversibility of hydrogen absorption/resorption. *Acta Mater.* **2013**, *61*, 6882–6893.

19. Borgschulte, A.; Callini, E.; Probst, B.; Jain, A.; Kato, S.; Friedrichs, O.; Remhof, A.; Bielmann, M.; Ramirez-Cuesta, A.; Züttel, A. Impurity gas analysis of the decomposition of complex hydrides. *J. Phys. Chem. C* **2011**, *115*, 17220–17226.

20. Stadie, N.P.; Callini, E.; Richter, B.; Jensen, T.R.; Borgschulte, A.; Züttel, A. Supercritical N$_2$ Processing as a Route to the Clean Dehydrogenation of Porous Mg(BH$_4$)$_2$. *J. Am. Chem. Soc.* **2014**, *136*, 8181–8184.

21. He, L.; Li, H.-W.; Tumanov, N.; Filinchuk, Y.; Akiba, E. Facile synthesis of anhydrous alkaline earth metal dodecaborates MB$_{12}$H$_{12}$ (M = Mg, Ca) from M(BH$_4$)$_2$. *Dalton Trans.* **2015**, *44*, 15882–15887.

22. Kim, Y.; Hwang, S.-J.; Shim, J.-H.; Lee, Y.-S.; Han, H.N.; Cho, Y.W. Investigation of the Dehydrogenation Reaction Pathway of Ca(BH$_4$)$_2$ and Reversibility of Intermediate Phases. *J. Phys. Chem. C* **2012**, *116*, 4330–4334.

23. Li, H.-W.; Akiba, E.; Orimo, S.-I. Comparative study on the reversibility of pure metal borohydrides. *J. Alloy. Compd.* **2013**, *580*, S292–S295.

24. Minella, C.B.; Pistidda, C.; Garroni, S.; Nolis, P.; Baró, M.D.; Gutfleisch, O.; Klassen, T.; Bormann, R.; Dornheim, M. Ca(BH$_4$)$_2$ + MgH$_2$: Desorption reaction and role of Mg on its reversibility. *J. Phys. Chem. C* **2013**, *117*, 3846–3852.

25. Sivaev, I.B.; Bregadze, V.I.; Sjöberg, S. Chemistry of *closo*-Dodecaborate Anion [B$_{12}$H$_{12}$]$^{2-}$: A Review. *Collect. Czechoslov. Chem. Commun.* **2002**, *67*, 679–727.

26. Chen, X.; Liu, Y.-H.; Alexander, A.-M.; Gallucci, J.C.; Hwang, S.-J.; Lingam, H.K.; Huang, Z.; Wang, C.; Li, H.; Zhao, Q.; *et al.* Desolvation and dehydrogenation of solvated magnesium salts of dodecahydrododecaborate: Relationship between structure and thermal decomposition. *Chem. A Eur. J.* **2014**, *20*, 7325–7333.

27. He, L.; Li, H.-W.; Hwang, S.-J.; Akiba, E. Facile Solvent-Free Synthesis of Anhydrous Alkali Metal Dodecaborate M$_2$B$_{12}$H$_{12}$ (M= Li, Na, K). *J. Phys. Chem. C* **2014**, *118*, 6084–6089.

28. He, L.; Li, H.-W.; Nakajima, H.; Tumanov, N.; Filinchuk, Y.; Hwang, S.-J.; Sharma, M.; Hagemann, H.; Akiba, E. Synthesis of a bimetallic dodecaborate LiNaB$_{12}$H$_{12}$ with outstanding superionic conductivity. *Chem. Mater.* **2015**, *27*, 5483–5486.

29. Hwang, S.-J.; Bowman Jr, R.C.; Kim, C.; Zan, J.A.; Reiter, J.W. Solid State NMR Characterization of Complex Metal Hydrides systems for Hydrogen Storage Applications. *J. Anal. Sci. Technol.* **2011**, *2*, A159–A162.

30. Heřmánek, S. B NMR spectra of boranes, main-group heteroboranes, and substituted derivatives: Factors influencing chemical shifts of skeletal atoms. *Chem. Rev.* **1992**, *92*, 325–362.

31. Verdal, N.; Her, J.-H.; Stavila, V.; Soloninin, A.V.; Babanova, O.A.; Skripov, A.V.; Udovic, T.J.; Rush, J.J. Complex high-temperature phase transitions in Li$_2$B$_{12}$H$_{12}$ and Na$_2$B$_{12}$H$_{12}$. *J. Solid State Chem.* **2014**, *212*, 81–91.

32. Remhof, A.; Yan, Y.; Rentsch, D.; Borgschulte, A.; Jensen, C.M.; Züttel, A. Solvent-free synthesis and stability of $MgB_{12}H_{12}$. *J. Mater. Chem. A* **2014**, *2*, 7244–7249.

33. Nakamori, Y.; Miwa, K.; Ninomiya, A.; Li, H.; Ohba, N.; Towata, S.-I.; Züttel, A.; Orimo, S.-I. Correlation between thermodynamical stabilities of metal borohydrides and cation electronegativites: First-principles calculations and experiments. *Phys. Rev. B* **2006**, *74*.

34. Guo, Y.; Jia, J.; Wang, X.-H.; Ren, Y.; Wu, H. Prediction of thermodynamically reversible hydrogen storage reactions in the KBH_4/M (M= Li, Na, Ca)$(BH_4)_n$ (n = 1, 2) system from first-principles calculation. *Chem. Phys.* **2013**, *418*, 22–27.

Efficient Synthesis of an Aluminum Amidoborane Ammoniate

Junzhi Yang, Paul R. Beaumont, Terry D. Humphries, Craig M. Jensen and Xingguo Li

Abstract: A novel species of metal amidoborane ammoniate, $[Al(NH_2BH_3)_6{}^{3-}]$ $[Al(NH_3)_6{}^{3+}]$ has been successfully synthesized in up to 95% via the one-step reaction of $AlH_3 \cdot OEt_2$ with liquid $NH_3BH_3 \cdot nNH_3$ (n = 1~6) at 0 °C. This solution based reaction method provides an alternative pathway to the traditional mechano-chemical ball milling methods, avoiding possible decomposition. MAS ^{27}Al NMR spectroscopy confirms the formulation of the compound as an $Al(NH_2BH_3)_6{}^{3-}$ complex anion and an $Al(NH_3)_6{}^{3+}$ cation. Initial dehydrogenation studies of this aluminum based M-N-B-H compound demonstrate that hydrogen is released at temperatures as low as 65 °C, totaling ~8.6 equivalents of H_2 (10.3 wt %) upon heating to 105 °C. This method of synthesis offers a promising route towards the large scale production of metal amidoborane ammoniate moieties.

Reprinted from *Energies*. Cite as: Yang, J.; Beaumont, P.R.; Humphries, T.D.; Jensen, C.M.; Li, X. Efficient Synthesis of an Aluminum Amidoborane Ammoniate. *Energies* **2015**, *8*, 9107–9116.

1. Introduction

A critical challenge facing the advancement of hydrogen fuel cells for automotive applications is the development of safe and energy efficient hydrogen storage materials. Metal amidoboranes (MNH_2BH_3, MAB) and metal borohydride ammoniates ($MBH_4 \cdot nNH_3$, MBA) are currently among the most promising candidate materials [1–6]. Recent demonstration of the regeneration of ammonia borane derivatives using hydrazine in liquid ammonia point to the feasibility of off-board reversibility [7–9]. Substitution of one protic H atom in the $[NH_3]$ of NH_3BH_3 by a metal atom leads to the formation of MAB complexes.

Aluminum amidoborane ($Al(NH_2BH_3)_3$, AlAB), first synthesized by Hawthorne *et al.* [10], possesses one of the highest theoretical hydrogen capacities among MABs (12.9 wt % H), capable of releasing 6 wt % H_2 at 190 °C and approximately 8 wt % H_2 in the presence of an ionic liquid at lower temperatures [10]. As such, this material has already experienced intensive explorations, although up to now only a few reports have identified its existence owing to its poor stability and spontaneous H_2 loss caused by the chemically vulnerable Lewis-acidic Al^{3+} center [10–12]. The improved dehydrogenation properties of AlAB (Al χ_p = 1.5), relative to ammonia borane [10], makes the Al-N-B-H systems attractive, albeit difficult to synthesize. Recently, Guo *et al.* reported on the stability of $[Al(NH_3)_6](BH_4)_3$ in air, which differs quite significantly from the analogous volatile liquid $Al(BH_4)_3$ [13,14]. Strong N-H$^{\delta+}\cdots^{-\delta}$H-B dihydrogen bonds contribute to the stability of this compound resulting in its long term stability in air. Another recently reported Al based amidoborane complex includes $Li_3AlH_6 \cdot n(NH_2BH_3)$ which releases 9 wt % H_2 at a temperature of 130 °C [15].

A variety of B-N amidoborane ammoniates, have previously been synthesized by reacting MAB and NH_3, including $LiNH_2BH_3 \cdot NH_3$ [16], $Mg(NH_2BH_3)_2 \cdot 3NH_3$ [17], and $Ca(NH_2BH_3)_2 \cdot NH_3$ [18]. However, to the best of our knowledge, there has been no prior report of the synthesis of an aluminum analog. Herein we report the first synthesis of aluminum amidoborane ammoniate, $[Al(NH_2BH_3)_6{}^{3-}][Al(NH_3)_6{}^{3+}]$.

2. Results and Discussion

The synthesis of $[Al(NH_2BH_3)_6{}^{3-}][Al(NH_3)_6{}^{3+}]$ (according to Equations (1) and (2)) was achieved using a specially-designed polytetrafluoroethylene reactor, which allowed the reactants to be stirred at low temperature under ammonia pressure. Under these conditions, ammonia borane reversibly absorbs up to 6 equivalents of NH_3, forming liquid $NH_3BH_3 \cdot nNH_3$ (n = 1–6) complexes [19]. $AlH_3 \cdot OEt_2$, which is insoluble in $NH_3BH_3 \cdot nNH_3$, was utilized as a highly reactive Al source [20,21]. Immersing the Al source in ammonia borane ammoniate complex permits the selective uptake of ammonia in a one-step synthesis of $Al(NH_2BH_3)_3 \cdot 3NH_3$ as a solid precipitate that can be isolated in up to 95% purity (based on $AlH_3 \cdot OEt_2$) by filtration (Equations (1) and (2), details in Section 3, Experimental Section). It should be emphasized that this method avoids the high-energy impact generally encountered in traditional ball milling methods and further prevents possible decomposition of components. This synthesis strategy may also be effective for other amidoborane ammoniates.

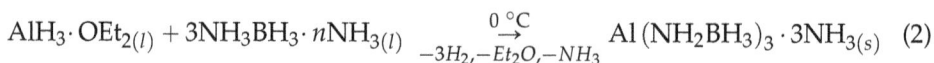

$$NH_3BH_{3(s)} + nNH_{3(g)} \underset{n=1\sim6}{\overset{0\,°C}{\rightleftharpoons}} NH_3BH_3 \cdot nNH_{3(l)} \tag{1}$$

$$AlH_3 \cdot OEt_{2(l)} + 3NH_3BH_3 \cdot nNH_{3(l)} \underset{-3H_2, -Et_2O, -NH_3}{\overset{0\,°C}{\rightarrow}} Al(NH_2BH_3)_3 \cdot 3NH_{3(s)} \tag{2}$$

Figure 1a illustrates the XRD pattern obtained for a sample of $Al(NH_2BH_3)_3 \cdot 3NH_3$ prepared via the method described above. The pattern does not index to any previously reported Al-N-B-H quaternary compound and contains at most, only very minor contributions from unreacted starting material. FTIR analysis of $Al(NH_2BH_3)_3 \cdot 3NH_3$ featured a N-B stretch at 875 cm^{-1} and peaks at 426 and 461 cm^{-1} which were assigned to Al-N lattice vibrations, (Figure S1,

Table S1). Attempts were also made to prepare $Al(NH_2BH_3)_3 \cdot 3NH_3$ using ball milling techniques. As shown in Figure 1c, no new species evolved from a mixture of $AlH_3 \cdot OEt_2 + 3NH_3BH_3$, which was ball milled under ammonia atmosphere at $0\,^\circ C$, at a speed of 150 rpm for at least 2 hours. Moreover, increasing the ball milling energy, such as higher rotational speed or temperature ($>40\,^\circ C$) during the synthesis causes dissociation of the ether adduct, which often leads to the production of γ-AlH_3 (Figure 1b). This alane polymorph incidentally shows much lower reactivity in liquid $NH_3BH_3 \cdot nNH_3$ than pure $AlH_3 \cdot OEt_2$, and inhibits the formation of an Al-N bond [22].

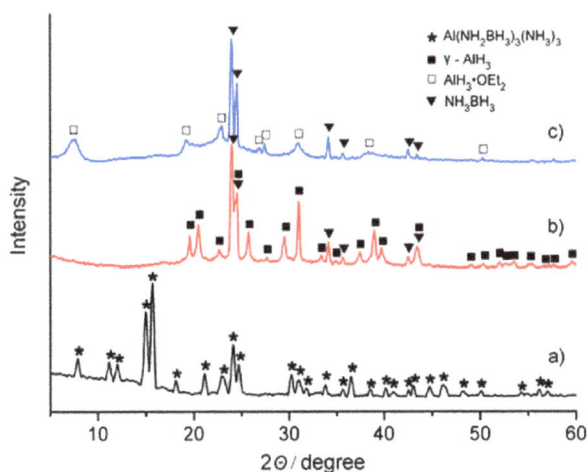

Figure 1. XRD patterns of (**a**) as-prepared $Al(NH_2BH_3)_3 \cdot 3NH_3$; (**b**) mixture of γ-$AlH_3 + 3NH_3BH_3$; (**c**) ball milled $AlH_3 \cdot OEt_2 + 3NH_3BH_3$. $\lambda = 1.5406$ Å.

The ^{27}Al MAS NMR spectrum (Figure 2b) verifies the formation of $Al(NH_2BH_3)_3 \cdot 3NH_3$ and provides key information about its molecular structure. After reaction, only traces of the characteristic resonance for $AlH_3 \cdot OEt_2$ at 109.9 ppm are observed [23]. Two major resonances at 65.5 ppm and 33.6 ppm dominate the spectrum, clearly indicating that $Al(NH_2BH_3)_3 \cdot 3NH_3$ contains equal amounts of aluminum in two very different coordination environments. The MAS ^{11}B NMR spectrum of the product contains a major resonance for $Al(NH_2BH_3)_3 \cdot 3NH_3$ at 19.6 ppm and a and a minor resonance at -38 ppm which is due the presence of $[(NH_3)_2BH_2^+][BH_4^-]$ DADB or a related decomposition product that was also observed in the starting material [24]. As seen in Figure 2a, the ^{11}B chemical shift of $Al(NH_2BH_3)_3 \cdot 3NH_3$ is 2.7 ppm upfield from the 22.3 ppm shift observed for $NH_3BH_3 \cdot nNH_3$. Similar upfield shifts have been observed for other metal amidoboranes and hence this observation confirms the substitution of an H atom by an Al atom in the ammonia borane molecule [25,26].

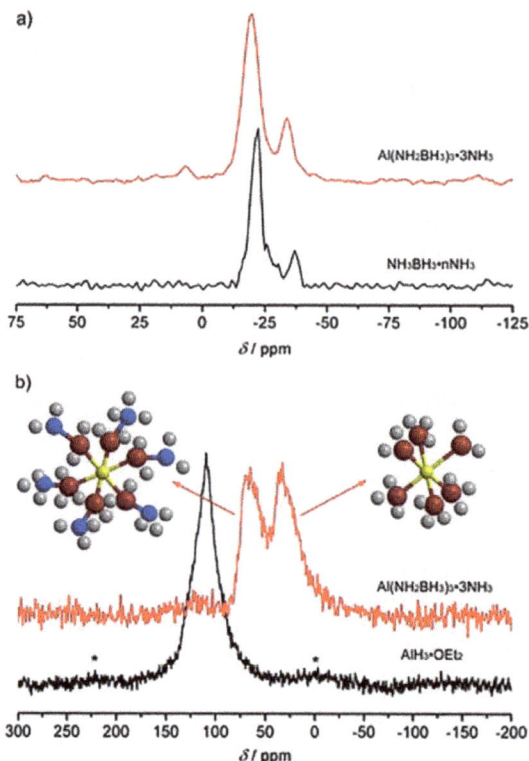

Figure 2. (a) ^{11}B MAS NMR spectra of $[\mathrm{Al(NH_2BH_3)_6}^{3-}][\mathrm{Al(NH_3)_6}^{3+}]$ and $\mathrm{NH_3BH_3 \cdot} n\mathrm{NH_3}$; (b) ^{27}Al MAS NMR spectra of $[\mathrm{Al(NH_2BH_3)_6}^{3-}][\mathrm{Al(NH_3)_6}^{3+}]$ and $\mathrm{AlH_3 \cdot OEt_2}$. The molecular structure of the octahedral Al complexes are also depicted (yellow balls represent Al, red for N, blue for B, and grey for H).

Al^{3+} generally has either tetrahedral or octahedral coordination. Thus *a priori* there are four possible coordination geometries for $\mathrm{Al(NH_2BH_3)_3 \cdot 3NH_3}$: 1) Al coordinates three $(\mathrm{NH_2BH_3})^-$ anions and three ammonia molecules to give a neutral $\mathrm{Al(NH_3)_3(NH_2BH_3)_3}$ complex; 2) Al coordinates octahedrally with only ammonia giving a hexamminealuminum cation [27], $\mathrm{Al(NH_3)_6}^{3+}$ and leaving three free $(\mathrm{NH_2BH_3})^-$ anions; 3) Al coordinates tetrahedrally with $(\mathrm{NH_2BH_3})^-$ anions, forming an $\mathrm{Al(NH_2BH_3)_4}^-$ anion and three of these anions pair with one $\mathrm{Al(NH_3)_6}^{3+}$; and 4) Al octahedrally coordinates with $(\mathrm{NH_2BH_3})^-$ anions giving a $\mathrm{Al(NH_2BH_3)_6}^{3-}$ complex anion and ion pairs with the $\mathrm{Al(NH_3)_6}^{3+}$ cation. The observation of two peaks with equal intensity in the ^{27}Al MAS NMR spectrum is consistent with only the $[\mathrm{Al(NH_2BH_3)_6}^{3-}][\mathrm{Al(NH_3)_6}^{3+}]$ formulation and as such the ^{27}Al NMR resonances are assigned as follows: $\mathrm{Al(NH_2BH_3)_6}^{3-}$ at 65.5 ppm and $\mathrm{Al(NH_3)_6}^{3+}$ at 33.6 ppm (Figure 2b). This is quite similar to the reported structure of $\mathrm{Mg(NH_2BH_3)_2 \cdot 3NH_3}$

80

where Mg^{2+} exhibits both tetrahedral and octahedral coordination [17]. Elemental analysis (Table S2) also shows the ratio of Al:N:B is 1:6:3, and supports the $[Al(NH_2BH_3)_6{}^{3-}][Al(NH_3)_6{}^{3+}]$ formulation.

No apparent reaction was observed after exposure of a sample of $[Al(NH_2BH_3)_6{}^{3-}][Al(NH_3)_6{}^{3+}]$ to dry air for 3 days (Figure S2). The time-programmed-desorption/mass spectroscopy (TPD/MS) results reveal that the thermal decomposition of $[Al(NH_2BH_3)_6{}^{3-}][Al(NH_3)_6{}^{3+}]$ occurs in the temperature range of 65–180 °C, with the release corresponding to 7.5 wt % (Figure 3a,c). The desorbed gaseous product comprises of both H_2 and NH_3.

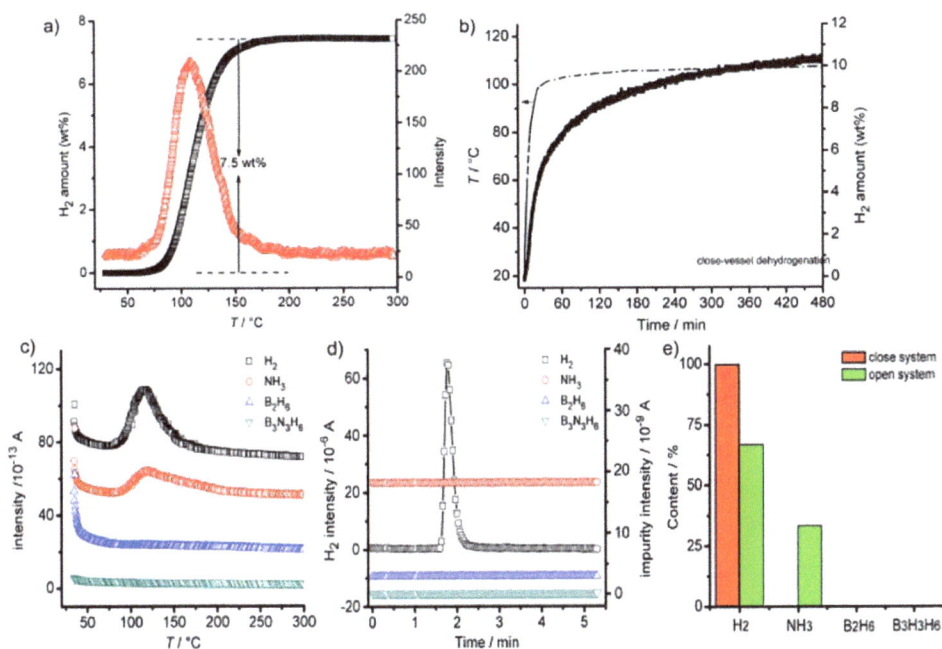

Figure 3. (a) TPD (Δ) and gas release ($\square$) profiles of $[Al(NH_2BH_3)_6{}^{3-}][Al(NH_3)_6{}^{3+}]$ at a heating rate of 5 °C min^{-1} under argon flow; (b) Isothermal desorption of $[Al(NH_2BH_3)_6{}^{3-}][Al(NH_3)_6{}^{3+}]$ in a closed vessel; the temperature ramping shown by dash dot line; (c) MS signals in (a): $\square$ H_2, $\bigcirc$ NH_3, $\triangle$ B_2H_6, $\triangledown$ $B_3N_3H_6$; (d) MS signals in (b) measured using Calibration Injection Mode: $\square$ H_2, $\bigcirc$ NH_3, $\triangle$ B_2H_6, $\triangledown$ $B_3N_3H_6$; (e) H_2 purity comparison between different systems.

Figure 4 shows the N 1s XPS results of AlAB· 3NH$_3$ before and after thermal decomposition (experimental details described in Supplemental Information). The peaks at ~396.6, ~398.0 and ~399.6 eV are attributed to N-Al, N-B and NH_3, respectively. After decomposition, the evolution of NH_3 and the corresponding peak at ~189.8 eV in B 1s XPS (Figure S3) suggests the formation of an Al-N-B matrix. Combined with the remaining B-H vibrations in micro-FTIR (Figure S4, Table S3), the reaction under dynamic inert gas flow can be described by Equation (3). The anticipated borazine-derived structure is illustrated in Figure S5 representing $AlN_3B_3H_6$.

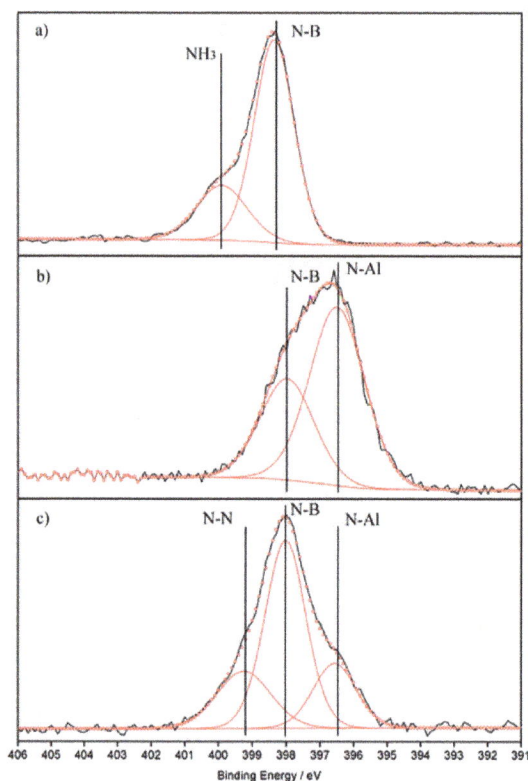

$$[Al(NH_2BH_3)_6{}^{3-}][Al(NH_3)_6{}^{3+}] \rightarrow 2AlN_3B_3H_3 + 6NH_3 + 12H_2 \qquad (3)$$

Figure 4. The N 1s XPS results of $Al(NH_2BH_3)_3 \cdot 3NH_3$ (**1**) before (**a**) and after thermal decomposition in an open system (**b**) and in a closed system (**c**). The experiment data are in black, while the fitted ones are in red.

The isothermal desorption in a closed vessel was examined using a Sieverts method at 105 °C (experimental details described in Supplemental Information). The gas evolved is calculated to be 10.3 wt % (Figure 3b), while only 0.05% NH_3 is detectable (Figure 3d). Obviously, the mass difference (Figure 3e) indicates that a significant partial pressure of NH_3 in a closed system suppresses further NH_3 desorption. This phenomenon is in accordance with the decomposition pathways of other metal amidoborane ammoniates [17,28,29]. Element analysis shows that the composition of Al, N, B and H are 15.83%, 48.18%, 18.36% and 3.83%, respectively, indicating an empirical formula of $AlN_6B_3H_{6.5}{}^*$. Similarly, N-H or B-H vibrations are not observed in the micro-FTIR spectrum, while Al-N stretching vibrations and weak H wagging vibrations are apparent (Figure S4 and Table S3). Meanwhile, the peaks at 1367 and 1627 cm^{-1} are typical of N-B stretching in h-BN [30]. The N 1s XPS peak ~396.4 eV (Figure 4c) and is attributed to the formation of an Al-N bond, while the two overlapped B 1s XPS peaks in Figure S3 suggests that the decomposed product comprises of not only [AlNBH] but also another B moiety. On the other hand, the ^{11}B MAS NMR spectrum presents at least two overlapping resonances at 6.3 ppm and 18.3 ppm (Figure S6), which is due to the second-order quadrupolar interaction. Thus, the B atoms are likely in a BN_3 or BN_2H environment [31,32]. The N 1s peak at ~399.1 eV is possibly a N-N bond. Clearly, the decomposition mechanism of $[Al(NH_2BH_3)_6{}^{3-}][Al(NH_3)_6{}^{3+}]$ in a closed system is much more complicated than that of an open system. On the basis of 8.6 equivalents of H_2, the dehydrogenation process can be briefly described by Equation (4).

$$[Al(NH_2BH_3)_6{}^{3-}][Al(NH_3)_6{}^{3+}] \rightarrow AlN_6B_3H_{6.5}{}^* + (8.5\sim9.0)H_2 \qquad (4)$$

3. Experimental Section

All starting materials, LiAlH$_4$ 99% (Sigma-Aldrich, Shanghai, China), AlCl$_3$ 99.99%, NH$_3$BH$_3$ 99% (Sigma-Aldrich), and NH$_3$ (Alfa Aesar, Shanghai, China), were obtained commercially and used without further purification. All manipulations were carried out under inert atmosphere conditions, either in an argon-filled glovebox or using standard Schlenk line techniques under a nitrogen atmosphere. The organometallic synthesis of AlH$_3 \cdot$ Et$_2$O is a chemically simple process, but a brief summary is presented. Generally, AlCl$_3$ was reacted with LiAlH$_4$ in diethyl ether with the LiCl precipitate being removed by filtration [21,33]. The excess diethyl ether was then removed under dynamic vacuum. AlH$_3 \cdot$ Et$_2$O was ground in a mortar with excess NH$_3$BH$_3$ and then sealed in a self-designed polytetrafluoroethylene (PTFE) reactor. The reactor was attached to the gas/vacuum manifold and rapidly evacuated/backfilled with 0.3–0.5 MPa NH$_3$. The system was cooled to $-70\,°C$ using acetone and dry ice, and gradually warmed to $0\,°C$ in an ice bath. At this temperature and under the NH$_3$ atmosphere, ammonia borane reversibly absorbed up to at least 6 equivalents of NH$_3$, forming liquid NH$_3$BH$_3 \cdot n$NH$_3$ (n = 1–6) complexes. AlH$_3 \cdot$ OEt$_2$ was dissolved in liquid NH$_3$BH$_3 \cdot n$NH$_3$, and the solution stirred for 2 h until the reaction was complete. The internal temperature and pressure of the reactor and manifold were recorded for the duration of the experiment. The ammonia and reaction produced hydrogen were then removed *in vacuo* at room temperature. Anhydrous diethyl ether was then added to the remaining products, thereby dissolving the excess NH$_3$BH$_3$ of which was removed by filtration. The residual solid $[Al(NH_2BH_3)_6{}^{3-}][Al(NH_3)_6{}^{3+}]$ was then heated to 45 °C for 12h to remove residual solvent to yield a solid white powder.

4. Conclusions

In summary, a novel aluminum amidoborane ammoniate, $[Al(NH_2BH_3)_6{}^{3-}]$ $[Al(NH_3)_6{}^{3+}]$, has been successfully synthesized. A reaction vessel has been designed that allows a one-step synthesis from the reaction of $AlH_3 \cdot OEt_2$ with liquid $NH_3BH_3 \cdot nNH_3$ (n = 1~6) at 0 °C. MAS ^{27}Al NMR spectroscopy confirms the formulation of the compound as an $Al(NH_2BH_3)_6{}^{3-}$ complex anion and a $Al(NH_3)_6{}^{3+}$ cation. This aluminum based M-N-B-H compound begins to release hydrogen at 65 °C, amounting to ~8.6 equivalents of H_2 (10.3 wt %) upon heating to 105 °C. This method of synthesis offers a promising route towards the large scale production of metal amidoborane ammoniate moieties.

Supplementary Materials: Supplementary materials can be accessed at: http://www.mdpi.com/1996-1073/8/9/9107/s1.

Acknowledgments: The authors would like to acknowledge the financial support given by the MOST of China (No. 2010CB631301, 2009CB939902 and 2012CBA01207) and the NSFC (No. U1201241 and 51071003).

Conflicts of Interest: The authors declare no conflict of interest.

References

1. Staubitz, A.; Robertson, A.P.M.; Manners, I. Ammonia-borane and related compounds as dihydrogen sources. *Chem. Rev.* **2010**, *110*, 4079–4124.
2. Chua, Y.S.; Chen, P.; Wu, G.; Xiong, Z. Development of amidoboranes for hydrogen storage. *Chem. Commun.* **2011**, *47*, 5116–5129.
3. Zhang, Y.S.; Wolverton, C. Crystal structures, phase stabilities, and hydrogen storage properties of metal amidoboranes. *J. Phys. Chem. C* **2012**, *116*, 14224–14231.
4. Zheng, X.L.; Wu, G.T.; Li, W.; Xiong, Z.T.; He, T.; Guo, J.P.; Chen, H.; Chen, P. Releasing 17.8 wt% H_2 from lithium borohydride ammoniate. *Energy Environ. Sci.* **2011**, *4*, 3593–3600.
5. Guo, Y.H.; Wu, H.; Zhou, W.; Yu, X.B. Dehydrogenation tuning of ammine borohydrides using double-metal cations. *J. Am. Chem. Soc.* **2011**, *133*, 4690–4693.
6. Jepsen, L.H.; Ley, M.B.; Lee, Y.-S.; Cho, Y.W.; Dornheim, M.; Jensen, J.O.; Filinchuk, Y.; Jørgensen, J.E.; Besenbacher, F.; Jensen, T.R.; *et al.* Boron-nitrogen based hydrides and reactive composites for hydrogen storage. *Mater. Today* **2014**, *17*, 129–135.
7. Sutton, A.D.; Burrell, A.K.; Dixon, D.A.; Garner, E.B.; Gordon, J.C.; Nakagawa, T.; Ott, K.C.; Robinson, P.; Vasiliu, M. Regeneration of ammonia borane spent fuel by direct reaction with hydrazine and liquid ammonia. *Science* **2011**, *331*, 1426–1429.
8. Tang, Z.W.; Tan, Y.B.; Chen, X.W.; Yu, X.B. Regenerable hydrogen storage in lithium amidoborane. *Chem. Commun.* **2012**, *48*, 9296–9298.

9. Li, S.; Tang, Z.; Gong, Q.; Yu, X.; Beaumont, P.R.; Jensen, C.M. Phenyl introduced ammonium borohydride: Synthesis and reversible dehydrogenation properties. *J. Mater. Chem.* **2012**, *22*, 21017–21023.

10. Hawthorne, M.F.; Jalisatgi, S.S.; Safronov, A.V.; Lee, H.B.; Wu, J. Chemical Hydrogen Storage Using Polyhedral Borane Anions and Aluminum-Ammonia-Borane Complexes. Available online: http://www.osti.gov/scitech//servlets/purl/990217-xUxbgx/ (accessed on 28 June 2015).

11. Harder, S.; Spielmann, J. Unprecedented reactivity of an aluminium hydride complex with $ArNH_2BH_3$: Nucleophilic substitution *versus* deprotonation. *Chem. Commun.* **2011**, *47*, 11945–11947.

12. Dou, D.; Ketchum, D.R.; Hamilton, E.J.M.; Florian, P.A.; Vermillion, K.E.; Grandinetti, P.J.; Shore, S.G. Reactions of aluminum hydride derivatives with ammonia-borane: A new approach toward AlN/BN materials. *Chem. Mater.* **1996**, *8*, 2839–2842.

13. Guo, Y.; Yu, X.; Sun, W.; Sun, D.; Yang, W. The hydrogen-enriched Al-B-N system as an advanced solid hydrogen-storage candidate. *Angew. Chem. Int. Ed.* **2011**, *50*, 1087–1091.

14. Guo, Y.H.; Jiang, Y.X.; Xia, G.L.; Yu, X.B. Ammine aluminium borohydrides: An appealing system releasing over 12 wt% pure H_2 under moderate temperature. *Chem. Commun.* **2012**, *48*, 4408–4410.

15. Xia, G.; Tan, Y.; Chen, X.; Guo, Z.; Liu, H.; Yu, X. Mixed-metal (Li, Al) amidoborane: Synthesis and enhanced hydrogen storage properties. *J. Mater. Chem. A* **2013**, *1*, 1810–1820.

16. Xia, G.L.; Yu, X.B.; Guo, Y.H.; Wu, Z.; Yang, C.Z.; Liu, H.K.; Dou, S.X. Ammine lithium amidoborane $Li(NH_3)NH_2BH_3$: A new coordination compound with favorable dehydrogenation characteristics. *Chem. Eur. J.* **2010**, *16*, 3763–3769.

17. Kang, X.D.; Wu, H.; Luo, J.H.; Zhou, W.; Wang, P. A simple and efficient approach to synthesize amidoborane ammoniates: Case study for $Mg(NH_2BH_3)_2(NH_3)_3$ with unusual coordination structure. *J. Mater. Chem.* **2012**, *22*, 13174–13179.

18. Chua, Y.S.; Wu, H.; Zhou, W.; Udovic, T.J.; Wu, G.T.; Xiong, Z.T.; Wong, M.W.; Chen, P. Monoammoniate of calcium amidoborane: Synthesis, structure, and hydrogen-storage properties. *Inorg. Chem.* **2012**, *51*, 1599–1603.

19. Gao, L.; Fang, H.C.; Li, Z.H.; Yu, X.B.; Fan, K.N. Liquefaction of solid-state BH_3NH_3 by gaseous NH_3. *Inorg. Chem.* **2011**, *50*, 4301–4306.

20. Graetz, J.; Chaudhuri, S.; Wegrzyn, J.; Celebi, Y.; Johnson, J.R.; Zhou, W.; Reilly, J.J. Direct and reversible synthesis of AlH_3-triethylenediamine from Al and H_2. *J. Phys. Chem. C* **2007**, *111*, 19148–19152.

21. Brower, F.M.; Matzek, N.E.; Reigler, P.F.; Rinn, H.W.; Roberts, C.B.; Schmidt, D.L.; Snover, J.A.; Terada, K. Preparation and properties of aluminum-hydride. *J. Am. Chem. Soc.* **1976**, *98*, 2450–2453.

22. Giannasi, A.; Colognesi, D.; Fichtner, M.; Rohm, E.; Ulivi, L.; Ziparo, C.; Zoppi, M. Temperature behavior of the AlH_3 polymorph by *in situ* investigation using high resolution raman scattering. *J. Phys. Chem. A* **2011**, *115*, 691–699.

23. Humphries, T.D.; Munroe, K.T.; Decken, A.; McGrady, G.S. Lewis base complexes of AlH$_3$: Prediction of preferred structure and stoichiometry. *Dalton Trans.* **2013**, *42*, 6965–6978.

24. Stowe, A.C.; Shaw, W.J.; Linehan, J.C.; Schmid, B.; Autrey, T. *In situ* solid state 11B MAS-NMR studies of the thermal decomposition of ammonia borane: Mechanistic studies of the hydrogen release pathways from a solid state hydrogen storage material. *Phys. Chem. Chem. Phys.* **2007**, *9*, 1831–1836.

25. Shimoda, K.; Zhang, Y.; Ichikawa, T.; Miyaoka, H.; Kojima, Y. Solid state NMR study on the thermal decomposition pathway of sodium amidoborane NaNH$_2$BH$_3$. *J. Mater. Chem.* **2011**, *21*, 2609–2615.

26. Shimoda, K.; Doi, K.; Nakagawa, T.; Zhan, Y.; Miyaoka, H.; Ichikawa, T.; Tansho, M.; Shimizu, T.; Burrell, A.K.; Kojima, Y. Comparative study of structural changes in NH$_3$BH$_3$, LiNH$_2$BH$_3$, and KNH$_2$BH$_3$ during dehydrogenation process. *J. Phys. Chem. C* **2012**, *116*, 5957–5964.

27. Semenenko, K.N.; Shilkin, S.P.; Polyakova, V.B. Vibrational spectra and structure of the di- and tetraammoniate of aluminum borohydride. *Bull. Acad. Sci. USSR Div. Chem. Sc.* **1978**, *27*, 859–864.

28. Chua, Y.S.; Wu, G.T.; Xiong, Z.T.; Karkamkar, A.; Guo, J.P.; Jian, M.X.; Wong, M.W.; Autrey, T.; Chen, P. Synthesis, structure and dehydrogenation of magnesium amidoborane monoammoniate. *Chem. Commun.* **2010**, *46*, 5752–5754.

29. Chua, Y.S.; Wu, G.T.; Xiong, Z.T.; He, T.; Chen, P. Calcium amidoborane ammoniate-synthesis, structure, and hydrogen storage properties. *Chem. Mater.* **2009**, *21*, 4899–4904.

30. Geick, R.; Perry, C.; Rupprecht, G. Normal modes in hexagonal boron nitride. *Phys. Rev.* **1966**, *146*, 543–547.

31. Marchetti, P.S.; Kwon, D.K.; Schmidt, W.R.; Interrante, L.V.; Maciel, G.E. High-field B11 magic angle spinning NMR characterization of boron nitrides. *Chem. Mater.* **1991**, *3*, 482–486.

32. Jeschke, G.; Hoffbauer, W.; Jansen, M. A comprehensive NMR study of cubic and hexagonal boron nitride. *Solid State Nucl. Magn. Reson.* **1998**, *12*, 1–7.

33. Humphries, T.D.; Munroe, K.T.; DeWinter, T.M.; Jensen, C.M.; McGrady, G.S. NMR spectroscopic and thermodynamic studies of the etherate and the α, α', and γ phases of AlH$_3$. *Int. J. Hydrog. Energy* **2013**, *38*, 4577–4586.

Melting Behavior and Thermolysis of $NaBH_4-Mg(BH_4)_2$ and $NaBH_4-Ca(BH_4)_2$ Composites

Morten B. Ley, Elsa Roedern, Peter M. M. Thygesen and Torben R. Jensen

Abstract: The physical properties and the hydrogen release of $NaBH_4-Mg(BH_4)_2$ and $NaBH_4-Ca(BH_4)_2$ composites are investigated using *in situ* synchrotron radiation powder X-ray diffraction, thermal analysis and temperature programmed photographic analysis. The composite, $xNaBH_4-(1-x)Mg(BH_4)_2$, $x = 0.4$ to 0.5, shows melting/frothing between 205 and 220 °C. However, the sample does not become a transparent molten phase. This behavior is similar to other alkali-alkaline earth metal borohydride composites. In the $xNaBH_4-(1-x)Ca(BH_4)_2$ system, eutectic melting is not observed. Interestingly, eutectic melting in metal borohydrides systems leads to partial thermolysis and hydrogen release at lower temperatures and the control of sample melting may open new routes for obtaining high-capacity hydrogen storage materials.

Reprinted from *Energies*. Cite as: Ley, M.B.; Roedern, E.; Thygesen, P.M.M.; Jensen, T.R. Melting Behavior and Thermolysis of $NaBH_4-Mg(BH_4)_2$ and $NaBH_4-Ca(BH_4)_2$ Composites. *Energies* **2015**, *8*, 2701–2713.

1. Introduction

In order to create a new sustainable energy economy, the storage of renewable energy is essential, e.g., directly as electricity in a battery or indirectly as hydrogen in a solid state metal hydride [1–4]. Metal borohydrides can store considerable amounts of energy as hydrogen in the solid state, but tend to exhibit poor thermodynamic and kinetic properties, which hamper their technological utilization [5,6]. In order to improve the properties for reversible solid-state hydrogen storage, continued research within energy storage materials science is required.

The structural flexibility observed for metal borohydrides is highlighted by magnesium borohydride with seven structurally different polymorphs: α-, β-, β'-, γ-, ε-, δ- and ζ-$Mg(BH_4)_2$ [7–12]. $Mg(BH_4)_2$ is among the more promising materials for hydrogen storage applications with a high gravimetric hydrogen content of 14.9 wt% H_2 and possible reformation from the decomposition products [13,14]. Calcium borohydride also exists in several structural polymorphs, α/α'-, β- and γ-$Ca(BH_4)_2$ [15,16]. $Ca(BH_4)_2$ ($\rho_m = 11.56$ wt% H_2) decomposes at 370 °C to CaH_2 and CaB_6, which can also be directly rehydrogenated to $Ca(BH_4)_2$ [17,18]. The finding of other bi- and tri-metallic borohydrides, like $K_2Mg(BH_4)_4$ or $LiKMg(BH_4)_5$,

87

illustrates the structural flexibility of complex metal hydrides [19–21], which can lead to unexpected properties, such as lithium ion conductivity, as observed in, e.g., $LiCe(BH_4)_3Cl$ [22,23]. Recently, a series of 30 new complex hydride perovskite-type materials with new photophysical, electronic and hydrogen storage properties was presented [24]. Combinations of metal borohydrides and metal hydrides in reactive hydride composite systems can influence thermodynamics and tune the gas release [25–27].

Composites of alkali and alkaline earth metal borohydrides may form eutectic mixtures with a lower melting point than the two individual components or any other composition of the two metal borohydrides [28]. $0.62LiBH_4–0.38NaBH_4$ melts at $T_m \sim 220\ °C$, as compared to pristine $LiBH_4$ at $T_m = 280\ °C$ and $NaBH_4$ at $T_m > 500\ °C$, and produces a uniform clear molten phase [28,29]. $0.725LiBH_4–0.275KBH_4$ has the lowest eutectic melting temperature at $T_m = 105\ °C$ (KBH_4, $T_m > 600\ °C$) [30]. The system $xLiBH_4–(1 - x)Mg(BH_4)$, $x = 0.5$ to 0.6, melts at $T_m \sim 180\ °C$ ($Mg(BH_4)_2$, $T_m > 280\ °C$) and shows improved thermodynamics and kinetics, as decomposition proceeds immediately after melting and releases 7 wt% H_2 already at $T = 270\ °C$ [11,31]. $0.68LiBH_4–0.32Ca(BH_4)_2$ has a eutectic melting temperature at $T_m = 200\ °C$ ($Ca(BH_4)_2$, $T_m = 370\ °C$), releases ~10 wt% H_2 at $T < 400\ °C$ and also shows partial reversibility with respect to hydrogen storage [28,32,33]. Interestingly, the systems composed of only alkali metal borohydrides produce transparent molten phases at their melting point, while mixtures of alkali and alkaline earth metal borohydrides showed frothing/bubbling at the melting point without producing a transparent molten phase [28]. These observations have prompted the present investigation of two other metal borohydride composites, $NaBH_4–Mg(BH_4)_2$ and $NaBH_4–Ca(BH_4)_2$.

2. Experimental Section

2.1. Synthesis

Magnesium borohydride, $\gamma-Mg(BH_4)_2$, was synthesized using a previously published method [8]. Sodium borohydride, $NaBH_4$ (Sigma-Aldrich, 98%), and calcium borohydride, $Ca(BH_4)_2$ (Sigma-Aldrich, 95%), were used as received. The samples $xNaBH_4–(1 - x)Mg(BH_4)_2$, $x = 0.1, 0.2, 0.3, 0.4, 0.5, 0.6, 0.7, 0.8$ and 0.9, and $xNaBH_4–(1 - x)Ca(BH_4)_2$, $x = 0.335, 0.375, 0.429, 0.445, 0.5$ and 0.665, were prepared by manual mixing using a mortar and pestle. The $0.5NaBH_4-0.5Mg(BH_4)_2$, $0.665NaBH_4–0.335Mg(BH_4)_2$ and $0.5NaBH_4–0.5Ca(BH_4)_2$ samples were prepared by ball-milling (BM) for 240 min and applying 2 min BM and 2-min pauses (120 repetitions) using a Fritsch Pulverisette 4 planetary mill under inert conditions (argon atmosphere) in 80-mL tungsten carbide containers with tungsten carbide balls (outer diameter (o.d.) 10 mm, sample to balls mass ratio 1:40, speed of main disk 200 rpm, speed of planetary disks 560 rpm). All preparation and manipulation of the samples were performed in a glove box with a circulation purifier maintained under an argon atmosphere with less than 1 ppm of O_2 and H_2O.

2.2. In Situ Time-Resolved Synchrotron Radiation Powder X-ray Diffraction

Synchrotron radiation powder X-ray diffraction (SR-PXD) data were collected at beamline I711 at the synchrotron MAX-II in the MAX IV laboratory Lund, Sweden, with a MAR165 CCD detector system, X-ray exposure time of 30 s and selected wavelengths of 0.999991 or 1.00355 Å [34,35]. The powdered sample was mounted in a sapphire (Al_2O_3) single-crystal tube (o.d. 1.09 mm, inner diameter (i.d.) 0.79 mm) in an argon-filled glove box $p(O_2, H_2O) < 1$ ppm. The temperature was controlled with a thermocouple placed in the sapphire tube 1 mm from the sample. All obtained raw images were calibrated against a standard NIST LaB_6 sample and transformed to 2D-powder patterns using the FIT2D program [36].

2.3. Thermal Analysis

All samples, including the reactants $Mg(BH_4)_2$ and $Ca(BH_4)_2$, were studied by simultaneous thermogravimetric analysis (TGA) and differential scanning calorimetry (DSC) using a PerkinElmer STA 6000 apparatus. Additionally, the $xNaBH_4$–$(1 - x)Mg(BH_4)_2$ samples were studied by mass spectrometry (MS) using a Hiden Analytical HPR-20 QMS sampling system. The samples (approximately 3 mg) were placed in an Al crucible and heated (5 °C/min) in an argon flow of 20 mL/min. The samples exhibit vigorous frothing above 400 °C, preventing further heating of the samples during the thermal analysis experiment.

2.4. Temperature Programmed Photographic Analysis

Temperature programmed photographic analysis (TPPA) was performed using a previously described setup [28]. Photographs were collected using a digital camera whilst heating the samples from RT to 400 °C ($\Delta T/\Delta t = 5$ °C/min). The samples (approximately 15 mg) were sealed under argon in a glass vial connected to an argon-filled balloon to maintain an inert atmosphere. A thermocouple was in contact with the sample within the glass vial to monitor the temperature during thermolysis. The glass vial was encased within an aluminum block with open viewing windows for photography, to provide near-uniform heating by rod heaters, interfaced with a temperature controller.

3. Results

3.1. Differential Scanning Calorimetry

Analysis of the melting in the samples is performed using DSC and TPPA from room temperature (RT) to 400 °C. The DSC data of the as-synthesized $Mg(BH_4)_2$

($x = 0$) reveal a single event with an onset temperature of 197 °C; see Figure 1. $NaBH_4$ is not affected by thermodynamic events below 400 °C [37]. The DSC data of the $xNaBH_4$–$(1 − x)Mg(BH_4)_2$ composites reveal two endothermic peaks with varying intensity and onset temperatures of 178 and 205 °C; see Figure 1 and Figure S1. The first event at 178 °C is most noticeably observed in the samples of $xNaBH_4$–$(1 − x)Mg(BH_4)_2$, $x = 0.1$ to 0.6; see Figure 1 and Figure S2. The second event at 205 °C is observed mainly in $xNaBH_4$–$(1 − x)Mg(BH_4)_2$, $x = 0.3$–0.6.

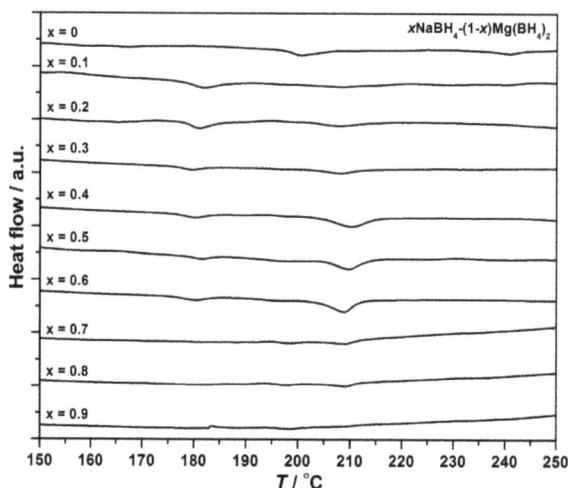

Figure 1. Normalized DSC curves of $Mg(BH_4)_2$ ($x = 0$) and $xNaBH_4$–$(1 − $ emphx$)Mg(BH_4)_2$, $x = 0.1$ to 0.9, in the temperature range of 150 to 250 °C, $\Delta T/\Delta t = 5$ °C/min.

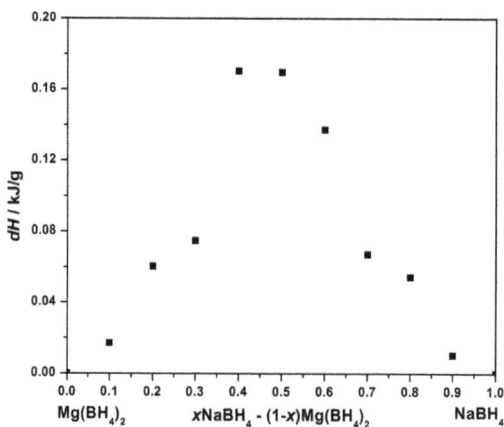

Figure 2. Integrated DSC signal in the temperature range 203 to 214 °C of the endothermic event per sample mass for $xNaBH_4$–$(1 − x)Mg(BH_4)_2$, $x = 0$ to 1.

The integrated area of the DSC peaks is proportional to the enthalpy change of the thermal events. The heat of reaction (dH) for the different sample compositions is extracted for the thermal events at 178 and 205 °C and shown in Figure S2 and Figure 2, respectively. The integrated area of the peak at 205 °C is largest for the samples $0.4NaBH_4$–$0.6Mg(BH_4)_2$ and $0.5NaBH_4$–$0.5Mg(BH_4)_2$.

The DSC data from the $xNaBH_4$–$(1-x)Ca(BH_4)_2$ composites are shown in the Supplementary Materials; see Figure S3. The DSC data for $Ca(BH_4)_2$ ($x = 0$) reveal an event at 345 °C. The other samples, $xNaBH_4$–$(1-x)Ca(BH_4)_2$, $x = 0.335, 0.375, 0.429,$ 0.429, 0.445 and 0.5, all have an endothermic peak at 280 °C followed by multiple thermal events above ~350 °C, which coincides with the single event observed for $Ca(BH_4)_2$. Furthermore, in $0.665NaBH_4$–$0.335Ca(BH_4)_2$, there are several small endothermic events above ~350 °C.

3.2. Temperature Programmed Photographic Analysis

In the samples $0.1NaBH_4$–$0.9Mg(BH_4)_2$ and $0.2NaBH_4$–$0.8Mg(BH_4)_2$, frothing/bubbling is observed above 280 °C; see Figure 3. In $0.4NaBH_4$–$0.6Mg(BH_4)_2$, $0.5NaBH_4$–$0.5Mg(BH_4)_2$ (BM) and $0.6NaBH_4$–$0.4Mg(BH_4)_2$, a decrease in the volume of the powder followed by melting/frothing are observed between 200 and 220 °C. Melting/frothing becomes more visible above 240 °C. The samples, $x = 0.8$–0.9, do not show visible changes below 400 °C, since these samples contain a majority of $NaBH_4$ with higher thermal stability. TPPA was conducted for two samples of $0.5NaBH_4$–$0.5Mg(BH_4)_2$ produced by ball milling and hand mixing, and the samples were found to behave similarly.

Figure 3. Temperature programmed photographic analysis (TPPA) sequence for (**a**) $0.2NaBH_4$-$0.8Mg(BH_4)_2$; (**b**) $0.4NaBH_4$-$0.6Mg(BH_4)_2$; (**c**) $0.5NaBH_4$-$0.5Mg(BH_4)_2$ (BM) and (**d**) $0.6NaBH_4$-$0.4Mg(BH_4)_2$, $0.8NaBH_4$-$0.2Mg(BH_4)_2$ at six selected temperatures between RT and 340 °C, $\Delta T / \Delta t = 5$ °C/min, Ar atmosphere.

The TPPA sequence for $0.5NaBH_4$-$0.5Ca(BH_4)_2$ is shown in Figure S4. A color change from white to light brown occurs in the sample above 290 °C. A similar behavior was observed for pure $Ca(BH_4)_2$ [28]. At 350 °C, the sample becomes partly molten. However, this also occurs for $Ca(BH_4)_2$ during thermolysis, and consequently, this is not interpreted as eutectic melting [28].

3.3. Thermogravimetric and Mass Spectrometry Analysis

The composites of $NaBH_4$ and $Mg(BH_4)_2$ are all destabilized and release hydrogen at lower temperatures as compared to $Mg(BH_4)_2$. The largest effect is observed for samples with $x = 0.4$–0.6; see Figure 4. In the samples with a majority of $NaBH_4$, $x = 0.8$–0.9, the amount of released hydrogen until 400 °C is minor, since $NaBH_4$ decomposes above 500 °C. The MS data show significant changes in the hydrogen release profile of the composites compared to $Mg(BH_4)_2$, in particular at $T < 300$ °C. The MS data for $xNaBH_4$–$(1 - x)Mg(BH_4)_2$, $x = 0.4$ and 0.6, show two hydrogen release reactions beginning at ~180 and 230 °C, which are not observed for $Mg(BH_4)_2$, where only a slight increase in the hydrogen signal is observed at 200 °C.

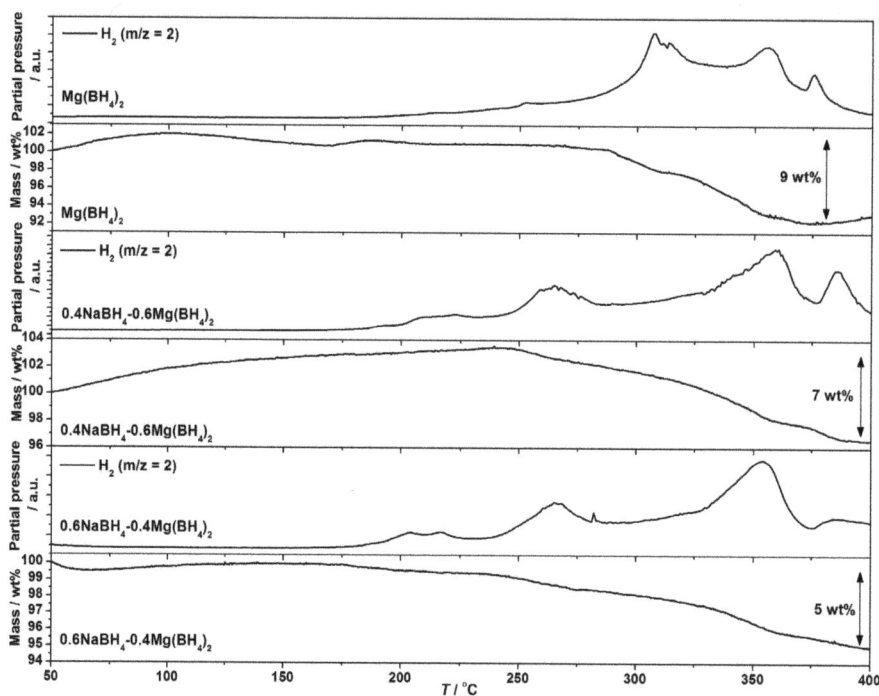

Figure 4. TGA and MS data for $Mg(BH_4)_2$, $0.4NaBH_4$–$0.6Mg(BH_4)_2$ and $0.6NaBH_4$–$0.4Mg(BH_4)_2$ from RT to 400 °C, $\Delta T / \Delta t = 5$ °C/min.

The TGA registers a beginning mass loss at $T \sim 240 \, °C$ for the composites and $T \sim 280 \, °C$ for $Mg(BH_4)_2$. A total mass loss of 9 wt% is recorded by TGA for $Mg(BH_4)_2$ below 380 °C, which suggest partial thermolysis, $\rho_m(Mg(BH_4)_2) = 14.9 \, wt\%$ H_2. Mass losses of 7 and 5 wt% are observed for $0.4NaBH_4$–$0.6Mg(BH_4)_2$ and $0.6NaBH_4$–$0.4Mg(BH_4)_2$ between 200 and 400 °C, which also suggests partial thermolysis. The theoretical hydrogen content for these samples is 14.0 and 12.7 wt% H_2.

The TGA data from the $xNaBH_4$–$(1 - x)Ca(BH_4)_2$ composites are shown in the Supplementary Materials; see Figure S5. The calculated hydrogen contents for the samples $Ca(BH_4)_2$, $0.4NaBH_4$–$0.6Ca(BH_4)_2$ and $0.665NaBH_4$–$0.335Ca(BH_4)_2$, are $\rho_m = 11.6$, 11.3 and 11.1 wt% H_2, respectively. The mass loss from $Ca(BH_4)_2$ amounts to 6.3 wt% from 360 to 400 °C. The mass loss observed in the samples, $xNaBH_4$–$(1 - x)Ca(BH_4)_2$, $x = 0.4$ and 0.665, are 6 and 3.7 wt%, respectively. The mass loss decreases with the added amount of $NaBH_4$.

3.4. Decomposition Mechanisms Observed by in Situ SR-PXD

The *in situ* SR-PXD data for the decomposition of $0.665NaBH_4$-$0.335Mg(BH_4)_2$ is shown in Figure S6. Normalized diffracted intensities of selected Bragg peaks of the compounds are extracted as a function of temperature; see Figure 5. The diffraction pattern measured at RT has a broad hump in the background in the range $9 < 2\theta < 13°$, originating from amorphous $Mg(BH_4)_2$ [11,38]. The porous structure of γ-$Mg(BH_4)_2$ may have collapsed during the ball-milling, as the characteristic Bragg peaks from γ-$Mg(BH_4)_2$ are not observed at RT in the *in situ* SR-PXD experiment. Bragg peaks from $NaBH_4$ are observed at RT, indicating that $NaBH_4$ and $Mg(BH_4)_2$ do not react during ball-milling. At $T \sim 110 \, °C$, α-$Mg(BH_4)_2$ crystallizes, and at $T \sim 180 \, °C$ the polymorphic phase change from the α- to β-$Mg(BH_4)_2$ occurs.

Crystalline β-$Mg(BH_4)_2$ disappears from the sample simultaneously with decreasing diffracted intensity from $NaBH_4$ at $T \sim 235 \, °C$. At $T \sim 240 \, °C$, an unknown compound, denoted **1**, appears, identified by three broad Bragg peaks at $2\theta = 5.95°$, 11.9° and 23.5°, which did not allow indexing. Compound **1** disappears at $T \sim 325 \, °C$ simultaneous with a further decrease in intensity of Bragg peaks belonging to $NaBH_4$. At $T \sim 300 \, °C$, two broad peaks assigned either MgO or MgB_2 appear. At $T \sim 400 \, °C$, Bragg peaks from MgH_2 are observed, and the diffracted intensity from $NaBH_4$ increases again. MgH_2 decomposes at $T \sim 465 \, °C$, followed by the formation of Mg metal. At $T \sim 525 \, °C$, the Bragg peak at $2\theta = 27.5°$ previously assigned to MgO shows a small shift to lower 2θ values, possibly due to an increased formation of MgB_2, which overlaps in peak position with MgO. At $T \sim 565 \, °C$, Bragg peaks from $NaBH_4$ vanish, while diffraction peaks from Mg metal, MgO and MgB_2 remain until $T = 600 \, °C$.

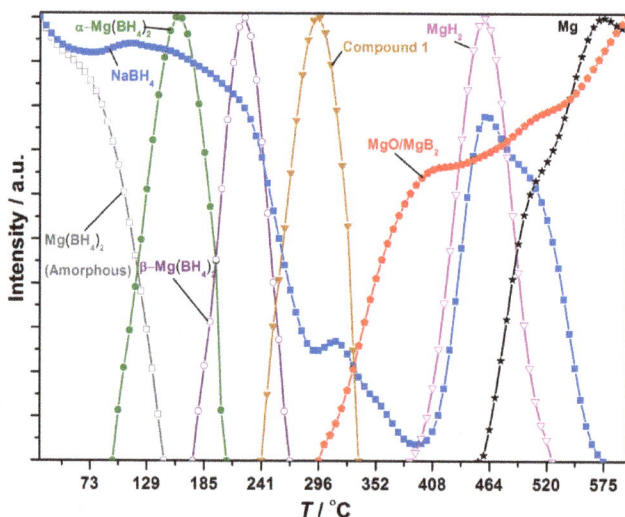

Figure 5. Normalized diffracted intensities of selected Bragg peaks from the compounds observed during the *in situ* synchrotron radiation powder X-ray diffraction (SR-PXD) experiment of $0.665NaBH_4$-$0.335Mg(BH_4)_2$. Legend: $NaBH_4$ (blue square), $Mg(BH_4)_2$ amorphous (grey square), α-$Mg(BH_4)_2$ (green circle), β-$Mg(BH_4)_2$ (purple circle), Compound **1** (orange triangle), MgH_2 (magenta triangle), MgO/MgB_2 (red pentagon), Mg (black star).

4. Discussion

4.1. Discussion of the $xNaBH_4$-$(1-x)Mg(BH_4)_2$ Composite

Twelve samples of $NaBH_4$ and $Mg(BH_4)_2$ with varying compositions have been studied. An endothermic DSC event with an onset temperature of 178 °C was observed in all of the samples of $xNaBH_4$-$(1-x)Mg(BH_4)_2$. This thermal event likely corresponds to the polymorphic transition of α- to β-$Mg(BH_4)_2$. Additionally, the area of the peak is larger for samples with higher $Mg(BH_4)_2$ content; see Figure 1 and Figure S2. The polymorphic transition is also observed by *in situ* SR-PXD at 180 °C; see Figure 5. The transition temperature may be affected by the addition of $NaBH_4$, as the event occurs at a lower temperature compared to the pristine sample; see Figure 1. A similar effect is observed in the $LiBH_4$-$LiCl$ system, where the onset temperature for the orthorhombic to hexagonal transition for $LiBH_4$ is lowered by the addition of $LiCl$ [39].

The *in situ* SR-PXD, TPPA and DSC experiments suggest that melting occurs in the samples between 205 and 240 °C. The DSC and TPPA experiments reveal that the effect is more pronounced in the samples $xNaBH_4$-$(1-x)Mg(BH_4)_2$, $x = 0.4$-0.6. The integrated peak area of the endothermic event with an onset temperature $T \sim 205$ °C is largest in 0.4NaBH$_4$–0.6Mg(BH$_4$)$_2$ and 0.5NaBH$_4$–0.5Mg(BH$_4$)$_2$, while the peak area is slightly smaller for 0.6NaBH$_4$–0.4Mg(BH$_4$)$_2$; see Figure 2. Additionally, the TPPA experiments reveal that vigorous frothing occurs above 280 °C in samples with a majority of Mg(BH$_4$)$_2$, 0.2NaBH$_4$–0.8Mg(BH$_4$)$_2$ and 0.4NaBH$_4$–0.6Mg(BH$_4$)$_2$; see Figure 3 [11].

The *in situ* SR-PXD experiment of 0.665NaBH$_4$–0.335Mg(BH$_4$)$_2$ shows decreasing Bragg peak intensity of NaBH$_4$ already at $T > 200$ °C. However, a major decrease in intensity only occurs at $T > 235$ °C, about 15 °C later than the frothing/melting was observed in the xNaBH$_4$–$(1 - x)$Mg(BH$_4$)$_2$, $x = 0.4$–0.6, by DSC and TPPA. This may be due to the higher amount of NaBH$_4$ [28,30].

Consequently, the xNaBH$_4$–$(1 - x)$Mg(BH$_4$)$_2$, $x = 0.4$–0.5, system is proposed to be a eutectic melting system with $T_m \sim 205$ °C. This is the first eutectic system within mixtures of alkali and alkaline metal borohydrides, which may have an excess of the alkaline earth metal borohydride. However, the melting point of Mg(BH$_4$)$_2$ is lower than that of NaBH$_4$ [11,28]. The lower melting metal borohydride also makes up the larger part in other eutectic metal borohydride systems [28,30]. Like in other eutectic alkali-alkaline earth metal borohydride systems, the molten phase is not a transparent liquid. This may be due to partial thermolysis and gas release of the alkaline earth metal borohydrides during melting/frothing [28].

Weak Bragg peaks assigned to Compound **1** were observed during the *in situ* SR-PXD experiment after the disappearance of β-Mg(BH$_4$)$_2$. Compound **1** does not show similarities to any of the recently discovered Mg(BH$_4$)$_2$ polymorphs [8,9,12,38]. The compound appears to crystallize from the molten phase, possibly due to excess sodium borohydride. A minor increase in the diffracted intensity from NaBH$_4$ occurs during the decomposition of **1**, as well as the formation of MgO. Compound **1** may be analogous to compounds in the LiBH$_4$–Ca(BH$_4$)$_2$ system, *i.e.*, Ca$_3$(BH$_4$)$_3$(BO$_3$) and LiCa$_3$(BH$_4$)(BO$_3$)$_2$ [40,41]. *In situ* SR-PXD shows that MgH$_2$ forms and decomposes at $T > 465$ °C. MgH$_2$ should form after the decomposition of Mg(BH$_4$)$_2$ at $T > 300$ °C. However, the melting/frothing in the system may make it difficult to observe Bragg peaks from MgH$_2$ before $T > 465$ °C.

The composites of xNaBH$_4$–$(1 - x)$Mg(BH$_4$)$_2$ are destabilized, and hydrogen release occurs at lower temperatures, as compared to Mg(BH$_4$)$_2$; and a larger amount of hydrogen is released in between 180 and 300 °C. However, larger amounts of NaBH$_4$ decrease the total amount of released hydrogen in the temperature range of RT to 400 °C. The MS data for hydrogen release from 0.4NaBH$_4$–0.6Mg(BH$_4$)$_2$ and 0.6NaBH$_4$–0.4Mg(BH$_4$)$_2$ looks very similar to 0.55LiBH$_4$-0.45Mg(BH$_4$)$_2$ [28,31]. The decomposition mechanism might be similar for the systems, as both melt/froth and contain Mg(BH$_4$)$_2$. However, because of the melting/frothing in the samples, the detailed decomposition mechanism is difficult to establish by *in situ* SR-PXD. Nanoconfinement has been used to improve the kinetics of other eutectic metal borohydride systems and may also improve the xNaBH$_4$–$(1 - x)$Mg(BH$_4$)$_2$ system, whereby more hydrogen can be collected at the eutectic melting point [33,42–44]. Increased amounts of hydrogen could also be harvested by adding a catalyst [45].

4.2. Discussion of the xNaBH$_4$-$(1 - x)$Ca(BH$_4$)$_2$ Composite

The DSC data for each of the xNaBH$_4$–$(1 - x)$Ca(BH$_4$)$_2$, x = 0.335, 0.375, 0.429, 0.429, 0.445 and 0.5, samples reveal an endothermic event at ~280 °C, which may correspond to the formation of Ca$_3$(BH$_4$)$_3$(BO$_3$); see Figure S3, Figure S7 and Figure S8. The endothermic peaks observed at ~350 °C in all of the samples of xNaBH$_4$–$(1 - x)$Ca(BH$_4$)$_2$ are assigned to the decomposition of Ca(BH$_4$)$_2$; see Figure S3. However, partial melting was observed by TPPA in the 0.5NaBH$_4$–0.5Ca(BH$_4$)$_2$ sample above 350 °C; see Figure S4.

NaBH$_4$ may react with either β-Ca(BH$_4$)$_2$ or Ca$_3$(BH$_4$)$_3$(BO$_3$) and produce Compound **2**, observed in the *in situ* SR-PXD (see Figure S7), which can explain the decreasing diffracted intensity from NaBH$_4$ above ~300 °C. Furthermore, the decomposition products from Ca(BH$_4$)$_2$ appear along with an increase in the diffracted intensity from NaBH$_4$. NaBH$_4$ recrystallizes after the decomposition of Compound **2**, and **2** may be analogous to LiCa$_3$(BH$_4$)(BO$_3$)$_2$ [41].

The mass loss observed from the samples containing NaBH$_4$ and Ca(BH$_4$)$_2$ remains below the theoretical content of hydrogen in all samples; see Figure S5. The hydrogen release is lower, since NaBH$_4$ only decomposes above ~500 °C. Therefore, the behavior of the composite xNaBH$_4$–$(1 - x)$Ca(BH$_4$)$_2$ during thermolysis appears to resemble that of the individual compounds with the exception of the formation of Compound **2**. Furthermore, the mixing of NaBH$_4$ and Ca(BH$_4$)$_2$ does not lead to destabilization and hydrogen release at a lower temperatures, as observed in the xNaBH$_4$–$(1 - x)$Mg(BH$_4$)$_2$ composite.

5. Conclusions

The composites, xNaBH$_4$–$(1 - x)$Mg(BH$_4$)$_2$ and xNaBH$_4$–$(1 - x)$Ca(BH$_4$)$_2$, were studied by *in situ* SR-PXD, temperature programmed photographic analysis and thermal analysis combined with mass spectrometry. The composite 0.4NaBH$_4$–0.6Mg(BH$_4$)$_2$ shows eutectic melting with T_m ~ 205 °C. However, the sample is not a transparent molten phase after the melting point, behaving similarly to other eutectic mixtures of alkali and alkaline earth metal borohydride mixtures. The 0.4NaBH$_4$–0.6Mg(BH$_4$)$_2$ mixture is destabilized compared to pristine Mg(BH$_4$)$_2$ and releases hydrogen at a lower temperature. Only partial melting is observed for the xNaBH$_4$–$(1 - x)$Ca(BH$_4$)$_2$ composite, and it is not related to a destabilization of the system. Eutectic melting changes the decomposition mechanisms significantly. Surprisingly, the addition of a more stable metal borohydride, NaBH$_4$, to a less stable metal borohydride, Mg(BH$_4$)$_2$, leads to partial thermolysis and hydrogen release at lower temperatures than observed for the individual components. Indeed, the control of sample melting may open new routes for obtaining high-capacity hydrogen storage materials by tailoring the conditions for the release and uptake of hydrogen.

Supplementary Materials: Supplementary materials can be accessed at: http://www.mdpi.com/1996-1073/8/4/2701/s1.

Acknowledgments: We thank the Danish Natural Science Research Council for funding the research program DanScatt and the Danish Council for Strategic Research the project HyFillFast. We thank the MAX IV laboratory for the allocated beam time. We also thank the Center for Material Crystallography (CMC) funded by The Danish National Research Foundation for support, as well as the Carlsberg Foundation.

Author Contributions: All authors contributed to this work. Peter M. M. Thygesen and Morten Brix Ley performed the synthesis and characterization for the xNaBH$_4$-(1 − x)Mg(BH$_4$)$_2$ samples. Elsa Roedern and Morten Brix Ley performed the synthesis and characterization for the xNaBH$_4$-(1 − x)Ca(BH$_4$)$_2$ composites. Elsa Roedern and Morten Brix Ley analyzed the experimental data from all of the samples. Morten Brix Ley and Torben R. Jensen wrote the manuscript.

Conflicts of Interest: The authors declare no conflict of interest.

References

1. Ley, M.B.; Jepsen, L.H.; Lee, Y.-S.; Cho, Y.W.; Bellosta von Colbe, J.M.; Dornheim, M.; Rokni, M.; Jensen, J.O.; Sloth, M.; Filinchuk, Y.; *et al.* Complex hydrides for hydrogen storage—New perspectives. *Mater. Today* **2014**, *17*, 122–128.

2. Jepsen, L.H.; Ley, M.B.; Lee, Y.-S.; Cho, Y.W.; Dornheim, M.; Jensen, J.O.; Filinchuk, Y.; Jørgensen, J.E.; Besenbacher, F.; Jensen, T.R. Boron-nitrogen based hydrides and reactive composites for hydrogen storage. *Mater. Today* **2014**, *17*, 129–135.

3. Fichtner, M. Conversion materials for hydrogen storage and electrochemical applications—Concepts and similarities. *J. Alloys Compd.* **2011**, *509*, 529–534.

4. Goodenough, J.B.; Kim, Y. Challenges for Rechargeable Li Batteries. *Chem. Mater.* **2009**, *22*, 587–603.

5. Orimo, S.; Nakamori, Y.; Eliseo, J.R.; Züttel, A.; Jensen, C.M. Complex hydrides for hydrogen storage. *Chem. Rev.* **2007**, *107*, 4111–4132.

6. Rude, L.H.; Nielsen, T.K.; Ravnsbæk, D.B.; Bösenberg, U.; Ley, M.B.; Richter, B.; Arnbjerg, L.M.; Dornheim, M.; Filinchuk, Y.; Besenbacher, F.; *et al.* Tailoring properties of borohydrides for hydrogen storage: A review. *Phys. Status Solidi* **2011**, *208*, 1754–1773.

7. Amieiro-Fonseca, A.; Ellis, S.R.; Nuttall, C.J.; Hayden, B.E.; Guerin, S.; Purdy, G.; Soulié, J.-P.; Callear, S.K.; Culligan, S.D.; David, W.I.F.; *et al.* A multidisciplinary combinatorial approach for tuning promising hydrogen storage materials towards automotive applications. *Faraday Discuss.* **2011**, *151*, 369–384.

8. Filinchuk, Y.; Richter, B.; Jensen, T.R.; Dmitriev, V.; Chernyshov, D.; Hagemann, H. Porous and dense magnesium borohydride frameworks: Synthesis, stability, and reversible absorption of guest species. *Angew. Chem.* **2011**, *123*, 11358–11362.

9. David, W.I.F.; Callear, S.K.; Jones, M.O.; Aeberhard, P.C.; Culligan, S.D.; Pohl, A.H.; Johnson, S.R.; Ryan, K.R.; Parker, J.E.; Edwards, P.P.; *et al.* The structure, thermal properties and phase transformations of the cubic polymorph of magnesium tetrahydroborate. *Phys. Chem. Chem. Phys.* **2012**, *14*, 11800–11807.

10. Her, J.H.; Stephens, P.W.; Gao, Y.; Soloveichik, G.L.; Rijssenbeek, J.; Andrus, M.; Zhao, J.C. Structure of unsolvated magnesium borohydride $Mg(BH_4)_2$. *Acta Crystallogr. B* **2007**, *63*, 561–568.

11. Paskevicius, M.; Pitt, M.P.; Webb, C.J.; Sheppard, D.A.; Filsø, U.; Gray, E.M.; Buckley, C.E. *In-Situ* X-ray Diffraction Study of γ-$Mg(BH_4)_2$ Decomposition. *J. Phys. Chem. C* **2012**, *116*, 15231–15240.

12. Richter, B.; Ravnsbæk, D.B.; Tumanov, N.; Filinchuk, Y.; Jensen, T.R. Manganese borohydride; synthesis and characterization. *Dalt. Trans.* **2015**, *44*, 3988–3996.

13. Pistidda, C.; Garroni, S.; Dolci, F.; Bardají, E.G.; Khandelwal, A.; Nolis, P.; Dornheim, M.; Gosalawit, R.; Jensen, T.; Cerenius, Y.; *et al.* Synthesis of amorphous $Mg(BH_4)_2$ from MgB_2 and H2 at room temperature. *J. Alloys Compd.* **2010**, *508*, 212–216.

14. Severa, G.; Rönnebro, E.; Jensen, C.M. Direct hydrogenation magnesium boride to magnesium borohydride: Demonstration of >11 weight percent reversible hydrogen storage. *Chem. Commun.* **2010**, *46*, 421–423.

15. Filinchuk, Y.; Ronnebro, E.; Chandra, D. Crystal structures and phase transformations in $Ca(BH_4)_2$. *Acta Mater.* **2009**, *57*, 732–738.

16. Aeberhard, P.C.; Refson, K.; Edwards, P.P.; David, W.I.F. High-pressure crystal structure prediction of calcium borohydride using density functional theory. *Phys. Rev. B* **2011**, *83*, 174102.

17. Riktor, M.D.; Sørby, M.H.; Chłopek, K.; Fichtner, M.; Hauback, B.C. The identification of a hitherto unknown intermediate phase CaB_2H_x from decomposition of $Ca(BH_4)_2$. *J. Mater. Chem.* **2009**, *19*, 2754–2759.

18. Barkhordarian, G.; Jensen, T.R.; Doppiu, S.; Bösenberg, U.; Borgschulte, A.; Gremaud, R.; Cerenius, Y.; Dornheim, M.; Klassen, T.; Bormann, R. Formation of $Ca(BH_4)_2$ from hydrogenation of CaH_2+MgB_2 composite. *J. Phys. Chem. C* **2008**, *112*, 2743–2749.

19. Schouwink, P.; D'Anna, V.; Ley, M.B.; Lawson Daku, L.M.; Richter, B.; Jensen, T.R.; Hagemann, H.; Černý, R. Bimetallic borohydrides in the system $M(BH_4)_2$—KBH_4 (M = Mg, Mn): On the structural diversity. *J. Phys. Chem. C* **2012**, *116*, 10829–10840.

20. Schouwink, P.; Ley, M.B.; Jensen, T.R.; Smrčok, L.; Černý, R. Borohydrides: From sheet to framework topologies. *Dalton Trans.* **2014**, *43*, 7726–7733.

21. Nickels, E.A.; Jones, M.O.; David, W.I.F.; Johnson, S.R.; Lowton, R.L.; Sommariva, M.; Edwards, P.P. Tuning the Decomposition Temperature in Complex Hydrides: Synthesis of a Mixed Alkalali Metal Borohydride. *Angew. Chem. Int. Ed.* **2008**, *47*, 2817–2819.

22. Ley, M.B.; Ravnsbæk, D.B.; Filinchuk, Y.; Lee, Y.-S.; Janot, R.; Cho, Y.W.; Skibsted, J.; Jensen, T.R. $LiCe(BH_4)_3Cl$, a New Lithium-Ion Conductor and Hydrogen Storage Material with Isolated Tetranuclear Anionic Clusters. *Chem. Mater.* **2012**, *24*, 1654–1663.

23. Ley, M.B.; Boulineau, S.; Janot, R.; Filinchuk, Y.; Jensen, T.R. New li ion conductors and solid state hydrogen storage materials: $LiM(BH_4)_3Cl$, M = La, Gd. *J. Phys. Chem. C* **2012**, *116*, 21267–21276.

24. Schouwink, P.; Ley, M.B.; Tissot, A.; Hagemann, H.; Jensen, T.R.; Smrčok, L.; Černý, R. Structure and properties of complex hydride perovskite materials. *Nat. Commun.* **2014**, *5*, 5706.

25. Vajo, J.J.; Skeith, S.L.; Mertens, F. Reversible storage of hydrogen in destabilized $LiBH_4$. *J. Phys. Chem. B* **2005**, *109*, 3719–3722.

26. Dornheim, M.; Doppiu, S.; Barkhordarian, G.; Bösenberg, U.; Klassen, T.; Gutfleisch, O.; Bormann, R. Hydrogen storage in magnesium-based hydrides and hydride composites. *Scr. Mater.* **2007**, *56*, 841–846.

27. Roedern, E.; Jensen, T.R. Thermal decomposition of $Mn(BH_4)_2-M(BH_4)_x$ and $Mn(BH_4)_2-MH_x$ composites with M = Li, Na, Mg, and Ca. *J. Phys. Chem. C* **2014**, *118*, 23567–23574.

28. Paskevicius, M.; Ley, M.B.; Sheppard, D.A.; Jensen, T.R.; Buckley, C.E. Eutectic melting in metal borohydrides. *Phys. Chem. Chem. Phys.* **2013**, *15*, 19774–19789.

29. Semenenko, K.N.; Chavgun, A.P.; Surov, V.N. Interaction of sodium tetrahydroborate with potassium and lithium tetrahydroborates. *Russ. J. Inorg. Chem.* **1971**, *16*, 271–273.

30. Ley, M.B.; Roedern, E.; Jensen, T.R. Eutectic melting of $LiBH_4$-KBH_4. *Phys. Chem. Chem. Phys.* **2014**, *16*, 24194–24199.

31. Bardaji, E.G.; Zhao-Karger, Z.; Boucharat, N.; Nale, A.; van Setten, M.J.; Lohstroh, W.; Rohm, E.; Catti, M.; Fichtner, M. $LiBH_4$-$Mg(BH_4)_2$: A physical mixture of metal borohydrides as hydrogen storage material. *J. Phys. Chem. C* **2011**, *115*, 6095–6101.

32. Lee, J.Y.; Ravnsbæk, D.; Lee, Y.-S.; Kim, Y.; Cerenius, Y.; Shim, J.-H.; Jensen, T.R.; Hur, N.H.; Cho, Y.W. Decomposition reactions and reversibility of the $LiBH_4$-$Ca(BH_4)_2$ composite. *J. Phys. Chem. C* **2009**, *113*, 15080–15086.

33. Lee, H.-S.; Lee, Y.-S.; Suh, J.-Y.; Kim, M.; Yu, J.-S.; Cho, Y.W. Enhanced Desorption and Absorption properties of eutectic $LiBH_4$-$Ca(BH_4)_2$ infiltrated into mesoporous carbon. *J. Phys. Chem. C* **2011**, *115*, 20027–20035.

34. Cerenius, Y.; Stahl, K.; Svensson, L.A.; Ursby, T.; Oskarsson, A.; Albertsson, J.; Liljas, A. The crystallography beamline I711 at MAX II. *J. Synchrotron Radiat.* **2000**, *7*, 203–208.

35. Jensen, T.R.; Nielsen, T.K.; Filinchuk, Y.; Jørgensen, J.E.; Cerenius, Y.; Gray, E.M.; Webb, C.J. Versatile *in situ* powder X-ray diffraction cells for solid–gas investigations. *J. Appl. Crystallogr.* **2010**, *43*, 1456–1463.

36. Hammersley, A.P.; Svensson, S.O.; Hanfland, M.; Fitch, A.N.; Hausermann, D. Two-dimensional detector software: From real detector to idealised image or two-theta scan. *High Press. Res.* **1996**, *14*, 235–248.

37. Martelli, P.; Caputo, R.; Remhof, A.; Mauron, P.; Borgschulte, A.; Züttel, A. Stability and decomposition of $NaBH_4$. *J. Phys. Chem. C* **2010**, *114*, 7173–7177.

38. Ban, V.; Soloninin, A.V.; Skripov, A.V.; Hadermann, J.; Abakumov, A.; Filinchuk, Y. Pressure-collapsed amorphous $Mg(BH_4)_2$: An Ultradense complex hydride showing a reversible transition to the porous framework. *J. Phys. Chem. C* **2014**, *118*, 23402–23408.

39. Arnbjerg, L.M.; Ravnsbæk, D.B.; Filinchuk, Y.; Vang, R.T.; Cerenius, Y.; Besenbacher, F.; Jørgensen, J.-E.; Jakobsen, H.J.; Jensen, T.R. Structure and dynamics for $LiBH_4$—LiCl solid solutions. *Chem. Mater.* **2009**, *21*, 5772–5782.

40. Riktor, M.D.; Filinchuk, Y.; Vajeeston, P.; Bardají, E.G.; Fichtner, M.; Fjellvåg, H.; Sørby, M.H.; Hauback, B.C. The crystal structure of the first borohydride borate, $Ca_3(BD_4)_3(BO_3)$. *J. Mater. Chem.* **2011**, *21*, 7188–7193.

41. Lee, Y.-S.; Filinchuk, Y.; Lee, H.-S.; Suh, J.-Y.; Kim, J.W.; Yu, J.-S.; Cho, Y.W. On the formation and the structure of the first bimetallic borohydride borate, $LiCa_3(BH_4)(BO_3)_2$. *J. Phys. Chem. C* **2011**, *115*, 10298–10304.

42. Zhao-Karger, Z.; Witter, R.; Bardají, E.G.; Wang, D.; Cossement, D.; Fichtner, M. Altered reaction pathways of eutectic $LiBH_4$-$Mg(BH_4)_2$ by nanoconfinement. *J. Mater. Chem. A* **2013**, *1*, 3379.

43. Javadian, P.; Jensen, T.R. Enhanced hydrogen reversibility of nanoconfined $LiBH_4$–$Mg(BH_4)_2$. *Int. J. Hydrog. Energy* **2014**, *39*, 9871–9876.

44. Javadian, P.; Sheppard, D.A.; Buckley, C.E.; Jensen, T.R. Hydrogen storage properties of nanoconfined $LiBH_4$–$Ca(BH_4)_2$. *Nano Energy* **2015**, *11*, 96–103.

45. Zavorotynska, O.; Saldan, I.; Hino, S.; Humphries, T.D.; Deledda, S.; Hauback, B.C. Hydrogen cycling in γ-$Mg(BH_4)_2$ with cobalt-based additives. *J. Mater. Chem. A* **2015**, *3*, 6592–6602.

Dehydriding Process and Hydrogen–Deuterium Exchange of LiBH$_4$–Mg$_2$FeD$_6$ Composites

Guanqiao Li, Motoaki Matsuo, Katsutoshi Aoki, Tamio Ikeshoji and Shin-ichi Orimo

Abstract: The dehydriding process and hydrogen–deuterium exchange (H–D exchange) of xLiBH$_4$ + (1 − x)Mg$_2$FeD$_6$ (x = 0.25, 0.75) composites has been studied in detail. For the composition with x = 0.25, only one overlapping mass peak of all hydrogen and deuterium related species was observed in mass spectrometry. This implied the simultaneous dehydriding of LiBH$_4$ and Mg$_2$FeD$_6$, despite an almost 190 °C difference in the dehydriding temperatures of the respective discrete complex hydrides. *In situ* infrared spectroscopy measurements indicated that H–D exchange between [BH$_4$]$^-$ and [FeD$_6$]$^{4-}$ had occurred during ball-milling and was promoted upon heating. The extent of H–D exchange was estimated from the areas of the relevant mass signals: immediately prior to the dehydriding, more than two H atoms in [BH$_4$]$^-$ was replaced by D atoms. For x = 0.75, H–D exchange also occurred and about one to two H atoms in [BH$_4$]$^-$ was replaced by D atoms immediately before the dehydriding. In contrast to the situation for x = 0.25, firstly LiBH$_4$ and Mg$_2$FeD$_6$ dehydrided simultaneously with a special molar ratio = 1:1 at x = 0.75, and then the remaining LiBH$_4$ reacted with the Mg and Fe derived from the dehydriding of Mg$_2$FeD$_6$.

Reprinted from *Energies*. Cite as: Li, G.; Matsuo, M.; Aoki, K.; Ikeshoji, T.; Orimo, S.-I. Dehydriding Process and Hydrogen–Deuterium Exchange of LiBH$_4$–Mg$_2$FeD$_6$ Composites. *Energies* **2015**, *8*, 5459–5466.

1. Introduction

The complex hydride LiBH$_4$, consisting of Li$^+$ cations and [BH$_4$]$^-$ complex anions, has a high gravimetric hydrogen density of 18.4 mass% and a volumetric hydrogen density of 121 kg H$_2$/m^3 [1]. The main issues to be resolved for developing LiBH$_4$ as a hydrogen storage material are lowering of its high dehydriding temperature of >420 °C and moderating the harsh rehydriding conditions of 35 MPa H$_2$ and high temperatures above 600 °C [2]. Many attempts have been made to improve the dehydriding properties of LiBH$_4$ by incorporating various additives, confining within nanoporous materials, or by preparing reactive composites with metal hydrides, and so on [3–12].

101

Recently, we reported that the dehydriding temperature of $LiBH_4$ is distinctly decreased upon combination with the complex hydride Mg_2FeH_6 composed of Mg^{2+} cations and $[FeH_6]^{4-}$ complex anions [13]. For example, the dehydriding temperature of $LiBH_4$ in $xLiBH_4 + (1 - x)Mg_2FeH_6$ composite with $x = 0.5$ is 350 °C, which is 100 °C lower than that of pure $LiBH_4$.

Besides this decreased dehydriding temperature of $LiBH_4$, a unique dehydriding process was identified. In thermal gravimetry–mass spectrometry measurements (TG–MS) of $xLiBH_4 + (1 - x)Mg_2FeH_6$ ($0.1 \leqslant x \leqslant 0.83$) composites, when $x \leqslant 0.5$, only one MS peak was observed. When $x > 0.5$, more than two MS peaks were observed. Moreover, over the entire composition range, the dehydriding temperature of $LiBH_4$ decreased almost linearly with the proportion of Mg_2FeH_6. These results suggested that when $x \leqslant 0.5$, Mg_2FeH_6 and $LiBH_4$ dehydrided simultaneously, despite the almost 190 °C difference in the dehydriding temperatures of the respective discrete complex hydrides. Conversely, when $x > 0.5$, firstly Mg_2FeH_6 dehydrided to Mg and Fe, and then $LiBH_4$ dehydrided by reacting with the Mg and/or Fe formed.

Several studies on $LiBH_4$-rich compositions of $xLiBH_4 + (1 - x)Mg_2FeH_6$ ($x > 0.65$) composites have been published. The dehydriding process of $0.8LiBH_4 + 0.2Mg_2FeH_6$ was investigated by Langmi et al. [14] and Deng et al. [15]. Both of them reported that the dehydriding temperature of $LiBH_4$ was decreased by combining with Mg_2FeH_6 by the following two-step dehydriding process. Firstly, Mg_2FeH_6 dehydrided to Mg and Fe, and then $LiBH_4$ reacted with the Mg and Fe formed. The boron in $LiBH_4$ was stabilized by Mg and Fe to form MgB_2 and FeB. Ghaani et al. [16] reported the destabilized thermodynamics of $2/3LiBH_4 + 1/3Mg_2FeH_6$ composites when compared with that of the respective discrete complex hydrides. In these studies, neither the linear variation in the dehydriding temperature with changing composition nor the evidence of simultaneous dehydriding of $LiBH_4$ and Mg_2FeH_6 has apparently been noticed. Therefore, clarifying the unique dehydriding process is important to gain a deep knowledge of such composites of complex hydrides.

Because the hydrogen released from $LiBH_4$ and Mg_2FeH_6 cannot be differentiated in MS measurements, the evidence for simultaneous dehydriding when $x \leqslant 0.5$ was not conclusive and the assignment of the multiple MS peaks observed to the dehydriding of $LiBH_4$ and Mg_2FeH_6 when $x > 0.5$ was not unequivocal. In this study, instead of Mg_2FeH_6 we have used Mg_2FeD_6 to prepare $xLiBH_4 + (1 - x)Mg_2FeD_6$ composites with two compositions, $x = 0.25$ and 0.75, in the expectation of distinguishing the dehydriding processes of the respective components by the MS signals of H_2 and D_2. During the actual measurement, H–D exchange between $LiBH_4$ and Mg_2FeD_6 was observed, and the relationship between this H–D exchange and the simultaneous dehydriding processes is discussed herein.

2. Results and Discussion

2.1. Dehydriding Property of $0.25LiBH_4 + 0.75Mg_2FeD_6$

The TG–MS profile of $0.25LiBH_4 + 0.75Mg_2FeD_6$ is shown in Figure 1. The dehydriding profile was almost the same as that of $0.25LiBH_4 + 0.75Mg_2FeH_6$, in which only one MS peak was observed, implying that both $LiBH_4$ and Mg_2FeH_6 were dehydrided, as explained in the Introduction. Taking the purity of Mg_2FeD_6 into consideration, the experimental weight loss of 9.5 wt% was in reasonable agreement with the theoretical weight loss (10.5 wt%) for full dehydriding of $LiBH_4$ and Mg_2FeD_6 and the general reaction equation may be as follows:

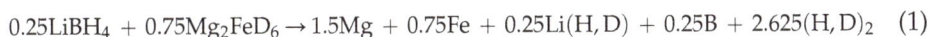

$$0.25LiBH_4 + 0.75Mg_2FeD_6 \rightarrow 1.5Mg + 0.75Fe + 0.25Li(H,D) + 0.25B + 2.625(H,D)_2 \quad (1)$$

Actually, XRD analysis of the sample collected after TG measurement (400 °C), as illustrated in Figure S1 in the Supplementary Material, showed that the dehydrided products contained Mg, Fe, Fe_2B, and LiH(D).

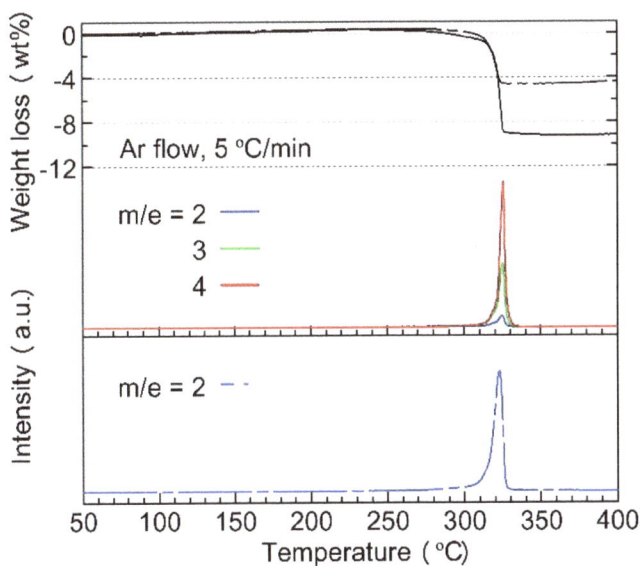

Figure 1. TG–MS profile of $0.25LiBH_4 + 0.75Mg_2FeD_6$ (solid line) and $0.25LiBH_4 + 0.75Mg_2FeH_6$ (dash line). m/e = 2, 3, and 4 are signals of H_2, HD, and D_2, respectively.

In the MS measurement, peaks due to H_2 and D_2 were observed at the same temperature of 325 °C. This proved our expectation that $LiBH_4$ and Mg_2FeD_6

dehydrided simultaneously. Besides the H_2 and D_2 peaks, a peak due to HD was also observed, indicative of H–D exchange between $LiBH_4$ and Mg_2FeD_6.

To prove this H–D exchange, *in situ* IR spectra were recorded, and the results are shown in Figure 2. As shown in the spectrum at 25 °C, H–D exchange has already occurred during ball-milling process. Referring to theoretical data, for isotopically pure $LiBH_4$, a BH stretching peak at $v \approx 2350$ cm^{-1} and bending peaks at $v \approx 1300$ and 1100 cm^{-1} should be observed [17]; for isotopically pure Mg_2FeD_6, an FeD stretching peak should appear at $v \approx 1260$ cm^{-1} [18]. The experimentally observed peak at $v = 2340$ cm^{-1} at 25 °C was assigned to the BH stretching mode, and that at $v = 1310$ cm^{-1} was assigned to the FeD stretching mode. The missing BH bending peak and the broadened peak of the BH stretching mode indicate that the symmetry of $[BH_4]^-$ had been broken owing to the part replacement of H atoms by D atoms [19]. The peak at $v \approx 1840$ cm^{-1} was assigned as the FeH stretching mode in Mg_2FeD_5H [18]. A very broad peak in the region $v = 1600–1900$ cm^{-1} was considered to be due to merged FeH and BD stretching peaks [18,20]. It was difficult to distinguish the FeH stretching peak at $v = 1794$ cm^{-1} and the BD stretching peak at $v = 1775$ cm^{-1} within the spectral resolution.

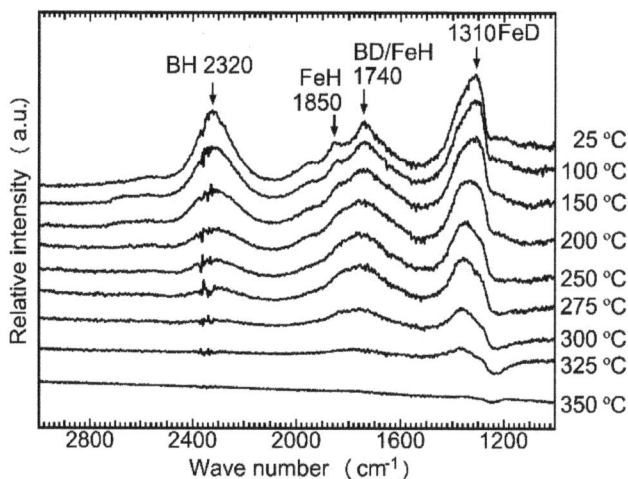

Figure 2. *In situ* IR spectra of ball-milled $0.25LiBH_4 + 0.75Mg_2FeD_6$. The heating rate was 5 °C/min. The atmosphere in the sample holder was 0.1 MPa Ar without gas flow.

When we take a look at the IR spectra during the heating process, the area of the BD/FeH stretching peak did not change discernibly when compared to that of the steadily diminishing BH stretching peak, although the whole peak intensities weakened upon heating due to deterioration of the optical focusing by the thermally expanding sample. In addition, the FeD stretching peak at $v = 1310$ cm^{-1} was shifted to higher wavenumber and broadened significantly during the heating process. Referring to previous H–D exchange studies of LiBH$_4$ and Mg(BH$_4$)$_2$ [19–21], these results suggest that H–D exchange was promoted during the heating process. At 325 °C, all of the peaks faded as a result of the dehydriding reaction, consistent with the TG–MS measurements shown in Figure 1. The peaks disappeared and the spectrum did not change further at 350 °C, suggesting completion of the dehydriding reaction.

To further assess the extent of H–D exchange, we attempted a quantitative analysis based on the areas of MS signals that directly related to the amount of gas released. If it is supposed that H and D atoms are firstly released from H–D exchanged Mg$_2$FeH$_{y/3}$D$_{6-y/3}$ and LiBH$_{4-y}$D$_y$, and then mix and combine freely to form H$_2$, HD, and D$_2$ gas molecules, then statistically the area ratio of the MS signals of H$_2$, HD, and D$_2$ should be 25:12:1. In fact, the experimentally measured area ratio of these MS signals was 9:5:1, quite different from the statistical distribution. Thus, the processes of H–D exchange and dehydriding need to be interpreted differently. Here, we suppose that:

(a)   H–D exchange occurred during ball-milling and the heating process but stopped as soon as the dehydriding started;

(b)   H$_2$, HD, or D$_2$ molecules were directly released from either Mg$_2$FeH$_{y/3}$D$_{6-y/3}$ or LiBH$_{4-y}$D$_y$; H or D atoms derived from the two different complex hydrides cannot combine to form gas molecules.

Based on this assumption, the extent of H–D exchange was estimated from the area ratio of MS signals. The result shows that the extent of H–D exchange, y, was around 2.5 immediately prior to the onset of dehydriding:

$$0.25\text{LiBH}_4 + 0.75\text{Mg}_2\text{FeD}_6 \rightarrow 0.25\text{LiBH}_{4-y}\text{D}_y + 0.75\text{Mg}_2\text{FeH}_{y/3}\text{D}_{6-y/3} \; (y \approx 2.5) \quad (2)$$

According to the estimation, less than one D atom in [FeD$_6$]$^{4-}$ was replaced by H atom before the dehydriding and more than two H atoms in [BH$_4$]$^-$ were replaced by D atoms. This result is in good agreement with the IR spectra: even though a shift to higher wavenumber was observed, the area of the FeD stretching peak did not decrease markedly because the replacement in [FeD$_6$]$^{4-}$ was slight. Conversely, the BH stretching peak at $\nu \approx 2320$ cm^{-1} was weakened and the BD stretching peak at $\nu \approx 1740$ cm^{-1} was significantly intensified since the symmetry of [BH$_4$]$^-$ was severely disrupted.

2.2. Dehydriding Property of 0.75LiBH$_4$ + 0.25Mg$_2$FeD$_6$

In contrast to the situation for 0.25LiBH$_4$ + 0.75Mg$_2$FeD$_6$, multiple MS peaks were observed for the dehydriding process of $0.75LiBH_4 + 0.25Mg_2FeD_6$. The TG–MS profile is shown in Figure 3, together with that of $0.75LiBH_4 + 0.25Mg_2FeH_6$ as a reference. The total weight loss was 9.0 wt%, which suggests full dehydriding of both $LiBH_4$ and Mg_2FeD_6. The dehydrided products were confirmed as Mg, FeB_2, and LiH(D) by XRD analysis, as shown in Figure S2 in the Supplementary Material.

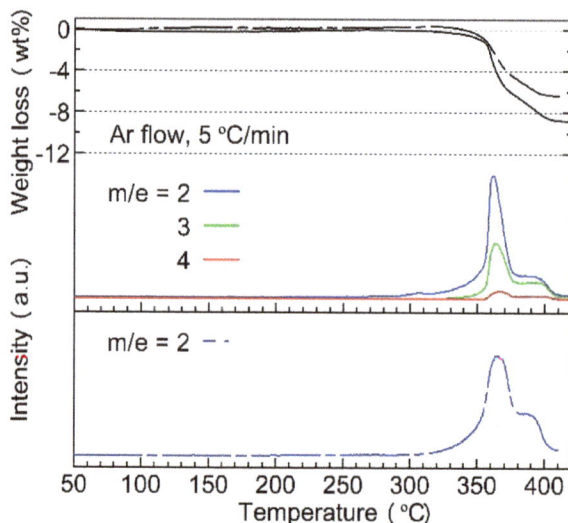

Figure 3. TG–MS profile of $0.75LiBH_4 + 0.25Mg_2FeD_6$ (solid line) and $0.75LiBH_4 + 0.25Mg_2FeH_6$ (dash line).

For the $0.25LiBH_4 + 0.75Mg_2FeD_6$ composition, *In situ* IR confirmed that H-D exchange between $LiBH_4$ and Mg_2FeD_6 occurred during ball-milling and was promoted during the heating process, as shown in Figure S3 in the Supplementary Material. With the same premises as in the case of the $0.25LiBH_4 + 0.75Mg_2FeD_6$ composition, the extent of H-D exchange immediately prior to dehydriding was estimated as $y \approx 4.5$ from the area ratio of the MS signals (sum of the two peaks) of H_2, HD, and D_2. This result shows that more than one H atom in $[BH_4]^-$ was replaced by D:

$$0.75LiBH_4 + 0.25Mg_2FeD_6 \rightarrow 0.75LiBH_{4-y/3}D_{y/3} + 0.25Mg_2FeH_yD_{6-y} \quad (y \approx 4.5) \quad (3)$$

The slope of the TG profile changed and the MS curves separated at around 370 °C. The weight loss was 5.4 wt% before the inflection temperature followed by 3.6 wt% until completion of the dehydriding. Considering the estimated extent of HD exchange, the weight loss indicated that $0.25LiBH_{4-y/3}D_{y/3} + 0.25Mg_2FeH_yD_{6-y}$ dehydrided before the inflection temperature and then the residue $0.5LiBH_{4-y/3}D_{y/3}$ dehydrided. The area ratio of the first MS peak was 14:7:1, and this changed to 8:6:1 for the second MS peak. This result supports the interpretation that the first MS peak corresponds to the simultaneous dehydrogenation of isotopically exchanged $LiBH_4$ and Mg_2FeD_6 and the second MS peak corresponds to the dehydriding of $LiBH_{4-y/3}D_{y/3}$. Therefore, it is evident that even though multiple MS peaks were observed, $LiBH_4$ and Mg_2FeD_6 were still dehydrided simultaneously. Following on from our previous report on the dehydriding properties of other compositions of $xLiBH_4 + (1-x)Mg_2FeH_6$ composites, when $x \leqslant 0.5$ (molar ratio of $LiBH_4:Mg_2FeD_6 = 1:1$), only one MS peak is observed [13]. It can be surmised that the molar ratio 1:1 is a special composition: $LiBH_4$ can dehydride simultaneously with Mg_2FeH_6/Mg_2FeD_6 up to this molar ratio; if there is more $LiBH_4$ in the composite, the residual $LiBH_4$ will subsequently dehydride by reacting with Mg and Fe derived from the dehydriding of Mg_2FeH_6/Mg_2FeD_6.

3. Experimental Section

Mg_2FeD_6 was synthesized by pressing a $2Mg + Fe$ mixture into pellets and subjecting it to heat treatment at 400 °C for 20 h under 3 MPa D_2. The product yield was 91% according to TG measurement and the isotopic purity was almost 100% according to the MS measurement. Mg_2FeD_6 was then mixed with $LiBH_4$ (95%, Aldrich, St. Louis, MO, USA) and $xLiBH_4 + (1-x)Mg_2FeD_6$ composites with compositions $x = 0.25$ and 0.75 were prepared by planetary ball-milling (Fritsch P-5, Fritsch, Idar-Oberstein, Germany) for 5 h under argon.

The dehydriding properties were examined by TG–MS measurements (TG8120, Rigaku, Tokyo, Japan, Ar flow of 150 mL/min, heating rate of 5 °C/min). Powder X-ray diffraction (XRD) measurements were conducted on an X'Pert-Pro diffractometer (Cu-K_α radiation, PANalytical, Almelo, The Netherlands). *In situ* infrared spectroscopy measurements were performed on a iZ10 infrared spectrometer (diffuse-reflectance mode, heating rate 5 °C/min, resolution 4 cm^{-1}, Thermo Nicolet, Thermo Fisher Scientific, Waltham, MA, USA). The samples were always handled in a glove box filled with purified argon.

4. Conclusions

We have investigated the dehydriding processes of $xLiBH_4 + (1-x)Mg_2FeD_6$ ($x = 0.25, 0.75$) composites in detail. For both of these compositions, H–D exchange between $LiBH_4$ and Mg_2FeD_6 occurred during ball-milling and was promoted during the heating process, as confirmed by *in situ* infrared spectroscopy and mass spectrometry measurements. The extent of H–D exchange immediately prior to the dehydriding reaction was estimated from the area ratio of MS signals. For the composition with $x = 0.25$, more than two H atoms in $[BH_4]^-$ were replaced by D atoms and for that with $x = 0.75$, one to two H atoms in $[BH_4]^-$ were replaced by D atoms.

For the composition with $x = 0.25$, only one MS peak was observed, which resulted from the simultaneous dehydriding of isotopically exchanged $LiBH_4$ and Mg_2FeD_6. For the composition with $x = 0.75$, two MS peaks were observed, which resulted from partial simultaneous dehydriding of isotopically exchanged $LiBH_4$ and Mg_2FeD_6, and subsequent dehydriding of the residue isotopically exchanged $LiBH_4$. A special molar ratio of 1:1 has been identified as the limit for simultaneous dehydriding of $LiBH_4$ with Mg_2FeH_6/Mg_2FeD_6. Experiments aimed at delineating the detailed thermodynamics of the dehydriding process based on pressure–composition isotherm analysis is underway and the kinetics of the H–D exchange is being investigated by *in situ* Raman spectroscopy.

Supplementary Materials: Supplementary materials can be accessed at: http://www.mdpi.com/1996-1073/8/6/5459/s1.

Acknowledgments: The authors would like to thank Naoko Warifune for her technical support, and Stefano Deledda and Olena Zavorotynska for their precious advice. This research is funded by JSPS KAKENHI Grant Numbers 25220911, 26820311, JSPS Fellows and Cooperative Research and Development Center for Advanced Materials of Institute for Institute for Materials Research, Tohoku University.

Author Contributions: All of the authors contributed to this work. Guanqiao Li and Motoaki Matsuo designed and conducted the experiments, and wrote the paper. Katsutoshi Aoki, Tamio Ikeshoji and Shin-ichi Orimo helped analyze the data and revise the paper.

Conflicts of Interest: The authors declare no conflict of interest.

References

1. Züttel, A.; Borgschulte, A.; Orimo, S. Tetrahydroborates as new hydrogen storage materials. *Scr. Mater.* **2007**, *56*, 823–828.
2. Orimo, S.; Nakamori, Y.; Kitahara, G.; Miwa, K.; Ohba, N.; Towata, S.; Züttel, A. Dehydriding and rehydriding reactions of $LiBH_4$. *J. Alloy Compd.* **2005**, *404*, 427–430.
3. Züttel, A.; Wenger, P.; Rentsch, S.; Sudan, P.; Mauron, P.; Emmenegger, C. $LiBH_4$ a new hydrogen storage material. *J. Power Sources* **2003**, *118*, 1–7.
4. Gross, A.F.; Vajo, J.J.; Van Atta, S.L.; Olson, G.L. Enhanced hydrogen storage kinetics of $LiBH_4$ in nanoporous carbon scaffolds. *J. Phys. Chem. C* **2008**, *112*, 5651–5657.
5. Liu, X.F.; Peaslee, D.; Jost, C.Z.; Majzoub, E.H. Controlling the decomposition pathway of $LiBH_4$ via confinement in highly ordered nanoporous carbon. *J. Phys. Chem. C* **2010**, *114*, 14036–14041.
6. Ngene, P.; Adelhelm, P.; Beale, A.M.; de Jong, K.P.; de Jongh, P.E. $LiBH_4$/SBA-15 nanocomposites prepared by melt infiltration under hydrogen pressure: Synthesis and hydrogen sorption properties. *J. Phys. Chem. C* **2010**, *114*, 6163–6168.
7. Vajo, J.J.; Skeith, S.L.; Mertens, F. Reversible storage of hydrogen in destabilized $LiBH_4$. *J. Phys. Chem. B* **2005**, *109*, 3719–3722.

8. Pinkerton, F.E.; Meyer, M.S.; Meisner, G.P.; Balogh, M.P.; Vajo, J.J. Phase boundaries and reversibility of $LiBH_4/MgH_2$ hydrogen storage material. *J. Phys. Chem. C* **2007**, *111*, 12881–12885.

9. Nakagawa, T.; Ichikawa, T.; Hanada, N.; Kojima, Y.; Fujii, H. Thermal analysis on the Li-Mg-B-H systems. *J. Alloy Compd.* **2007**, *446*, 306–309.

10. Jin, S.A.; Lee, Y.S.; Shim, J.H.; Cho, Y.W. Reversible hydrogen storage in $LiBH_4$-MH_2 (M = Ce, Ca) composites. *J. Phys. Chem. C* **2008**, *112*, 9520–9524.

11. Fu, H.; Yang, J.Z.; Wang, X.J.; Xin, G.B.; Zheng, J.; Li, X.G. Preparation and dehydrogenation properties of lithium hydrazidobis(borane) $(LiNH(BH_3)NH_2BH_3)$. *Inorg. Chem.* **2014**, *53*, 7334–7339.

12. Au, M.; Jurgensen, A.; Zeigler, K. Modified lithium borohydrides for reversible hydrogen storage. *J. Phys. Chem. B* **2006**, *110*, 26482–26487.

13. Li, G.; Matsuo, M.; Deledda, S.; Sato, R.; Hauback, B.C.; Orimo, S. Dehydriding property of $LiBH_4$ combined with Mg_2FeH_6. *Mater. Trans.* **2013**, *54*, 1532–1534.

14. Langmi, H.W.; McGrady, G.S.; Newhouse, R.; Rönnebro, E. Mg_2FeH_6–$LiBH_4$ and Mg_2FeH_6–$LiNH_2$ composite materials for hydrogen storage. *Int. J. Hydrog. Energy* **2012**, *37*, 6694–6699.

15. Deng, S.S.; Xiao, X.Z.; Han, L.Y.; Li, Y.; Li, S.Q.; Ge, H.W.; Wang, Q.D.; Chen, L.X. Hydrogen storage performance of $5LiBH_4+Mg_2FeH_6$ composite system. *Int. J. Hydrog. Energ.* **2012**, *37*, 6733–6740.

16. Ghaani, M.R.; Catti, M.; Nale, A. Thermodynamics of dehydrogenation of the $2LiBH_4$–Mg_2FeH_6 composite. *J. Phys. Chem. C* **2012**, *116*, 26694–26699.

17. Zavorotynska, O.; Corno, M.; Damin, A.; Spoto, G.; Ugliengo, P.; Baricco, M. Vibrational properties of MBH_4 and MBF_4 crystals (M = Li, Na, K): A combined DFT, infrared, and Raman Study. *J. Phys. Chem. C* **2011**, *115*, 18890–18900.

18. Parker, S.F.; Williams, K.P.J.; Bortz, M.; Yvon, K. Inelastic neutron scattering, infrared, and Raman spectroscopic studies of Mg_2FeH_6 and Mg_2FeD_6. *Inorg. Chem.* **1997**, *36*, 5218–5221.

19. Borgschulte, A.; Jain, A.; Ramirez-Cuesta, A.J.; Martelli, P.; Remhof, A.; Friedrichs, O.; Gremaud, R.; Züttel, A. Mobility and dynamics in the complex hydrides $LiAlH_4$ and $LiBH_4$. *Faraday Discuss.* **2011**, *151*, 213–230.

20. Andresen, E.R.; Gremaud, R.; Borgschulte, A.; Ramirez-Cuesta, A.J.; Züttel, A.; Hamm, P. Vibrational dynamics of $LiBH_4$ by infrared pump-probe and 2D spectroscopy. *J. Phys. Chem. A* **2009**, *113*, 12838–12846.

21. Hagemann, H.; D'Anna, V.; Rapin, J.P.; Yvon, K. Deuterium-hydrogen exchange in solid $Mg(BH_4)_2$. *J. Phys. Chem. C* **2010**, *114*, 10045–10047.

The improved Hydrogen Storage Performances of the Multi-Component Composite: $2Mg(NH_2)_2$–$3LiH$–$LiBH_4$

Han Wang, Hujun Cao, Guotao Wu, Teng He and Ping Chen

Abstract: $2Mg(NH_2)_2$–$3LiH$–$LiBH_4$ composite exhibits an improved kinetic and thermodynamic properties in hydrogen storage in comparison with $2Mg(NH_2)_2$–$3LiH$. The peak temperature of hydrogen desorption drops about 10 K and the peak width shrinks about 50 K compared with the neat $2Mg(NH_2)_2$–$3LiH$. Its isothermal dehydrogenation and re-hydrogenation rates are respectively 2 times and 18 times as fast as those of $2Mg(NH_2)_2$–$3LiH$. A slope desorption region with higher equilibrium pressure is observed. By means of X-ray diffraction (XRD), Fourier transform infrared spectroscopy (FTIR) and nuclear magnetic resonance (NMR) analyses, the existence of Li_2BNH_6 is identified and its roles in kinetic and thermodynamic enhancement are discussed.

Reprinted from *Energies*. Cite as: Wang, H.; Cao, H.; Wu, G.; He, T.; Chen, P. The improved Hydrogen Storage Performances of the Multi-Component Composite: $2Mg(NH_2)_2$–$3LiH$–$LiBH_4$. *Energies* **2015**, *8*, 6898–6909.

1. Introduction

Owing to the worldwide demand for the renewable energy sources, hydrogen will be an ideal energy carrier if it can be stored safely, efficiently and conveniently [1]. Numerous solid-state hydrogen storage materials have been developed to store hydrogen in an energy or volume efficient way [2]. Complex hydrides, e.g., alanates [3–5], borohydrides [6–8], and amide-hydride systems [9–20] are promising to fulfill the on-board hydrogen storage requirements. In particular, a number of amide–hydride systems, such as Li–N–H [10], Li–Mg–N–H [11,13,17–20], Li–Ca–N–H [14], Li–Al–N–H [15], Mg–N–H [12], Ca–N–H [16], and so on [9,14,21,22] have been investigated since 2002. As for the Li–Mg–N–H system, $Li_2Mg(NH)_2$ can be obtained by releasing hydrogen from MgH_2–$2LiNH_2$ composite or $Mg(NH_2)_2$–$2LiH$ composite, which has attracted considerable attention due to its favorable thermodynamics ($\Delta H \approx 40$ kJ/mol-H_2), relatively high hydrogen capacity (5.6 wt%) and good reversibility.

Hydrogenation/dehydrogenation processes of $Li_2Mg(NH)_2$ were probed under the pressure composition isotherms conditions by *ex-situ* X-ray diffraction (XRD) [17] and *in-situ* neutron measurements [23]. During the hydrogenation process, $Li_2Mg(NH)_2$ is first converted to $Li_2Mg_2(NH)_3$, $LiNH_2$ and LiH at a low pressure slope, and then form $Mg(NH_2)_2$ and 2LiH at the high pressure plateau. In the dehydrogenation, $Mg(NH_2)_2$ and $2LiH$ take the reverse reaction route and go back to $Li_2Mg(NH)_2$.

The reactions can be described as follows:

$$2Mg(NH_2)_2 + 3LiH \rightleftharpoons Li_2Mg_2(NH)_3 + LiNH_2 + 2H_2 \tag{1}$$

$$Li_2Mg_2(NH)_3 + LiNH_2 + LiH \rightleftharpoons Li_2Mg(NH)_2 + 2H_2 \tag{2}$$

Considerable works have been carried out to lower down the operating temperature and improve the dehydrogenation kinetics of $Mg(NH_2)_2$–$2LiH$ composite by doping additives, such as alkali metal compounds [24–27] and metal borohydrides [28,29]. The addition of KH significantly reduced the operating temperature from *ca.* 523 to 380 K, while the equilibrium pressure was *ca.* 0.2 MPa. Rb based compounds were found to be more effective for improving the kinetic properties of $Mg(NH_2)_2$–$2LiH$ composite than the potassium based additives, and the active species was identified to be RbH [24,30].Yang *et al.* [31] reported that a multi-component composite of $LiNH_2$, MgH_2 and $LiBH_4$ with a molar ratio of 2:1:1 exhibited enhanced low-temperature desorption kinetics and a significant reduction in ammonia liberations. As a kind of self-catalyzing hydrogen storage material, $LiBH_4$ catalyzed the dehydrogenation of $2LiNH_2$–MgH_2 composite to form $Li_2Mg(NH)_2$, furthermore $Li_2Mg(NH)_2$ could react with $LiBH_4$ to release more hydrogen. Following this work, Hu *et al.* [29] doped 10 mol% $LiBH_4$ in $Mg(NH_2)_2$–$2LiH$ composite and found the doping of $LiBH_4$ not only improved the kinetics but also reduced the heat of dehydrogenation from 40 kJ/mol-H_2 of $Mg(NH_2)_2$–$2LiH$ composite to 36.5 kJ/mol-H_2 of $Mg(NH_2)_2$–$2LiH$–$0.1LiBH_4$ composite. Other borohydrides, such as $Mg(BH_4)_2$ and $Ca(BH_4)_2$ [32–34], have similar effects as $LiBH_4$ in $Mg(NH_2)_2$–$2LiH$ composite because they convert to $LiBH_4$ after a metathesis reaction. Li *et al.* [35] introduced 5 mol% LiBr to $2LiNH_2$–MgH_2 composite and found certain effects on its thermodynamics and kinetics. In Cao's recentwork [36], thermodynamic properties of the dehydrogenation of $2Mg(NH_2)_2$–$3LiH$ composite were improved by stabilizing the dehydrogenated product, $LiNH_2$, in Reaction (1). $LiNH_2$ reacts with $LiBH_4$, LiI and LiBr exothermically and forms more stable compounds, *i.e.*, $Li_4(NH_2)_3BH_4$, $Li_3(NH_2)_2I$, and $Li_2(NH_2)Br$; the reactions are as follows:

$$2Mg(NH_2)_2 + 3LiH + 1/2LiI \rightleftharpoons Li_2Mg_2(NH)_3 + 1/2Li_3(NH_2)_2I + 3H_2 \tag{3}$$

$$2Mg(NH_2)_2 + 3LiH + LiBr \rightleftharpoons Li_2Mg_2(NH)_3 + Li_2NH_2Br + 3H_2 \tag{4}$$

$$2Mg(NH_2)_2 + 3LiH + 1/3LiBH_4 \rightleftharpoons Li_2Mg_2(NH)_3 + 1/3Li_4BN_3H_{10} + 3H_2 \tag{5}$$

The additions of $LiBH_4$, LiI and $LiBr$ noticeably reduced the heat of dehydrogenation from 40 kJ/mol-H_2 of $2Mg(NH_2)_2$–3LiH composite to 35.8, 33.3 and 31.9 kJ/mol-H_2, respectively, which suggests that hydrogen release at 0.1 MPa equilibrium pressure can be thermodynamically allowed at 337, 333 and 320 K, respectively. Quaternary complex hydrides, such as Li_2BNH_6 [37] and $Li_4BN_3H_{10}$ [38], were found by ball milling or heating the mixtures of $LiNH_2$ and $LiBH_4$ with corresponding molar ratios. If with more $LiBH_4$ to facilitate the formation of Li_2BNH_6, Reaction (5) can be rewritten as Reaction (6):

$$2Mg(NH_2)_2 + 3LiH + LiBH_4 \rightleftharpoons Li_2Mg_2(NH)_3 + Li_2BNH_6 + 3H_2 \qquad (6)$$

Herein we report the modification of $2Mg(NH_2)_2$–3LiH system by adding a different ratio of $LiBH_4$. The dehydrogenation and re-hydrogenation performances of $Mg(NH_2)_2$–2LiH–1LiBH$_4$ composite were investigated by Thermogravimetry coupled differential thermal analysis (TG-DTA) measurement, temperature programmed desorption-mass spectroscopy (TPD-MS), isothermal volumetric release, and soak. We found that the increase in the amount of $LiBH_4$ improved the kinetics of $2Mg(NH_2)_2$–3LiH composite.

An interesting phenomenon was observed that there was a higher pressure slope in pressure composition isotherm (PCI) measurements of $2Mg(NH_2)_2$–3LiH–LiBH$_4$ composite than $2Mg(NH_2)_2$–3LiH–1/3LiBH$_4$ composite [36], which reveals the change of thermodynamic properties. Samples at this stage were analyzed by means of X-ray diffraction (XRD), Fourier transform infrared spectroscopy (FTIR), and magic angle spinning nuclear magnetic resonance (MAS NMR).

2. Results and Discussion

2.1. The hydrogen Desorption Properties of $2Mg(NH_2)_2$–3LiH–LiBH$_4$ Composite

TG curve of $2Mg(NH_2)_2$–3LiH–LiBH$_4$ composite (T_2) is shown in Figure 1a. The thermal desorption of this sample exhibits two main steps, one is from 400 K to 450 K and the other is from 475 K to 575 K. The weight losses of the first step and the second step are 3.63 wt% and 6.24 wt%, respectively. Moreover, the first-step weight loss is near the theoretical loss of H_2 (3.7 wt%) deduced from Reaction (6).

The DTA curve of T_2 gives two main endothermic signals corresponding to the two desorption steps in TG curve. The small endothermic peak in Figure 1 at 363 K can be assigned to the phase transformation of $LiBH_4$ in starting materials.

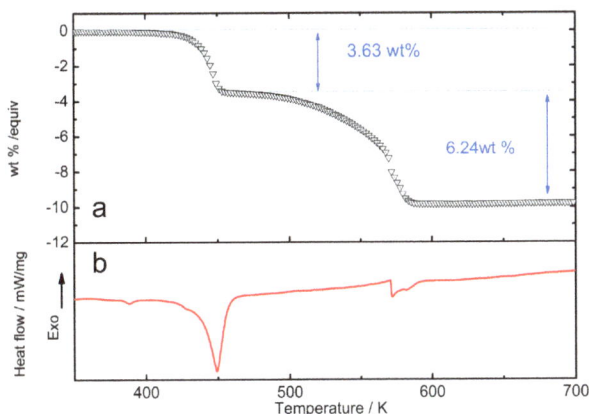

Figure 1. TG-DTA curves of $2Mg(NH_2)_2$–$3LiH$–$LiBH_4$ composite (T_2) (**a**) TG and (**b**) DTA.

The kinetic desorption behaviors of $2Mg(NH_2)_2$–$3LiH$ (T_p) and $2Mg(NH_2)_2$–$3LiH$–$LiBH_4$ (T_2) composites were investigated by TPD-MS too. H_2 and NH_3 signals are shown in Figure 2a,b, respectively. As shown in Figure 2b, the byproduct ammonia can be effectively inhibited in T_2 sample during desorption process, and the two-step dehydrogenation of T_2 sample is corresponding to the result of TG. In the first dehydrogenation step of T_2 sample, 3.63 wt% hydrogen can be released according to the thermogravimetry analysis, which is near the weight loss of 3.7 wt% deduced from Reaction (6). The first dehydrogenation step of T_2 sample occurred at the temperature of 400 K, and finished at 460 K. The first dehydrogenation peak temperature of T_2 sample lowers *ca.* 10 K compared with that of T_p sample (462 K), and the peak width shrinks *ca.* 50 K from that of T_p samples. Such a sharp dehydrogenation peak reflects a fast rate of dehydrogenation near the peak temperature. Furthermore, the dehydrogenation activation energies (E_a) of T_2 sample and T_p sample were determined by the Kissinger's method [36]. Through the calculation, the activation energy (E_a) of T_2 sample is 109 kJ/mol, which is lower than that of T_p (127 kJ/mol).

Isothermal dehydrogenation and re-hydrogenation of T_2 and T_p samples at 416 K were performed. Figure 3 (more detailed information is showed in Figure S1) shows that desorption and absorption kinetics of T_2 sample are accelerated remarkably in comparison with the T_p sample. In 100 min T_2 sample releases *ca.* 80% hydrogen, while T_p sample only releases 40% hydrogen, where the total amount of hydrogen is calculated according to Reaction (6) and Reaction (1), respectively. The tangent slopes of the initial linear parts also indicate that the dehydrogenation rate of the T_2 sample is *ca.* 2 times as fast as that of T_p sample. In the re-hydrogenation test, *ca.* 80% hydrogen can be charged back to T_2 sample within 50 min, while under

113

the same condition, only *ca.* 20% hydrogen can be soaked in T_p sample. And the tangent slopes of the initial linear parts also show that T_2 is *ca.* 18 times as fast as that of the T_p sample.

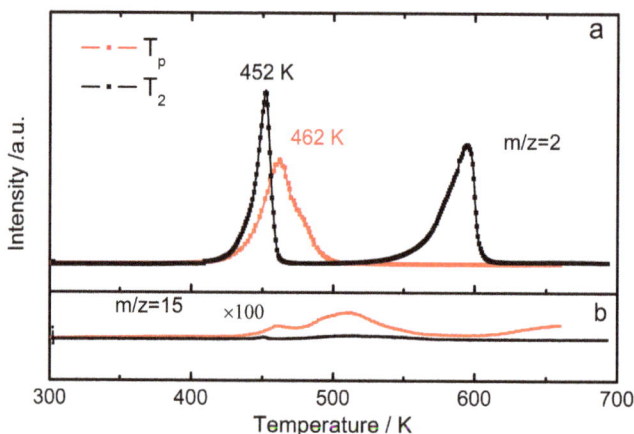

Figure 2. TPD-MS curves of $2Mg(NH_2)_2$–$3LiH$ composite (T_p), and $2Mg(NH_2)_2$–$3LiH$–$LiBH_4$ composite (T_2): (**a**) H_2 signals and (**b**) NH_3 signals.

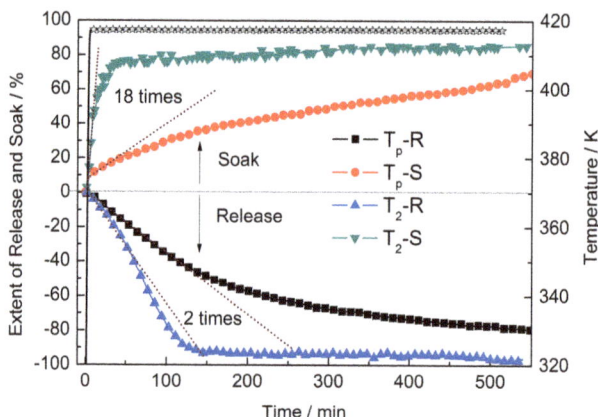

Figure 3. Isothermal hydrogen desorption and sorption of the $2Mg(NH_2)_2$–$3LiH$ composite (T_p), $2Mg(NH_2)_2$–$3LiH$–$LiBH_4$ composite (T_2) at 416 K under the pressure of 0.01 and 6 MPa, respectively.

XRD patterns of T_2 samples after dehydrogenation and re-hydrogenation at 460 K were collected for structural analyses. As shown in Figure 4a, the dehydrogenation products of T_2 sample are $Li_2Mg_2(NH)_3$ and $Li_4BN_3H_{10}$, which is somehow out of our initial expectation. Li_2BNH_6 has a hexagonal structure, and melts at ~365 K. $Li_4BN_3H_{10}$ has a body-centered cubic structure, and melts at

~465 K. Li_2BNH_6 is less stable than $Li_4BN_3H_{10}$ at temperatures above the melting temperature and decomposes to $Li_4BN_3H_{10}$ and $LiBH_4$ [39]. However, we did not observe crystalline $LiBH_4$ phase in XRD patterns in Figure 4a, which may become amorphous during this process. After the following re-hydrogenation, $Mg(NH_2)_2$ and LiH are regenarated and $Li_4BN_3H_{10}$ still remains (seen in Figure 4a). The melting of $Li_4BN_3H_{10}$ and/or Li_2BNH_6 may create a unique reaction environment, allowing interface reaction and mass transport of Reaction (1) to proceed at a faster rate [40]. Figure 4b shows the FTIR spectra of dehydrogenation and re-hydrogenation samples. The symmetric and asymmetric N–H stretching vibrations of $Li_4BN_3H_{10}$ are at 3243 and 3302 cm^{-1}, while the symmetric and asymmetric N–H stretching vibrations of $LiNH_2$ are at 3257 and 3310 cm^{-1}. Corresponding to the results of XRD characterization, the vibrations of $Li_4BN_3H_{10}$ can be found in the dehydrogenation and re-hydrogenation products of T_2 sample. Isothermal dehydrogenation/re-hydrogenation processes of T_2 sample are not fully reversible due to the kinetic and/or thermodynamic reasons. $Li_2Mg_2(NH)_3$ or MgNH is found in the re-hydrogenated sample, which exhibits the vibration at 3197 cm^{-1}.

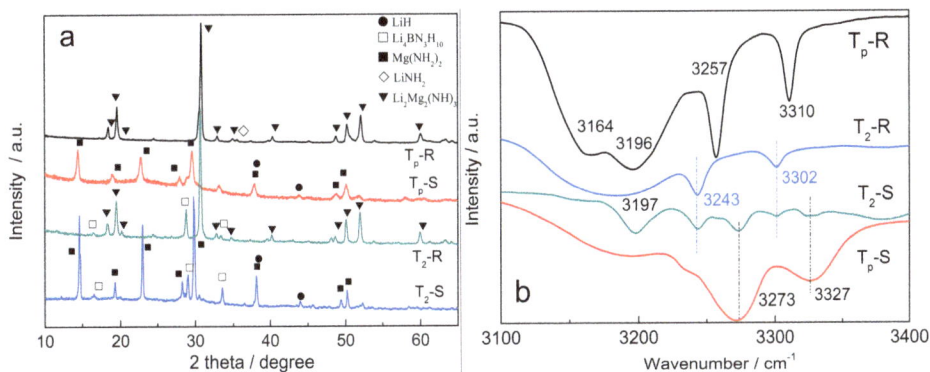

Figure 4. (a) XRD patterns and (b) FT-IR spectrum of $2Mg(NH_2)_2$–$3LiH$ (T_p) and $2Mg(NH_2)_2$–$3LiH$–$LiBH_4$ (T_2) samples release (R) and soak (S) at 460 K.

2.2. Pressure–Composition–Isotherm (PCI) Dehydrogenation at High Pressure

Pressure–Composition–Isotherm (PCI) dehydrogenation curves at 473 K of $2Mg(NH_2)_2$–$3LiH$ (T_p), $2Mg(NH_2)_2$–$3LiH$–$LiBH_4$ (T_2) and $2Mg(NH_2)_2$–$3LiH$–$1/3LiBH_4$ (T_4) samples are shown in Figure 5. It can be seen that the equilibrium plateau pressures of T_2 and T_4 composites at 473 K are almost the same. However, in the region where the amount of H_2 desorption is between 0 and 1.2 wt% (2 hydrogen atoms), the dehydrogenation curves of T_2 and T_4 at the higher pressure show certain differences (see the insert figure), i.e., there is a higher pressure slope of the T_2 ($2Mg(NH_2)_2$–$3LiH$–$LiBH_4$) composite than T_4 ($2Mg(NH_2)_2$–$3LiH$–$1/3LiBH_4$)

composite [36]. The sloping curve indicates that the constitution of the material changes along with the hydrogen desorption. T_2 samples with different amount of hydrogen desorption are characterized by means of X-ray diffraction (XRD), Fourier transform infrared spectroscopy (FTIR), and magic angle spinning nuclear magnetic resonance (MAS NMR).

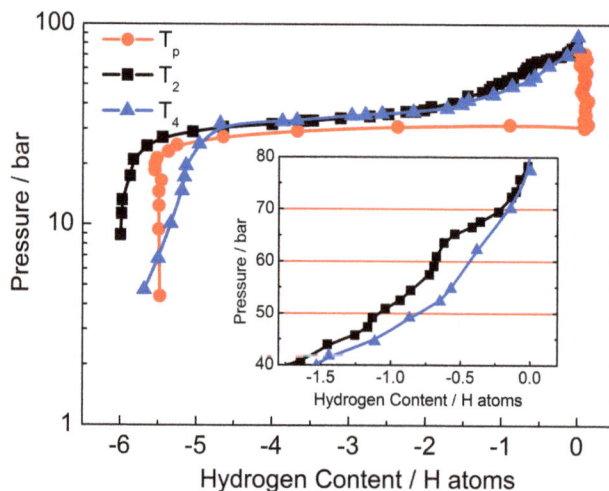

Figure 5. Dehydrogenation Pressure–Composition–Isotherm (PCI) curves of $2Mg(NH_2)_2$–$3LiH$ (T_p), $2Mg(NH_2)_2$–$3LiH$–$LiBH_4$ (T_2) and $2Mg(NH_2)_2$–$3LiH$–$1/3LiBH_4$ (T_4) composite at 473 K. The inset is the PCI curves at H/(mol of T_2 and T_4): 0 to 2.

XRD patterns of T_2 after ball milling and desorbing different amount of hydrogen at 460 K are showed in Figure 6, the pattern of fresh-made sample confirms the presence of starting materials $Mg(NH_2)_2$, LiH and $LiBH_4$. When T_2 desorbs 0.25 wt% hydrogen, the peaks of $LiBH_4$ disappear and new peaks assignable to $Li_4BN_3H_{10}$ are visible. As the amount of desorption is up to 0.58 wt%, peaks belonging to Li_2BNH_6 and $Li_4BN_3H_{10}$ appear at the same time. However, with the increasing amount of released hydrogen, the peaks of $Li_4BN_3H_{10}$ become stronger, the peaks of Li_2BNH_6 are getting weaker and finally disappear upon 1.4 wt% hydrogen is desorbed. Because diffraction peaks of Li_2BNH_6 are weak and some of them are overlapped with those of $Li_4BN_3H_{10}$, FTIR are preformed synchronously.

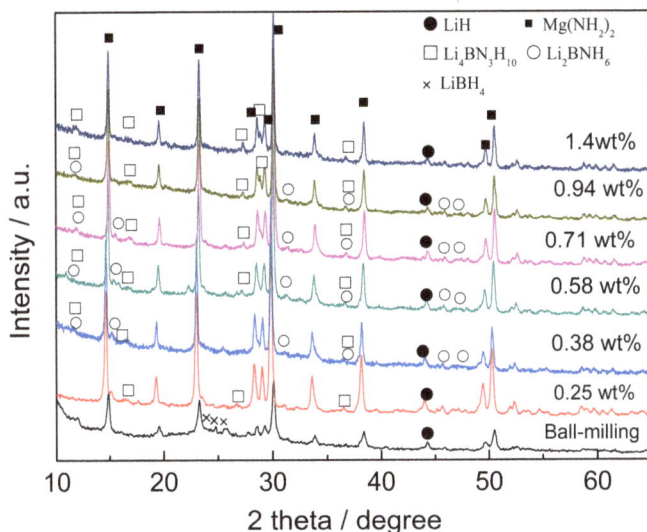

Figure 6. XRD patterns of T_2 (2Mg(NH$_2$)$_2$–3LiH–LiBH$_4$) samples collected at 460 K at different degree of dehydrogenation: 0 wt%, ball-milling stage; 0.25 wt%; 0.38 wt%; 0.58 wt%; 0.71 wt%; 0.94 wt%; and 1.4 wt%.

FTIR spectra of N–H stretching vibrations from the corresponding samples are shown in Figure 7. The B–H stretching frequencies (vibrations from 2000 to 2600 cm^{-1}) were also investigated (Figure S2), but due to the sensitivity of the equipment of FT-IR, there is not much useful information that can be gained. In Figure 7, the post-milled composite gives typical N–H stretching vibrations at 3327 and 3273 cm^{-1} assigned to Mg(NH$_2$)$_2$, which means Mg(NH$_2$)$_2$ does not react with LiBH$_4$ during the ball milling process. During the PCI dehydrogenation, LiNH$_2$ produced by Reaction (1) is immediately combined with the nearby LiBH$_4$, which causes that the typical N–H stretching vibrations at 3301 and 3258 cm^{-1} assigned to LiNH$_2$ are not observed. Furthermore, the spectral signals at 3243 and 3301 cm^{-1}, which are assigned to the characteristic vibrations of Li$_4$BN$_3$H$_{10}$, suggest that LiNH$_2$ reacts with LiBH$_4$ to form Li$_4$BN$_3$H$_{10}$ at the beginning of PCI dehydrogenation. It is likely that the LiNH$_2$–LiBH$_4$ interface reaction produces the Li$_2$BNH$_6$ layer and Li$_4$BN$_3$H$_{10}$ layer. However, the possibility of melting and decomposition of Li$_2$BNH$_6$ to Li$_4$BN$_3$H$_{10}$ and LiBH$_4$ may also exist during the PCI measures. The vibrations at 3246 cm^{-1} and 3294 cm^{-1} assigned to Li$_2$BNH$_6$ appear after the dehydrogenation of 0.58 wt%, which corresponds well with the XRD measurement. The solid-state phase of Li$_2$BNH$_6$ may originate from the quenching of the melts. The vibrations of Li$_2$BNH$_6$ are generally weak and broad, and companied with those of Li$_4$BN$_3$H$_{10}$. Li$_2$BNH$_6$ finally disappears upon desorption of 1.4 wt% hydrogen. We tentatively propose that the formation of Li_2BNH_6 and $Li_2Mg_2(NH)_3$ might be the reason for the high pressure slope in the initial dehydrogenation step.

As shown in Figure 8, ^{11}B MAS NMR spectra of T_2 samples at the different dehydrogenation degree can be fitted into three species, *i.e.*, $LiBH_4$ (−41.62 ppm), $Li_4BN_3H_{10}$ (−39.81 ppm) and Li_2BNH_6 (−37.88 ppm). Figure 8a shows that the peak of ^{11}B signal gradually becomes asymmetric and shifts to the lower chemical shift in pace with the increasing degree of dehydrogenation of T_2 sample. After the weight loss of hydrogen reached 0.58 wt%, the board peak starts to appear the signal of Li_2BNH_6 (−37.88 ppm), which exhibits strongest intensity when desorption weight of hydrogen comes up to 0.94 wt%, plotted in Figure 8b. To summarize the tendency of the peak shift, Figure 8c reveals that the signal area of $Li_4BN_3H_{10}$ increases and the signal peak of $LiBH_4$ decreases with the dehydrogenation. The signal of Li_2BNH_6 can be distinguished upon releasing 0.58 wt% hydrogen and it soon becomes invisible upon releasing 1.4 wt% hydrogen, which is consistent with the findings of XRD and FT-IR results.

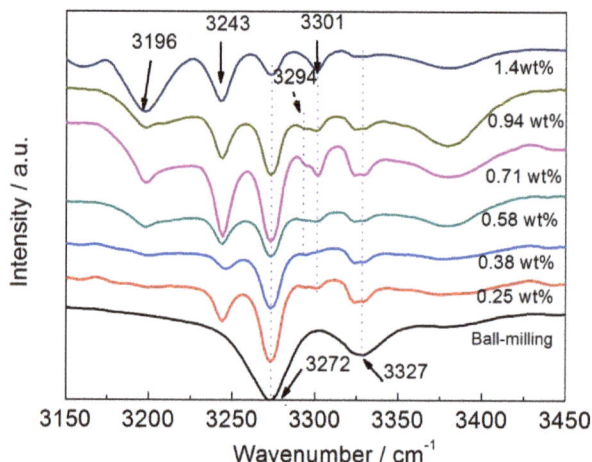

Figure 7. FT-IR spectra of N–H stretching vibrations from T_2 ($2Mg(NH_2)_2$–$3LiH$–$LiBH_4$) samples collected at 460 K at different degree of dehydrogenation: 0 wt%, ball-milling stage; 0.25 wt%; 0.38 wt%; 0.58 wt%; 0.71 wt%; 0.94 wt%; and 1.4 wt%.

Based on above analyses, the reaction of equimolar $LiNH_2$ and $LiBH_4$ is very likely to happen during the high-pressure slope of T_2 sample. The melted $LiNH_2$–$LiBH_4$ is cooled to room temperature, where $Li4BN_3H_{10}$, $LiBH_4$, and a small amount of solid Li_2BNH_6 can be found [39]. The existence of melted $LiNH_2$–$LiBH_4$ may result in a higher desorption pressure. Future work is still needed to make the mechanism clear.

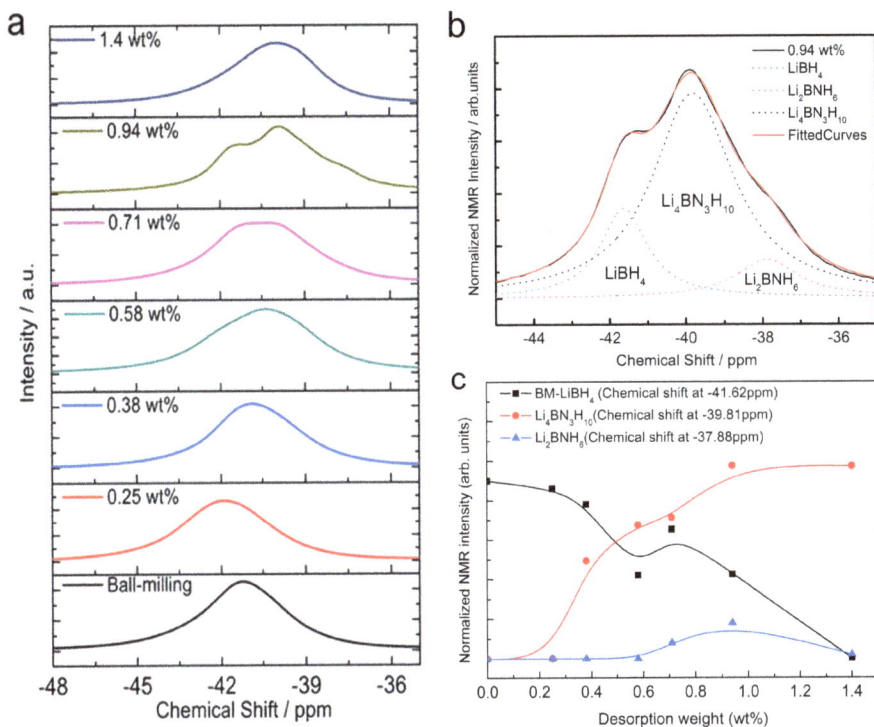

Figure 8. (**a**) [11]B magic angle spinning (MAS) NMR spectra of T_2 (2Mg(NH$_2$)$_2$–3LiH–LiBH$_4$) samples collected at 460 K at different degree of dehydrogenation: fresh-made, 0.25 wt%, 0.38 wt%, 0.58 wt%, 0.71 wt%, 0.94 wt%, and 1.4 wt%; (**b**) hydrogen desorption fitted NMR spectra of Figure 8a (0.94 wt%) in comparison with LiBH$_4$, Li$_2$BNH$_6$ and Li$_4$BN$_3$H$_{10}$; and (**c**) the normalized [11]B magic angle spinning (MAS) NMR peak intensities of LiBH$_4$ (−41.62 ppm), Li$_2$BNH$_6$ (−37.88 ppm) and Li$_4$BN$_3$H$_{10}$ (−39.81 ppm) in samples with different degree of hydrogen desorption.

3. Experimental Section

Lithium hydride (LiH, 99%) (Alfa Aesar, Ward Hill, MA, USA) and Lithium borohydride (LiBH$_4$, 95%) (Sigma Aldrich Fine Chemicals, St. Louis, MO, USA) were used without further purification. Magnesium amide (Mg(NH$_2$)$_2$) was synthesized by reacting metallic Mg powder (99%, Sigma) with NH$_3$.

A Retsch PM400 planetary (Haan, Germany) mill was used to ball mill the mixtures of Mg(NH$_2$)$_2$, LiH and LiBH$_4$ with different molar ratios, such as 2:3:1 and 2:3:1/3 (abbreviated as T_2 and T_4) at 200 rpm for 36 h. Pristine 2Mg(NH$_2$)$_2$–3LiH composite (short for T_p) without additives was also prepared under the same conditions for comparison. Li$_2$Mg$_2$(NH)$_3$ was synthesized by heating post-milled

$Mg(NH_2)_2$–LiH composite under vacuum for 24 h. Moreover, Li_2BNH_6 was synthesized by ball-milling the mixture of $LiBH_4$ and $LiNH_2$ with the same molar number for 24 h under the atmosphere of argon. To avoid oxygen and moisture contaminations, all the sample loadings were conducted inside a glove box that was filled with purified argon ($O_2 < 10$ ppm, $H_2O < 0.1$ ppm).

Thermal decomposition properties of samples were carried on a custom-built temperature programmed desorption (TPD)-mass spectrometer (MS, Hiden Analytical Limited, Warrington, UK) combined system. About 10 mg sample was tested each time at a ramping rate of 2 °C/min. H_2 and NH_3 signals were detected at the m/z ratios of 2 and 15, respectively. Thermogravimetry coupled differential thermal analysis (TG-DTA) measurements were conducted on a STA-449C (Netzsch, Wittelsbacherstraße, Germany), which was installed in the glove box as mentioned above. The heating rate was 2 °C/min and Ar was the carrier gas.

Pressure–Composition–Isotherm (PCI) measurements and dehydrogenation/ hydrogenation experiments were carried out on an automatic Sieverts-type apparatus (Hy-Energy scientific instruments PCT pro-2000, Newark, CA, USA). A sample of *ca.* 200 mg was used each time. Initial pressures in the sample chamber for dehydrogenation and re-hydrogenation experiments were 0.01 and 6 Mpa, respectively.

Isothermal dehydrogenation and re-hydrogenation experiments were carried out on an automatic Sieverts-type apparatus (Advanced Materials Co., PCT, Pittsburgh, PA, USA). Initial pressure in sample chamber for hydrogen desorption was 0.01 MPa, and for absorption was 6 MPa.

Powder X-ray diffraction (XRD) patterns were recorded over a 2θ range of 5°–80° on an X'Pert Pro (PANAnalytical) diffractometer (PANalytical, Almelo, The Netherlands) with $K\alpha$ radiation at 40 kV and 40 mA. The powder was placed on a self-made cell sealed with an air-tight hood. Fourier transform IR (FTIR) measurements were conducted on a Varian 3100 unit in DRIFT mode to detect the N–H vibration for metal amides or imides. Solid-state [11]B nuclear magnetic resonance (NMR) measurements were conducted on a Bruker Advance III 500 NMR spectrometer (Berlin, Germany) with a 4 mm MAS NMR probe working at a frequency of 128.28 MHz.

4. Conclusions

Increasing $LiBH_4$ amount in the $2Mg(NH_2)_2$–3LiH composite leads to a higher dehydrogenation pressure in the slope region. The reason for this change may be due to the melted reaction of $LiNH_2$–$LiBH_4$. Moreover, the kinetics of $2Mg(NH_2)_2$–3LiH–$LiBH_4$ composite are significantly improved in comparison with pristine $2Mg(NH_2)_2$–3LiH sample. Especially, it desorbs *ca.* 80% hydrogen in 100 min and re-hydrogenates 80% hydrogen in 50 min isothermally at 416 K, which is *ca.* 2 times and 18 times as fast as those of $2Mg(NH_2)_2$–3LiH composite.

Supplementary Materials: Supplementary materials can be accessed at: http://www.mdpi.com/1996-1073/8/7/6898/s1.

Acknowledgments: We acknowledge financial support from the project of National Natural Science Funds for Distinguished Young Scholar (51225206), projects of National Natural Science Foundation of China (Grant Nos 21273229, U1232120, 51301161, 21473181 and 51472237) and CAS-Helmholtz Association Collaborative Funding.

Author Contributions: Han Wang conceived and designed the experiments; Han Wang and Hujun Cao performed the experiments; Han Wang, Ping Chen and Guotao Wu analyzed the data; Teng He and Guotao Wu contributed analysis tools; Han Wang wrote the paper; Guotao Wu and Ping Chen revised it critically for intellectual content; Guotao Wu is the person for final approval of the version to be published.

Conflicts of Interest: The authors declare no conflict of interest.

References

1. Schlapbach, L.; Zuttel, A. Hydrogen-storage materials for mobile applications. *Nature* **2001**, *414*, 353–358.
2. Chen, P.; Zhu, M. Recent progress in hydrogen storage. *Mater. Today* **2008**, *11*, 36–43.
3. Bogdanovic, B.; Schwickardi, M. Ti-doped alkali metal aluminium hydrides as potential novel reversible hydrogen storage materials. *J. Alloy. Compd.* **1997**, *253*, 1–9.
4. Chen, J.; Kuriyama, N.; Xu, Q.; Takeshita, H.T.; Sakai, T. Reversible hydrogen storage via titanium-catalyzed $LiAlH_4$ and Li_3AlH_6. *J. Phys. Chem. B* **2001**, *105*, 11214–11220.
5. Fichtner, M.; Fuhr, O.; Kircher, O. Magnesium alanate—A material for reversible hydrogen storage. *J. Alloy. Compd.* **2003**, *356*, 418–422.
6. Chlopek, K.; Frommen, C.; Leon, A.; Zabara, O.; Fichtner, M. Synthesis and properties of magnesium tetrahydroborate, $Mg(BH_4)_2$. *J. Mater. Chem.* **2007**, *17*, 3496–3503.
7. Miwa, K.; Aoki, M.; Noritake, T.; Ohba, N.; Nakamori, Y.; Towata, S.; Zuttel, A.; Orimo, S. Thermodynamic stability of calcium borohydride $Ca(BH_4)_2$. *Phys. Rev. B* **2006**, *74*.
8. Zuttel, A.; Rentsch, S.; Fischer, P.; Wenger, P.; Sudan, P.; Mauron, P.; Emmenegger, C. Hydrogen storage properties of $LiBH_4$. *J. Alloy. Compd.* **2003**, *356*, 515–520.
9. Chen, P.; Xiong, Z. Metal–N–H systems for the hydrogen storage. *Scr. Mater.* **2007**, *56*, 817–822.
10. Chen, P.; Xiong, Z.T.; Luo, J.Z.; Lin, J.Y.; Tan, K.L. Interaction of hydrogen with metal nitrides and imides. *Nature* **2002**, *420*, 302–304.
11. Luo, W. ($LiNH_2$–MgH_2): A viable hydrogen storage system. *J. Alloy. Compd.* **2004**, *381*, 284–287.
12. Nakamori, Y.; Kitahara, G.; Orimo, S. Synthesis and dehydriding studies of Mg–N–H systems. *J. Power Sources* **2004**, *138*, 309–312.
13. Xiong, Z.; Wu, G.; Hu, J.J.; Chen, P. Ternary imides for hydrogen storage. *Adv. Mater.* **2004**, *16*, 1522–1525.

14. Wu, H. Structure of ternary imide $Li_2Ca(NH)_2$ and hydrogen storage mechanisms in amide–hydride system. *J. Am. Chem. Soc.* **2008**, *130*, 6515–6522.

15. Xiong, Z.; Wu, G.; Hu, J.; Liu, Y.; Chen, P.; Luo, W.; Wang, J. Reversible hydrogen storage by a Li–Al–N–H complex. *Adv. Funct. Mater.* **2007**, *17*, 1137–1142.

16. Hino, S.; Ichikawa, T.; Leng, H.Y.; Fujii, H. Hydrogen desorption properties of the Ca–N–H system. *J. Alloy. Compd.* **2005**, *398*, 62–66.

17. Hu, J.; Liu, Y.; Wu, G.; Xiong, Z.; Chen, P. Structural and compositional changes during hydrogenation/dehydrogenation of the Li–Mg–N–H system. *J. Phys. Chem. C* **2007**, *111*, 18439–18443.

18. Leng, H.; Ichikawa, T.; Fujii, H. Hydrogen storage properties of Li–Mg–N–H systems with different ratios of $LiH/Mg(NH_2)_2$. *J. Phys. Chem. B* **2006**, *110*, 12964–12968.

19. Luo, W.; Wang, J.; Stewart, K.; Clift, M.; Gross, K. Li–Mg–N–H: Recent investigations and development. *J. Alloy. Compd.* **2007**, *446*, 336–341.

20. Xiong, Z.T.; Wu, G.T.; Hu, J.J.; Chen, P.; Luo, W.F.; Wang, J. Investigations on hydrogen storage over Li–Mg–N–H complex—The effect of compositional changes. *J. Alloy. Compd.* **2006**, *417*, 190–194.

21. Chen, X.Y.; Guo, Y.H.; Yu, X.B. Enhanced dehydrogenation properties of modified $Mg(NH_2)_2$–$LiBH_4$ composites. *J. Phys. Chem. C* **2010**, *114*, 17947–17953.

22. Noritake, T.; Aoki, M.; Towata, S.; Ninomiya, A.; Nakamori, Y.; Orimo, S. Crystal structure analysis of novel complex hydrides formed by the combination of $LiBH_4$ and $LiNH_2$. *Appl. Phys. A* **2006**, *83*, 277–279.

23. Weidner, E.; Dolci, F.; Hu, J.J.; Lohstroh, W.; Hansen, T.; Bull, D.J.; Fichtner, M. Hydrogenation reaction pathway in $Li_2Mg(NH)_2$. *J. Phys. Chem. C* **2009**, *113*, 15772–15777.

24. Li, C.; Liu, Y.F.; Gu, Y.J.; Gao, M.X.; Pan, H.G. Improved hydrogen-storage thermodynamics and kinetics for an RbF-doped $Mg(NH_2)_2$–2LiH system. *Chem.Asian J.* **2013**, *8*, 2136–2143.

25. Liang, C.; Liu, Y.F.; Gao, M.X.; Pan, H.G. Understanding the role of K in the significantly improved hydrogen storage properties of a KOH-doped Li–Mg–N–H system. *J. Mater. Chem. A* **2013**, *1*, 5031–5036.

26. Liu, Y.F.; Li, C.; Li, B.; Gao, M.X.; Pan, H.G. Metathesis reaction-induced significant improvement in hydrogen storage properties of the KF-added $Mg(NH_2)_2$–2LiH system. *J. Phys. Chem. C* **2013**, *117*, 866–875.

27. Wang, J.H.; Liu, T.; Wu, G.T.; Li, W.; Liu, Y.F.; Araujo, C.M.; Scheicher, R.H.; Blomqvist, A.; Ahuja, R.; Xiong, Z.T.; *et al.* Potassium-modified $Mg(NH_2)_2/2$ LiH system for hydrogen storage. *Angew. Chem. Int. Ed.* **2009**, *48*, 5828–5832.

28. Hu, J.J.; Fichtner, M.; Chen, P. Investigation on the properties of the mixture consisting of $Mg(NH_2)_2$, LiH, and $LiBH_4$ as a hydrogen storage material. *Chem. Mater.* **2008**, *20*, 7089–7094.

29. Hu, J.J.; Liu, Y.F.; Wu, G.T.; Xiong, Z.T.; Chua, Y.S.; Chen, P. Improvement of hydrogen storage properties of the Li–Mg–N–H system by addition of $LiBH_4$. *Chem. Mater.* **2008**, *20*, 4398–4402.

30. Durojaiye, T.; Hayes, J.; Goudy, A. Rubidium hydride: An exceptional dehydrogenation catalyst for the lithium amide/magnesium hydride system. *J. Phys. Chem. C* **2013**, *117*, 6554–6560.

31. Yang, J.; Sudik, A.; Siegel, D.J.; Halliday, D.; Drews, A.; Carter, R.O., 3rd; Wolverton, C.; Lewis, G.J.; Sachtler, J.W.; Low, J.J.; *et al.* A self-catalyzing hydrogen-storage material. *Angew. Chem. Int. Ed.* **2008**, *47*, 882–887.

32. Li, B.; Liu, Y.F.; Gu, J.; Gao, M.X.; Pan, H.G. Synergetic effects of in situ formed CaH_2 and $LiBH_4$ on hydrogen storage properties of the Li–Mg–N–H system. *Chem. Asian J.* **2013**, *8*, 374–384.

33. Li, B.; Liu, Y.F.; Gu, J.; Gu, Y.J.; Gao, M.X.; Pan, H.G. Mechanistic investigations on significantly improved hydrogen storage performance of the $Ca(BH_4)_2$-added $2LiNH_2/MgH_2$ system. *Int. J. Hydrog. Energy* **2013**, *38*, 5030–5038.

34. Pan, H.G.; Shi, S.B.; Liu, Y.F.; Li, B.; Yang, Y.J.; Gao, M.X. Improved hydrogen storage kinetics of the Li–Mg–N–H system by addition of $Mg(BH_4)_2$. *Dalton Trans.* **2013**, *42*, 3802–3811.

35. Li, B.; Liu, Y.F.; Li, C.; Gao, M.X.; Pan, H.G. *In situ* formation of lithium fast-ion conductors and improved hydrogen desorption properties of the $LiNH_2$–MgH_2 system with the addition of lithium halides. *J. Mater. Chem. A* **2014**, *2*, 3155–3162.

36. Cao, H.J.; Wu, G.T.; Zhang, Y.; Xiong, Z.T.; Qiu, J.S.; Chen, P. Effective thermodynamic alteration to $Mg(NH_2)_2$–LiH system: Achieving near ambient-temperature hydrogen storage. *J. Mater. Chem. A* **2014**, *2*, 15816–15822.

37. Chater, P.A.; David, W.I.F.; Anderson, P.A. Synthesis and structure of the new complex hydride $Li_2BH_4NH_2$. *Chem. Commun.* **2007**, *45*, 4770–4772.

38. Wu, H.; Zhou, W.; Udovic, T.J.; Rush, J.J.; Yildirim, T. Structures and crystal chemistry of Li_2BNH_6 and $Li_4BN_3H_{10}$. *Chem. Mater.* **2008**, *20*, 1245–1247.

39. Borgschulte, A.; Jones, M.O.; Callini, E.; Probst, B.; Kato, S.; Zuttel, A.; David, W.I.F.; Orimo, S. Surface and bulk reactions in borohydrides and amides. *Energy Environ. Sci.* **2012**, *5*, 6823–6832.

40. Matsuo, M.; Remhof, A.; Martelli, P.; Caputo, R.; Ernst, M.; Miura, Y.; Sato, T.; Oguchi, H.; Maekawa, H.; Takamura, H.; *et al.* Complex hydrides with $(BH_4)^-$ and $(NH_2)^-$ anions as new lithium fast-ion conductors. *J. Am. Chem. Soc.* **2009**, *131*, 16389–16391.

Recent Advances in the Use of Sodium Borohydride as a Solid State Hydrogen Store

Jianfeng Mao and Duncan H. Gregory

Abstract: The development of new practical hydrogen storage materials with high volumetric and gravimetric hydrogen densities is necessary to implement fuel cell technology for both mobile and stationary applications. $NaBH_4$, owing to its low cost and high hydrogen density (10.6 wt%), has received extensive attention as a promising hydrogen storage medium. However, its practical use is hampered by its high thermodynamic stability and slow hydrogen exchange kinetics. Recent developments have been made in promoting H_2 release and tuning the thermodynamics of the thermal decomposition of solid $NaBH_4$. These conceptual advances offer a positive outlook for using $NaBH_4$-based materials as viable hydrogen storage carriers for mobile applications. This review summarizes contemporary progress in this field with a focus on the fundamental dehydrogenation and rehydrogenation pathways and properties and on material design strategies towards improved kinetics and thermodynamics such as catalytic doping, nano-engineering, additive destabilization and chemical modification.

Reprinted from *Energies*. Cite as: Mao, J.; Gregory, D.H. Recent Advances in the Use of Sodium Borohydride as a Solid State Hydrogen Store. *Energies* **2015**, *8*, 430–453.

1. Introduction

With concerning current trends in environmental pollution and depletion of fossil energy resources, there is an imperative to seek renewable and clean energy sources that can support the continued sustainable development of human society. Hydrogen is regarded as one of the best alternative sustainable energy carriers because of its abundance, high energy density and lack of adverse environmental impact (for example, when oxidized as water). However, an important challenge for the use of hydrogen for mobile (e.g., automotive) and small scale energy generation is how to achieve safe, cheap, high density storage [1]. Essentially, hydrogen can be stored either in a physical form (as a gas or liquid) or in a chemical form (e.g., within metal hydrides or so-called chemical hydrides). Compressed gas and liquid hydrogen storage technologies represent the current state-of-the art, but more compact (gravimetrically and/or volumetrically efficient) means of storing hydrogen are needed for mobile applications on a practical level. In principle, solid state hydrogen storage in metal hydrides is considered a more effective and safer way to handle hydrogen than its storage as either a compressed gas or cryogenic liquid.

The hydrides offer volumetric hydrogen densities substantially greater than that of compressed gas and comparable to or exceeding that of liquid hydrogen but without the requirement of very high pressure containment vessels or cryogenic tanks [2–4]. An ideal on-board hydrogen storage material will have a low molar weight, be inexpensive, have rapid kinetics for absorbing and desorbing H_2 in the 25–120 °C temperature range, and store large quantities of hydrogen reversibly [5]. Recently, light metal borohydrides such as $NaBH_4$ [6–8], $LiBH_4$ [9–12], $Mg(BH_4)_2$ [13–15], and $Ca(BH_4)_2$ [16–18] have attracted much attention as potential hydrogen storage media primarily due to their high gravimetric capacities. The physical and chemical properties of these borohydrides are shown in Table 1 [19,20]. Perhaps compared to the borohydrides of lithium, magnesium and calcium there has been little focus on $NaBH_4$ for hydrogen storage in the solid state due to its relatively much higher decomposition temperature. Given that the decomposition temperature of $NaBH_4$ at 1 bar of H_2 is in excess of 500 °C, the required operating temperature for a store would considerably exceed that required for practical application in hydrogen fuel cell vehicles [7]. This fact alone explains why most previous research has been conducted on the hydrolysis of $NaBH_4$ for hydrogen generation rather than its thermolysis as part of a solid state storage system [21]. However, the gravimetric hydrogen storage capacity of real hydrolysis-based storage systems will invariably be lower than the theoretical 10.6 wt% figure due to the excess water required to dissolve the $NaBH_4$ and its by-product, $NaBO_2$, as well as the added mass of the reaction and storage vessels. Hence, the U.S. Department of Energy (US DOE) issued a "No-Go" recommendation for the hydrolysis of $NaBH_4$ in 2007, and since then the approach has no longer been seriously considered for automotive applications [22].

Table 1. Physical and chemical properties of borohydrides.

Borohydride	Cost [a]/(USD/g)	Hydrogen density/wt%	T_d [b]/°C	Reaction	References
$NaBH_4$	6.47	10.6	505	$NaBH_4 \rightarrow Na + B + 2H_2$	[6–8]
$LiBH_4$	15.65	18.5	380	$LiBH_4 \rightarrow Li + B + 2H_2$	[9–12]
$Mg(BH_4)_2$	116.5	14.9	320	$Mg(BH_4)_2 \rightarrow MgB_2 + 4H_2$	[13–15]
$Ca(BH_4)_2$	142	11.6	367	$Ca(BH_4)_2 \rightarrow 2/3CaH_2 + 1/3CaB_6 + 10/3H_2$	[16–18]

[a] Prices from Sigma-Aldrich [23] for hydrogen storage grade materials; and
[b] dehydrogenation temperature.

However, use of solid $NaBH_4$ for hydrogen storage has many advantages. Compared to other borohydrides, $NaBH_4$ is cheaper and relatively stable in air [24]. For $NaBH_4$ to be suitable for practical applications, the desorption temperature must be reduced and appreciable cyclability must be demonstrated. Over the last several years some novel strategies such as catalysis, nano-engineering, additive destabilization and chemical modification have been employed to address the thermodynamic and kinetic limitations of the thermal decomposition of $NaBH_4$.

While $NaBH_4$ is not yet the solution to the problem of facile storage of hydrogen in the solid state, the progress of the various methodologies in improving both performance and understanding of this performance has been highly encouraging and hence we focus on these advances in this review. The primary purpose of this paper is to consider progress largely from 2009 onwards, comparing what is known regarding the decomposition behavior and mechanism of pristine $NaBH_4$ with materials modified using the approaches listed above and the prospects of such systems for practical exploitation.

2. Thermal Decomposition

$NaBH_4$ adopts a NaCl-type structure at ambient conditions in which four hydrogen atoms are covalently stabilized within the BH_4^- anion, which in turn is bonded essentially ionically to the counter-cation Na^+ [24]. The complete hydrogen desorption reaction of $NaBH_4$ can be expressed as follows:

$$NaBH_4 \rightarrow Na + B + 2H_2 \qquad 10.6\,wt\% \tag{1}$$

However, like many other borohydrides, the real decomposition process of $NaBH_4$ is likely to be more complex and involve intermediate phases such as NaH, $Na_2B_{12}H_{12}$, or even release impurity gases such as B_2H_6 [6]. Recent theoretical and experimental studies have provided insight into its decomposition behavior as well as identifying the extent of the challenges ahead in developing $NaBH_4$ as a viable hydrogen carrier.

First principles calculations suggest a scenario where BH_4^- ions decompose at the surface of $NaBH_4$ into H^- ions and BH_3 molecules [25]. The H^- ions remain in the lattice, locally converting $NaBH_4$ into NaH. The BH_3 molecules originating from the decomposition can escape to the gas phase and form B_2H_6 (diborane) molecules, for instance. Alternatively, they may decompose immediately to form hydrogen and B. However, there is no direct evidence from mass spectrometry data of B_2H_6 release during the decomposition of $NaBH_4$ [6]. Despite this lack of experimental evidence, it remains possible that $NaBH_4$ decomposition may involve diborane emission, as is seen in the thermal decomposition of less stable borohydrides [20]. Due to the high temperature necessary for decomposition on the one hand and the low thermal stability of diborane on the other, most of the diborane decomposes into the elements. At the same time, some of the gaseous species may react with remaining $NaBH_4$ to form $Na_2B_{12}H_{12}$.

Martelli *et al.* [7] investigated the stability and hydrogen desorption of $NaBH_4$ via dynamic pressure, composition, and temperature (PCT) measurements under constant hydrogen flows. It was found that only one plateau is visible in the isotherms, indicating that the decomposition occurs in one step (Figure 1). From the van't Hoff equation, the enthalpy and entropy of reaction are $-108 \pm 3 \, \text{kJ} \cdot \text{mol}^{-1}$ of H_2 and $133 \pm 3 \, \text{J} \cdot \text{K}^{-1} \cdot \text{mol}^{-1}$ of H_2 respectively. This corresponds to a decomposition temperature, $T_d = 534 \pm 10 \, °C$ at 1 bar of H_2. The high stability of $NaBH_4$ leads to a dehydrogenation temperature that is above the decomposition temperature of NaH [26].

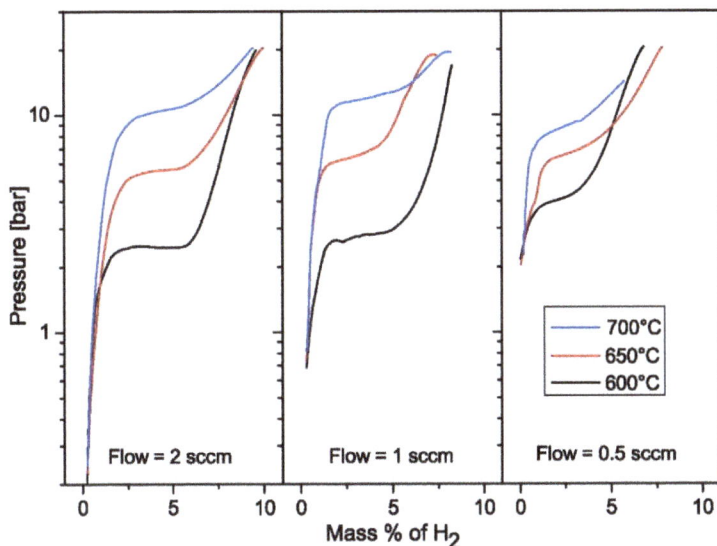

Figure 1. Pressure, composition, and temperature (PCT) isotherms measured on $NaBH_4$ at a constant hydrogen flow of 2, 1, and 0.5 cm^3 (STP) min^{-1} [7]. Reprinted with permission from [7], copyright 2010 The American Chemical Society.

Therefore, NaH is thermodynamically unstable under the decomposition conditions of $NaBH_4$ and will decompose into the corresponding elements without changing the observed equilibrium pressure. This rationalises the experimental analysis that Na is contained in the residue (as either Na or NaH). A second phase in the product was identified as either elemental boron or a boron-rich phase [7]. The Na:NaH ratio in the residue is determined by the reaction kinetics involved. The presence of traces of NaH in the product shows that $NaBH_4$ decays at least partially via NaH, which confirms the theoretical prediction [25].

Recently, the boron-containing intermediate phase $Na_2B_{12}H_{12}$ was found experimentally during the decomposition of $NaBH_4$ (and its composites). For example, Mao *et al.* [27] confirmed the formation of $Na_2B_{12}H_{12}$ by Fourier transform infrared spectroscopy (FTIR) in the decomposition of TiF_3-doped $NaBH_4$ or CaH_2-6$NaBH_4$ and $Ca(BH_4)_2$-4$NaBH_4$ composites. In contrast, Garroni *et al.* [28] detected amorphous $Na_2B_{12}H_{12}$ by nuclear magnetic resonance (NMR) in partially dehydrogenated 2$NaBH_4$-MgH_2 and in the final products of the decomposition reaction. More recently, Ngene *et al.* [29] detected $Na_2B_{12}H_{12}$ with ^{11}B solid state NMR after the dehydrogenation of a nanoconfined $NaBH_4$/porous carbon material. The formation mechanism of $Na_2B_{12}H_{12}$ is not yet clear; the borohydride may originate from the reaction of boranes with unreacted $NaBH_4$, which was proposed by the first principles calculations [25]. In fact, a similar decomposition route was proposed for $LiBH_4$, where Friedrichs *et al.* [30] suggested that the formation of $Li_2B_{12}H_{12}$ arises from the reaction of the borane evolving from $LiBH_4$ with the remaining starting material. First principles calculations suggest that $Na_2B_{12}H_{12}$ has significant ionic character and is relatively stable and if it was formed during the thermal decomposition of $NaBH_4$, thermodynamically one would not expect its existence to be fleeting [31]. Moreover, because of its anticipated low reactivity with hydrogen, when formed it might be expected to represent a limiting step in the reverse reaction to the fully hydrogenated $NaBH_4$. In this regard, further research is required to evaluate the effects of the formation of $Na_2B_{12}H_{12}$ on both the $NaBH_4$ dehydrogenation and its subsequent re-hydrogenation.

3. Strategies for Promoting H_2 Release from Solid-State Thermolysis of $NaBH_4$

From the point where $NaBH_4$ hydrolysis was no longer considered for automotive applications by the US DOE, solid-state thermolysis has become the only realistic option for the practical use of the borohydride in hydrogen storage applications. Before this can happen, however, the kinetic and thermodynamic limitations associated with the (de)hydrogenation of $NaBH_4$ must be removed. To this end, several strategies have recently been developed and proven effective in improving the thermally activated H_2 release from $NaBH_4$ and these are considered below.

3.1. Catalytic Doping

Catalysts play an important role in the hydrogen sorption processes in hydrides, since they improve the hydrogen uptake and release kinetics by reducing the activation barrier for diffusion and facilitating hydrogen dissociation. Therefore, it is of particular interest to use catalysts to promote hydrogen exchange reactions in $NaBH_4$ under moderate temperature and pressure conditions.

Mao *et al.* [32] investigated the effects of Ti-based additives, including Ti, TiH_2, and TiF_3, on the dehydrogenation of $NaBH_4$. It was revealed that all of the titanium-based additives were effective in improving the hydrogen desorption and absorption reactions of $NaBH_4$ and among them TiF_3 possessed the highest catalytic activity (Figure 2). Powder X-ray diffraction (PXD) and X-ray photoelectron spectroscopy (XPS) revealed that the dehydrogenation of TiF_3-doped $NaBH_4$ can be regarded as a two-step process: (i) the thermodynamically-favorable reaction between borohydride and fluoride at *ca.* 300 °C ($3NaBH_4 + TiF_3 \rightarrow 3NaF + TiB_2 +$

B + 6H$_2$); and (ii) the dehydrogenation of the remaining NaBH$_4$, catalysed by the NaF and TiB$_2$ formed *in situ* in step (i). The TiF$_3$-doped sample demonstrates good reversibility with *ca.* 4 wt% hydrogen absorbed below 500 °C at 5.5 MPa.

Ni-containing additives including Ni (20 nm), Ni$_3$B, NiCl$_2$, NiF$_2$, and Ni (65 wt%) supported on Si/Al$_2$O$_3$ reduce the dehydrogenation temperature of NaBH$_4$ by at least 60 °C (e.g., 65 wt% Ni on Si/Al$_2$O$_3$) [33]. PXD analysis has indicated that Ni reacts with B evolved during the thermal decomposition of NaBH$_4$ to form Ni$_x$B$_y$ species including Ni$_3$B, Ni$_2$B, and Ni$_3$B$_4$. The thermodynamically favorable formation of these species is likely one reason why the dehydrogenation temperature is reduced. The reversibility is poor however and re-hydrogenation forms NaH with a maximum hydrogen uptake of *ca.* 2 wt% and no activity to hydrogenation from the additives evident. The authors also conducted a catalyst screening study of NaBH$_4$ with a variety of metal nanoparticles, chlorides, borides, and mesoporous materials. The most effective catalysis was performed by Pd nanoparticles inducing a desorption temperature of 420 °C; a decrease of at least 85 °C compared to pristine NaBH$_4$. By analogy to the nickel additives above, the reduction in dehydrogenation temperature is probably enabled by the formation of Pd$_x$B$_y$ intermediate phases. The reversibility of hydrogen uptake and release in the system incorporating Pd (Pd$_x$B$_y$) has yet to be reported.

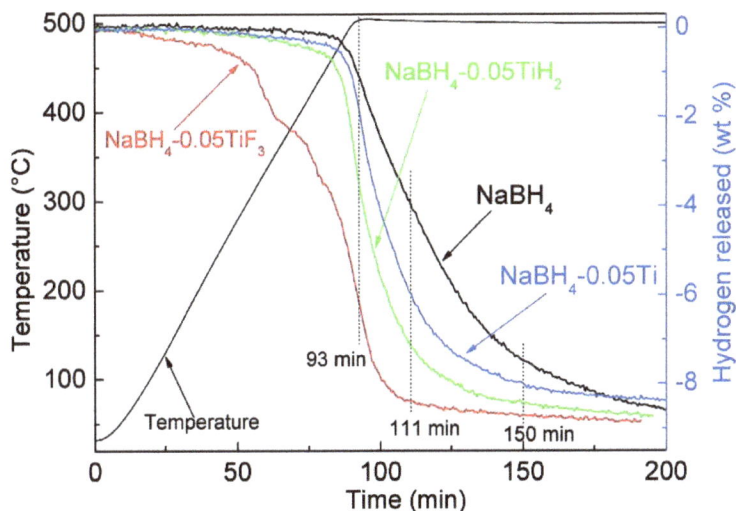

Figure 2. Temperature-programmed desorption (TPD) profiles of NaBH$_4$ with and without different titanium catalysts. The heating rate was 5 °C·min^{-1} [32]. Reprinted with permission from [32], copyright 2012 The American Chemical Society.

3.2. Nano-Engineering

It is well-documented that the physical and chemical properties of nanoparticles can be very different from those of the corresponding bulk materials [34]. Reducing the particle size of the metal hydride to the nanometer range can result in enhanced kinetics and in some cases, modified thermodynamics.

Metal hydride nanoparticles or nanocomposites are usually prepared by high-energy ball milling. However, the lower range of particle sizes obtained from milling is typically limited to hundreds of nanometers and the particle size distribution is usually non-uniform. For $NaBH_4$, Varin and Chiu [35] studied the variation of the cubic lattice parameter and crystallite (grain) size with milling times of up to 200 h. It was found that the lattice parameter of the compound varies only modestly during prolonged milling (maximum ~0.15% after 50 h) and the average crystallite (grain) size remains of the order of a few tens of nanometers. Therefore, it seems that ball milling even under these relatively extreme conditions is rather limited in its ability to nanostructure $NaBH_4$.

One sophisticated approach towards achieving genuinely nanoscale dimensions in the borohydride is to infiltrate the material into a mesoporous host matrix. Such approaches could improve the hydrogen uptake kinetics (and in some cases the thermodynamics) of hydrides significantly. Ampoumogli et al. [36] recently synthesized nanocomposites of $NaBH_4$/CMK-3 (an ordered mesoporous carbon) via the impregnation of the porous carbon with $NaBH_4$ dissolved in liquid ammonia and showed that the nanocomposite releases hydrogen at lower temperatures than bulk $NaBH_4$. Mass spectra however, showed that the released gases contained ammonia, which could either originate from solvent that is incorporated into the pores of the carbon or form a sodium borohydride ammine complex formed during the impregnation process. In contrast, by nano-confining $NaBH_4$ in a highly-ordered Si-based mesoporous scaffold (SBA-15) and its carbon (CMK-3) replica, respectively, through ammonia-free wet chemical impregnation, it was possible to avoid the formation of unwanted by-products [37]. Temperature-programmed desorption (TPD) highlighted a notable reduction in dehydrogenation temperature compared to bulk $NaBH_4$, but the details of the desorption pathway, associated structural evolution and reversibility in this system are not yet clear and require further study. Recently, Ngene et al. [29] synthesized $NaBH_4$/C nanocomposites in which the pores of the matrix were of 2–3 nm in diameter. The materials were prepared using pore volume impregnation either with an aqueous $NaBH_4$ solution (denoted SI) or via melt infiltration (MI). It was found that each method results in a lower dehydrogenation temperature compared to pristine $NaBH_4$ [29]. The onset of hydrogen release can be reduced from 470 °C for the bulk borohydride to less than 250 °C for the nanocomposites (Figure 3). In these cases the dehydrogenated nanocomposites could be partially re-hydrogenated with the absorption of about 43% of the initial hydrogen capacity under 60 bar H_2 at 325 °C. The loss of capacity in this system was directly connected to partial loss of Na during dehydrogenation and this loss could be ameliorated (to retention of 98% of initial capacity) by adding further Na to the nanocomposites.

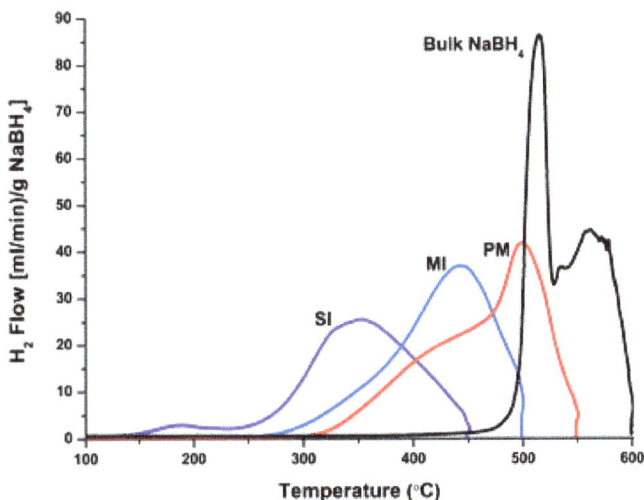

Figure 3. TPD experiments (5 °C· min^{-1} under Ar) showing hydrogen release from bulk $NaBH_4$ (black); a physical mixture of 25 wt% $NaBH_4$ and porous carbon (PM; red); solution impregnated 25 wt% $NaBH_4/C$ nanocomposites (SI; purple) and melt infiltration (MI; blue) 25 wt% $NaBH_4/C$ nanocomposites [29]. Reproduced from [29] with permission of The Royal Society of Chemistry.

Nano-confinement in porous matrices has become a well-traveled bridge connecting bulk and nanoscale hydrogen storage materials. In addition to the unique structures and size-specific chemistry of nanomaterials, the methodology is also expected to introduce a large number of defects. Moreover, the interactions between the M–H bond and the internal surface of the nanopores may also contribute a catalytic effect to the desorption process. All these effects would promote dehydrogenation at lower temperatures. However, the weight penalty of the supporting substrates (hosts) will always reduce the gravimetric hydrogen storage capacity of the system.

In an alternative approach, Christian and Aguey-Zinsou [38,39] synthesized $NaBH_4$ nanoparticles (<30 nm in diameter) by using an anti-solvent precipitation method. The procedure resulted in a decrease of the borohydride melting point and an initial release of hydrogen at 400 °C; *ca.* 100 °C lower than the bulk material. Encapsulation of these nanoparticles upon reaction with nickel chloride yielded core-shell nanostructures, $NaBH_4$@Ni. This core-shell material begins to release hydrogen at 50 °C with significant desorption from 350 °C. Even more remarkably, the core-shell configuration engenders full reversibility to NaBH$_4$ with hydrogen desorption/absorption occurring under 4 MPa at 350 °C (Figure 4). A consistent reversible hydrogen capacity of 5 wt% was achieved for NaBH$_4$@Ni, in which 80% of the hydrogen could be desorbed or absorbed in <60 min and full capacity could be achieved within 5 h. Although these conditions are still far from the ideal requirements for practical applications, this work suggests that the hydrogen storage performance of NaBH$_4$ can be altered dramatically by the integration of nano-engineering and catalysis concepts.

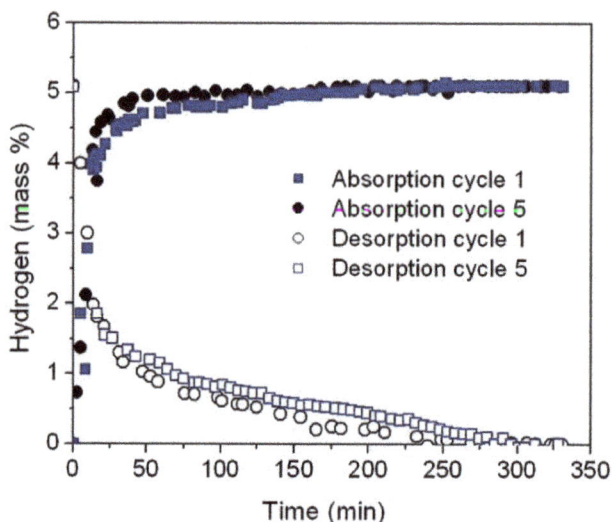

Figure 4. Kinetics of hydrogen desorption at 0.01 MPa and absorption under 4 MPa hydrogen pressure at 350 °C for NaBH$_4$@Ni [38]. Reprinted with permission from [38], copyright 2012 The American Chemical Society.

3.3. Destabilization Using Reactive Additives

Another possible way to lower the decomposition temperature of NaBH$_4$ and further tune its thermodynamic and kinetic characteristics is by the use of certain select additives. As opposed to the use of catalysts, the additive employed in these cases not only promotes the kinetics, but also tunes the thermodynamics through changing the reaction pathway. For example, the dehydrogenation thermodynamics and kinetics of NaBH$_4$ could be significantly improved by combining with fluorographite (FGi) according to the following reaction [40]:

$$x\text{NaBH}_4 + 4\text{CF}_x \rightarrow x\text{NaBF}_4 + 4\text{C} + 2x\text{H}_2 \ (x = 0.8 - 1) \tag{2}$$

The dehydrogenation onset temperature of ball-milled 55NaBH$_4$-45FGi composites can be decreased to 125 °C and approximately 4.8 wt% hydrogen can be released at 130 °C over a period of several seconds. Such additives are usually mixed with the hydrides by high-energy milling. Several additives such as hydrides and fluorides, have been found to destabilize NaBH$_4$ effectively and make the dehydrogenation or even hydrogenation possible at rather moderate pressures and temperatures.

3.3.1. Hydride Destabilization

An important potential advantage in using hydrides as a destabilizing additive over others, is that in addition to the tuning of the thermodynamics and kinetics of (de)hydrogenation, it is possible to maintain a high gravimetric capacity. For example, MgH$_2$ has been used successfully to modify the (de)hydrogenation thermodynamics relative to NaBH$_4$ by forming the compound MgB$_2$ upon dehydrogenation [41,42]. The main dehydrogenation reaction can either proceed to formation of NaH or sodium depending on the conditions:

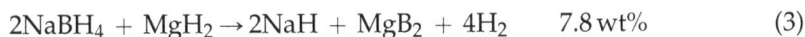

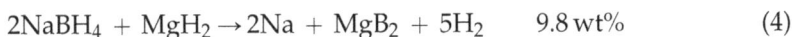

$$2NaBH_4 + MgH_2 \rightarrow 2NaH + MgB_2 + 4H_2 \qquad 7.8\,wt\% \qquad (3)$$

$$2NaBH_4 + MgH_2 \rightarrow 2Na + MgB_2 + 5H_2 \qquad 9.8\,wt\% \qquad (4)$$

This concept is called destabilization in a "reactive hydride composite (RHC)" in which two or more hydrides are combined in appropriate ratios to lower the dehydrogenation enthalpy of the system through forming a new hydrogen-free, thermodynamically stable compound as a by-product. Hence the desorption temperature is reduced and the reversibility of the system is improved [43].

By employing this strategy, the dehydrogenation of NaBH$_4$ can be facilitated by combining the borohydride with other metal hydrides such as LiAlH$_4$, Ca(BH$_4$)$_2$ and CaH$_2$ so as to form LiAl, AlB$_2$ and CaB$_6$ respectively upon dehydrogenation [27,44]:

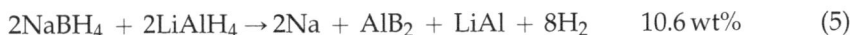

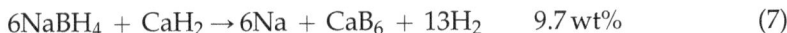

$$2NaBH_4 + 2LiAlH_4 \rightarrow 2Na + AlB_2 + LiAl + 8H_2 \qquad 10.6\,wt\% \qquad (5)$$

$$4NaBH_4 + Ca(BH_4)_2 \rightarrow 4Na + CaB_6 + 12H_2 \qquad 10.9\,wt\% \qquad (6)$$

$$6NaBH_4 + CaH_2 \rightarrow 6Na + CaB_6 + 13H_2 \qquad 9.7\,wt\% \qquad (7)$$

Similarly, addition of Mg$_2$NiH$_4$ to NaBH$_4$ leads to formation of the stable ternary boride phase MgNi$_{2.5}$B$_2$ and lowers the enthalpy of hydrogen desorption for NaBH$_4$ from 110 kJ$\cdot$mol^{-1} H$_2$ to 76 $\pm$ 5 kJ$\cdot$mol^{-1} H$_2$, according to the following reaction [45]:

$$4NaBH_4 + 5Mg_2NiH_4 \rightarrow 4NaH + 2MgNi_{2.5}B_2 + 8Mg + 16H_2 \qquad 4.5\,wt\% \quad (8)$$

As a consequence of this addition, the onset temperature of hydrogen desorption decreases from $ca.$ 500 °C for NaBH$_4$ to 360 °C for the NaBH$_4$/Mg$_2$NiH$_4$ composite mixture. When Mg$_2$FeH$_6$ is added to NaBH$_4$ the dehydrogenation can become quite complex [46]. A single dehydriding step is observed for xNaBH$_4$ + $(1 - x)$Mg$_2$FeH$_6$ when $x = 0.1$ and 0.125, but a multi-step process occurs when $x > 0.25$. Despite the different dehydriding process, PXD measurements maintain that NaH and MgB$_2$ are the dehydrogenation products over the entire composition range. The results also indicate that the dehydriding temperature of NaBH$_4$ is reduced by at least 150 °C when combined with Mg$_2$FeH$_6$.

As a model borohydride-hydride system, the NaBH$_4$-MgH$_2$ combination has been investigated extensively. For 2NaBH$_4$ + MgH$_2$, the dehydrogenation temperature is reduced by $ca.$ 40 °C compared to pure NaBH$_4$ [41]. The desorption was originally proposed to follow a two-step process:

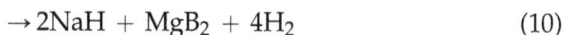

$$2NaBH_4 + MgH_2 \rightarrow 2NaBH_4 + Mg + H_2 \tag{9}$$

$$\rightarrow 2NaH + MgB_2 + 4H_2 \tag{10}$$

The dehydriding mechanism of the 2NaBH$_4$ + MgH$_2$ system was subsequently suggested to proceed in three steps under 1 bar of inert gas, by: (i) the dehydrogenation of MgH$_2$; (ii) the "disproportion" of NaBH$_4$; and (iii) the reaction of an intermediate borohydride compound, such as Na$_2$B$_{12}$H$_{12}$, with free Mg to give MgB$_2$, NaH and hydrogen [42,47]:

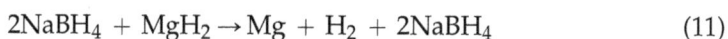

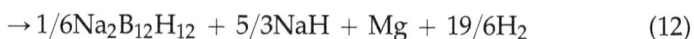

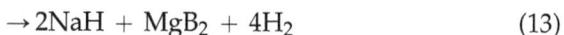

$$2NaBH_4 + MgH_2 \rightarrow Mg + H_2 + 2NaBH_4 \tag{11}$$

$$\rightarrow 1/6Na_2B_{12}H_{12} + 5/3NaH + Mg + 19/6H_2 \tag{12}$$

$$\rightarrow 2NaH + MgB_2 + 4H_2 \tag{13}$$

However, no direct observation of Na$_2$B$_{12}$H$_{12}$ was made. When the reaction is performed under static vacuum, however, the dehydrogenation of 2NaBH$_4$-MgH$_2$ appears to follow an alternative pathway (Figure 5) [48]:

$$2NaBH_4 + MgH_2 \rightarrow 2NaBH_4 + 1/2MgH_2 + 1/2Mg + 1/2H_2 \tag{14}$$

$$\rightarrow 3/2NaBH_4 + 1/4MgB_2 + 1/2NaH + 3/4Mg + 7/4H_2 \tag{15}$$

$$\rightarrow 2Na + B + 1/2Mg + 1/2MgB_2 + 5H_2 \tag{16}$$

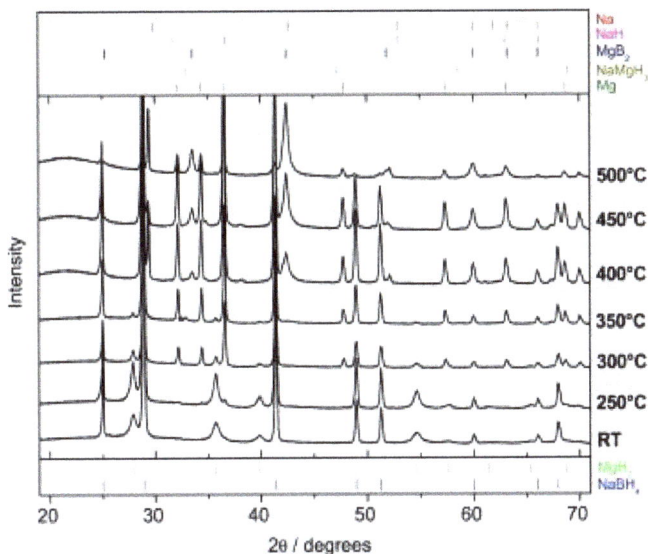

Figure 5. Powder X-ray diffraction (PXD) patterns of 2NaBH$_4$-MgH$_2$ with increasing temperature [48]. Reprinted from [48] with permission from the International Association of Hydrogen Energy.

The presence of the B$_{12}$H$_{12}$$^{2-}$ anion was confirmed experimentally by solid state NMR. Amorphous Na$_2$B$_{12}$H$_{12}$ was detected in a partially desorbed 2NaBH$_4$ + MgH$_2$ sample (following 2 h at 450 °C) and in the final products of the decomposition reaction by both direct comparison with the ^{11}B{^{1}H} NMR spectrum of pure Na$_2$B$_{12}$H$_{12}$ and by dynamic cross-polarization experiments [28].

Considering now the reverse hydrogenation reaction in the Na-Mg-B-H system, it has been suggested that the hydrogenation of 2NaH-MgB$_2$ proceeds according to the following reaction [41]:

$$2NaH + MgB_2 + 4H_2 \rightarrow 2NaBH_4 + MgH_2 \tag{17}$$

Further scrutiny of the process indicated that the absorption reaction does not occur in a single step. Nwakwuo *et al.* [49] and Pistidda *et al.* [50], respectively, characterized the uptake mechanism of ball-milled 2NaH-MgB$_2$ by using transmission electron microscopy (TEM) and *in situ* PXD. Under 50 bar of hydrogen, a new and unknown hydride phase was observed at *ca.* 280 °C. This phase remained present in diffraction patterns up to 325 °C followed by the formation of NaMgH$_3$ at about 330 °C. At 380 °C, crystals of NaBH$_4$ appeared and grew (Figure 6). The effect of the NaH:MgB$_2$ ratio on hydrogen uptake in the system has since become evident [51]. Unlike the 2:1 NaH:MgB$_2$ hydrogenation reaction, the only crystalline products of the hydrogenation of the 1:1 and 1:2 mixtures are NaBH$_4$ and MgH$_2$. Due to the reduced amount of NaH in the 1:2 system, the hydrogenation reaction proceeds towards the formation of $NaBH_4$ and MgH_2, completely consuming the $NaMgH_3$ formed and avoiding the formation of a molten NaH-$NaBH_4$ phase.

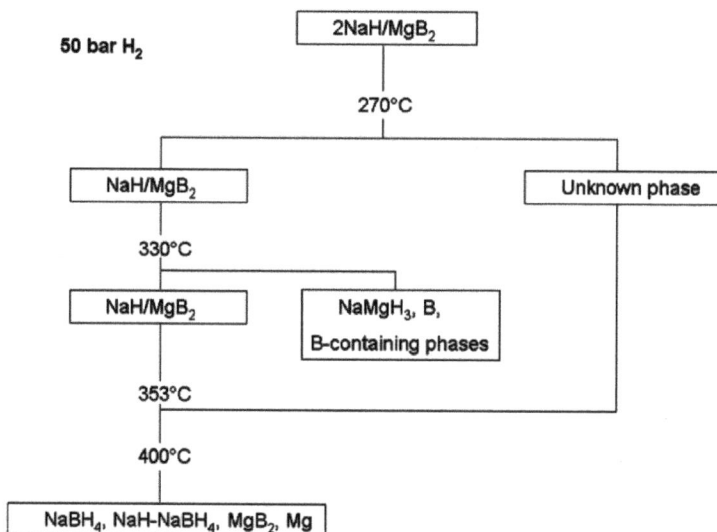

Figure 6. Scheme of the $2NaH + MgB_2$ absorption reaction performed under 50 bar H_2 [50]. Reprinted with permission from [50], copyright 2010 The American Chemical Society.

$NaBH_4$ can also be synthesized from NaH and MgB_2 under hydrogen by mechanochemical methods, but only partial hydrogenation is observed [52]. Although the formation of $NaBH_4$ was experimentally observed by *ex-situ*[11]B magic angle spinning (MAS) NMR under 1 bar of H_2 with a milling speed of 300 rpm, even at 120 bar H_2/550 rpm the yield of $NaBH_4$ was only 14 wt% (by PXD). Moreover, IR spectroscopy confirmed MgH_2 in the milling products.

Hence, the above examples demonstrate that the dehydrogenation thermodynamics of $NaBH_4$ are significantly improved by adding MgH_2 and that re-hydrogenation is possible. To obtain improvements in release kinetics and/or achieve dehydrogenation temperatures below 400 °C, approaches such as catalytic doping and nanoconfinement have been attempted for the Na-Mg-B-H system just as with $NaBH_4$ itself. 5 mol% TiF_3 doping reduces the dehydrogenation temperature of the $2NaBH_4$-MgH_2 system by 100 °C [41]. Moreover, TiF_3 doped $2NaBH_4$-MgH_2 can be rehydrogenated up to 5.89 wt% hydrogen within 12 h at 600 °C and 4 MPa H_2. $NaBH_4$ and MgH_2 are the clearly observed re-hydrogenation products by PXD. Of several other additives (fluorides, chlorides and hydroxides), the most promising would appear to be MgF_2 [8]. The MgF_2 reduces the $NaBH_4$

decomposition temperature by 30 °C and the desorption enthalpy by 2 kJ·mol^{-1}. The rate constant for desorption (fitted to a modified Avrami-Erofeev equation over the isothermal region) increases by a factor of 3.2 when MgF_2 is added.

Perhaps rather unexpectedly, short-term exposure to a moist atmosphere appears to have a positive effect on the desorption reaction of the $2NaBH_4 + MgH_2$ mixture [53]. The as-milled mixture desorbs 3.4 wt% of hydrogen at 450 °C, whereas 7.8 wt% of hydrogen is desorbed from the milled sample after 2 h of air exposure followed by drying. In this latter case, the final products are MgB_2 and NaH (in addition to some NaOH) whereas in the former, partial dehydrogenation yields $NaBH_4$ and Mg. Further investigation showed that the chemical state of the reactants is unchanged after exposure, but significant microstructural and morphological differences were revealed by Rietveld analysis and scanning electron microscope (SEM) characterization of the starting materials. It seems that the exposure of the 2:1 $NaBH_4$:MgH_2 system to moisture creates a scenario where the MgH_2 remains solid whereas the $NaBH_4$ forms a slurry that "wets" the surface of the MgH_2 particles (protecting the MgH_2 from reaction with air). The intimate interfacial contact is maintained in the solid state during the subsequent drying procedure facilitating dehydrogenation.

The effects of nanoconfinement have been evaluated against physically nanostructured mixtures by melt infiltration of $NaBH_4$-MgH_2 into mesoporous SBA-15 (NbF_5 was used as a catalyst in both cases) [54]. The thermal desorption profile of $2NaBH_4 + MgH_2$ shows two peaks at $ca.$ 300 °C and 410 °C, respectively. When 0.05 mol of NbF_5 is added, the desorption profile of the mixture displays three peaks centered at $ca.$ 200 °C, 300 °C, and 400 °C, respectively. The comparison of the two traces suggests that the NbF_5 additive actively alters the dehydrogenation process in the $2NaBH_4 + MgH_2$ system. If the same hydrides are nanoconfined ($2NaBH_4 + MgH_2 + 0.05$ mol NbF_5 confined into SBA-15), three desorption peaks occur at 134 °C, 323 °C and 354 °C, respectively. The changes in the desorption temperatures suggests that different reaction processes again occur in the nanoconfined-catalyzed material. Although Si-containing phases in the dehydrogenation product suggest that SBA-15 cannot be treated as an inert host, only hydrogen is observed as an evolved gas. The results indicate that favorable synergic effects between nanoconfinement and catalysis may exist for the Na-Mg-B-H system as have been observed for $NaBH_4$ itself. The reasons for these phenomena require further investigation.

3.3.2. Fluoride Destabilization

Despite improvements in performance over pristine $NaBH_4$, destabilized systems such as those in Section 3.3.1 have been unable to achieve gravimetric hydrogen capacities on a level with theoretical maxima. Given that the melting point of the decomposition product, Na, is ~371 K [55] and that molten Na may serve as an effective mass-transfer medium to promote atomic/ionic diffusion, one possible reason for this under-performance in the Na-Mg-B-H system is that a melting-induced phenomenon takes place during dehydrogenation. As a result, the local stoichiometry and homogeneity of the mixture may be disrupted, hindering rehydrogenation to $NaBH_4$. Alternatively, hydrogen released during dehydrogenation may transport Na away from the reaction mixture in the liquid and/or vapor phase. To prevent such eventualities, Na may be confined to the solid state (together with boron) by using certain additives such as metal fluorides. On the one hand, boron, the decomposition product of $NaBH_4$ can be stabilized by other metals to form borides while on the other, Na forms NaF with a significantly elevated melting point compared to Na metal (*i.e.*, a more than 3-fold increase; 1263 K) [56].

In addition, and also important, the thermodynamic and kinetic behaviour of $NaBH_4$ dehydrogenation itself may be tuned through the substitution of fluorine for hydrogen since H^- and F^- have similar ionic radii [57]. For example, a recent study on the $NaBH_4$–$NaBF_4$ system showed that hydrogen–fluorine exchange took place in a temperature range of 200–215 °C, leading to a new rock salt-type compound with idealized composition $NaBF_2H_2$ [58]. After further heating, the fluorine substituted compound becomes X-ray amorphous and decomposes to NaF at 310 °C. In particular, the $NaBH_4$–$NaBF_4$ composite decomposes at lower temperatures ($T = 300$ °C) compared to $NaBH_4$ ($T = 476$ °C) and retains 30% of the hydrogen storage capacity after three hydrogen release and uptake cycles compared to 6% for $NaBH_4$.

Two new systems based on $2NaF + MgB_2 + 0.05TiF_3$ (referred to here as the "Mg system") and $2NaF + AlB_2 + 0.05TiF_3$ ("Al system") were investigated by employing a fluorine-hydrogen substitution strategy [59]. The hydrogenation of the Mg system yielded $NaBH_4$ and MgF_2, which can be dehydrogenated to $NaMgF_3$ and MgB_2. In contrast, the hydrogenation of the Al system yielded $NaBH_4$ and Na_3AlF_6, which was dehydrogenated to NaF and AlB_2. These processes are therefore reversible and compared to pure $NaBH_4$ a significant kinetic and thermodynamic destabilisation with respect to the hydrogenation and dehydrogenation is achieved (Figure 7). The reversible hydrogen storage capacity reached 3.8 wt% and 2.5 wt% for the Mg and Al systems, respectively.

Combining $NaBH_4$ and ZnF_2 generates hydrogen by forming $NaBF_4$ with an onset temperature below 100 °C with favorable kinetics [60]. However, a small amount of B_2H_6 is released and the reversibility of the system is unknown. Studies of the effect of transition metal fluorides on the decomposition of $NaBH_4$ by reacting $NaBH_4$ with TiF_3 mechanochemically, MnF_3 or FeF_3 revealed that $NaBF_4$ was among the products in all cases [61]. Analysis of ^{11}B-NMR spectra gave $NaBF_4$:$NaBH_4$ ratios of 1:150, 1:40 and 1:10 for the Ti-, Mn- and Fe-containing systems respectively. The hydrogen release in the NaBH$_4$–MnF$_3$ system began at 130 °C while FeF$_3$ decreased the onset temperature to 161 °C and TiF$_3$ to 200 °C. TiF$_3$ reacted completely with NaBH$_4$ below 320 °C. All these 3d transition metal fluoride containing materials display negligible emissions of diborane species.

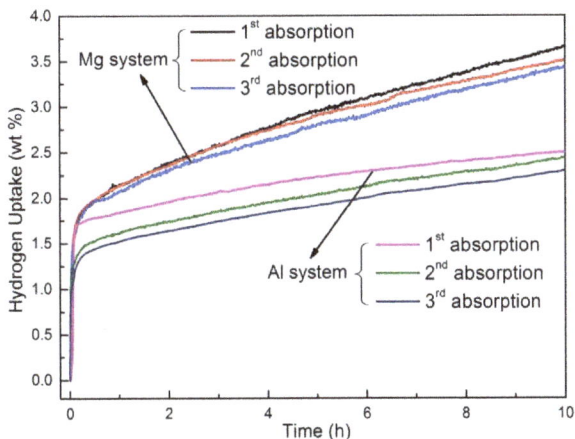

Figure 7. Hydrogenation and dehydrogenation curves for the 2NaF–MgB$_2$ (AlB$_2$)–0.05TiF$_3$ systems on cycling at 500 °C and 6 MPa hydrogen pressure [59]. Reproduced from [59] by permission of The Royal Society of Chemistry.

Both dehydrogenation and hydrogenation can be improved by adding rare earth fluorides (LnF$_3$, Ln = Nd, Y, La, Ho) [62–65]. The 3 NaBH$_4$/LnF$_3$ composites release hydrogen between 400 °C and 450 °C, which is lower than that of pure NaBH$_4$. Approximately 3 wt% hydrogen can be cycled in these systems. By analogy to the destabilization mechanisms proposed for other metal fluoride "composite" systems, the improvement can be attributed to the formation of borides according to the following reactions:

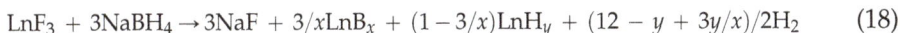

$$LnF_3 + 3NaBH_4 \rightarrow 3NaF + 3/xLnB_x + (1 - 3/x)LnH_y + (12 - y + 3y/x)/2H_2 \quad (18)$$

For example, dehydrogenation commences at 413 °C under 0.1 MPa Ar for the 3NaBH$_4$/NdF$_3$ system [62]. PXD revealed that NdB$_6$, Nd$_2$H$_5$ and NaF formed on decomposition. The process is pseudo-reversible, producing NaBH$_4$ and NaNdF$_4$ on hydrogenation. Similarly, dehydrogenation of 3NaBH$_4$/YF$_3$ starts at 423 °C but with a higher mass loss of 4.12 wt% (given the lower atomic mass of Y over Nd) [63]. PXD of the dehydrogenated products reveals NaF, YB$_4$ and YH$_2$ are formed and re-hydrogenation leads to NaBH$_4$ and NaYF$_4$ by analogy to the neodymium system.

The above shows that the products from NaBH$_4$ dehydrogenation can be stabilized simultaneously by introducing both fluorine and metals, hence effectively destabilizing $NaBH_4$. Although the conditions required for dehydrogenation and rehydrogenation in these systems are still too extreme for practical applications, the improvement in uptake and release provides the basis for a broader destabilization strategy.

3.4. Chemical Modification

3.4.1. Combination of Protic and Hydridic H Atoms

It is well known that hydrogen exists in a partially negatively charged state ($H^{\delta-}$) in complex hydrides such as $NaBH_4$, whereas it is partially positively charged ($H^{\delta+}$) in nitrogen containing compounds such as $LiNH_2$. Given the repulsive potential between two positively charged ($H^{\delta+}/H^{\delta+}$) or two negatively charged ($H^{\delta-}/H^{\delta-}$) species, there are relatively high energy barriers to the conversion of either $H^{\delta-}$ or $H^{\delta+}$ pairs to neutral H_2. This contrasts markedly with the ease of combining $H^{\delta-}$ and $H^{\delta+}$ [66]. These observations suggest that the reaction kinetics of H_2 formation can be enhanced by inducing H^+ and H^- mobility in compounds with suitable structures. Thus, it is of particular interest to consider the range of $H^{\delta+}$-rich compounds that could be combined with $NaBH_4$ to improve dehydrogenation.

Chater et al. [67] found a new cubic phase in the $NaNH_2$–$NaBH_4$ system with composition Na_2BNH_6 ($a \approx 4.7$ Å) at 190 °C. Hydrogen release initiates at ca. 290 °C in Na_2BNH_6 and peaks at ~350 °C. Ammonia is also released, constituting approximately 7 wt% of the total desorbed gas. The decomposition products are reported to be NaH, Na and an amorphous unidentified white solid. In fact, Na_2BNH_6 releases hydrogen while molten between 300 °C and 400 °C yielding NaH, Na and a grey amorphous powder [68]. On addition of excess amide ($\geqslant$2:1 $NaNH_2$:$NaBH_4$), Na_3BN_2 becomes the sole product.

Another new phase in the $NaNH_2$–$NaBH_4$ system, $Na_3(NH_2)_2BH_4$, is attainable by ball milling (at a molar ratio of 2:1) [69]. Thermal analysis shows that decomposition occurs in two main stages: (i) dehydrogenation below 400 °C to form Na_3BN_2 (6.85 wt%); and (ii) decomposition of Na_3BN_2 above 400 °C to produce Na, B, and N_2 according to the following reactions:

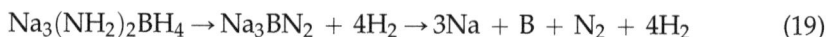

$$Na_3(NH_2)_2BH_4 \rightarrow Na_3BN_2 + 4H_2 \rightarrow 3Na + B + N_2 + 4H_2 \qquad (19)$$

Another example in terms of "protic" species that could be combined with borohydrides, is that of hydroxide. Drozd et al. [70] investigated the hydrogen-generating reaction between $NaBH_4$ and $Mg(OH)_2$, and found that reaction rate depends tremendously on the homogeneity and/or particle size of the reactants. PXD and Raman spectroscopy reveal that mechanically activated mixtures of $NaBH_4$ and $Mg(OH)_2$ react yielding MgO as the only crystalline phase between 240 °C and 318 °C. Ball milled $NaBH_4$-$2Mg(OH)_2$ mixtures release hydrogen in one exothermic reaction, with an onset temperature of 240 °C [71]. The estimated enthalpy for the reaction is 135.9 kJ·mol^{-1} and the dehydrogenation products contain $NaBO_2$ and MgO [70]. Therefore, the following dehydrogenation reaction was proposed:

$$NaBH_4 + 2Mg(OH)_2 \rightarrow NaBO_2 + 2MgO + 4H_2 \quad 5.18\,wt\%;\ \Delta H = -135.9\,kJ\cdot mol^{-1} \quad (20)$$

The various studies above show that the dehydrogenation of $NaBH_4$ can be improved by reaction with $H^{\delta+}$-containing starting materials (such as $NaNH_2$ or $Mg(OH)_2$) based on the premise of favorable H^+-H^- interactions. However, the dehydrogenation reaction of these systems can be exothermic, which clearly introduces substantial challenges in terms of cycling or regeneration.

3.4.2. Bimetallic Borohydrides

Nakamori *et al.* [72,73] theoretically and experimentally found that a clear correlation exists between the thermodynamic stability of metal borohydrides and the Pauling electronegativity of the respective metal cations (Figure 8). It was thus proposed that the dehydrogenation temperature of $M(BH_4)_n$, where M is a metal cation of valence n, decreases linearly with the increasing electronegativity of M. Hence one of the approaches to adjust the dehydrogenation thermodynamics of metal borohydrides is to substitute an alkali or alkaline earth metal, for example, by another metal with higher electronegativity.

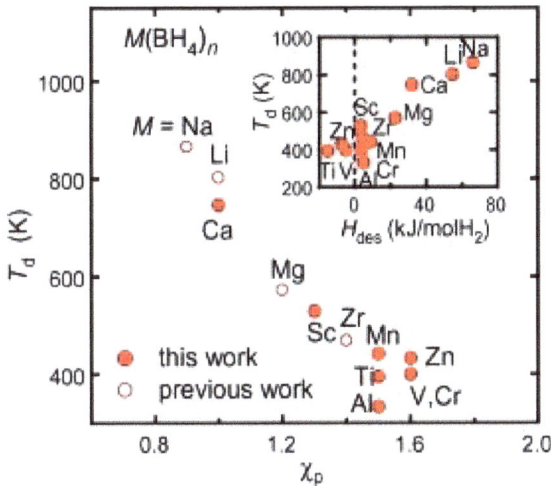

Figure 8. The dehydrogenation temperature, T_d as a function of the Pauling electronegativity χ_p for selected metals. The inset shows the correlation between T_d and estimated H_{des} for the desorption reaction [73]. Reprinted with permission from [73], copyright 2007 Elsevier.

NaK(BH$_4$)$_2$ was synthesized by mechanical milling of NaBH$_4$ and KBH$_4$ in a 1:1 ratio [74]. The new phase forms with a rhombohedral structure (tentatively space group $R3$), but *in situ* PXD indicated it was metastable, decomposing to the starting materials NaBH$_4$ and KBH$_4$ after 14 h at room temperature. A more common method for the synthesis of bimetallic sodium borohydrides is by solid or solution-state metathesis [75]:

$$x\text{NaBH}_4 + \text{MCl}_y \rightarrow \text{Na}_{x\text{-}y}\text{M(BH}_4)_x + y\text{NaCl} \tag{21}$$

where M is an alkali metal, alkaline earth metal, transition metal or lanthanide. For example, the new bimetallic borohydride NaSc(BH$_4$)$_4$ was synthesised by ball-milling mixtures of sodium borohydride and ScCl$_3$ [76]. The structure of NaSc(BH$_4$)$_4$ (orthorhombic space group $Cmcm$ a = 8.170(2) Å, b = 11.875(3) Å, c = 9.018(2) Å) consists of isolated scandium tetraborohydride tetrahedral anions, [Sc(BH$_4$)$_4$]$^-$, located inside slightly distorted trigonal Na$_6$ prisms (each second prism is empty) (Figure 9). Na$^+$ is surrounded by six BH$_4^-$ tetrahedra in almost regular octahedral coordination with a (6 + 12)-fold coordination of H to Na. NaSc(BH$_4$)$_4$ melts at ~137 °C subsequently releasing hydrogen in two steps between 167–217 °C and 222–267 °C. Scandium boride ScB$_x$ is tentatively identified as one of the decomposition products.

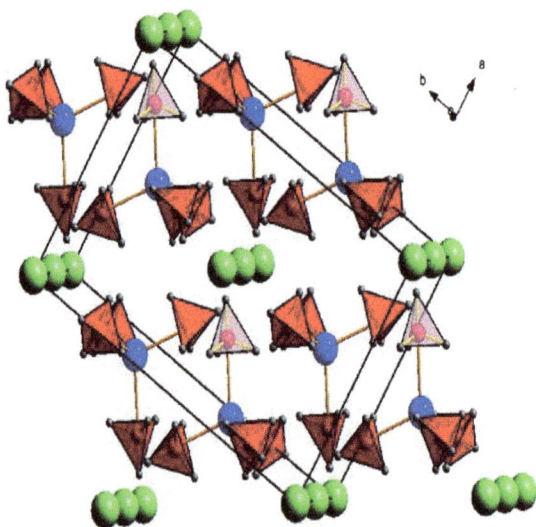

Figure 9. Crystal structure of NaSc(BH$_4$)$_4$ showing the coordination of Sc atoms (blue) by BH$_4$ tetrahedra (red); Na atoms are in green [76]. Reprinted with permission from [76], copyright 2010 The American Chemical Society.

NaZn(BH$_4$)$_3$ can be synthesized by ball milling NaBH$_4$ and ZnCl$_2$ in a 3:1 molar ratio and in fact if the starting ratios are varied (e.g., 2:1, 4:1) mixtures of NaZn$_2$(BH$_4$)$_5$ and NaZn(BH$_4$)$_3$ are obtained, indicating competitive reactions [77]. NaZn(BH$_4$)$_3$ (monoclinic space group $P2_1/c$; Figure 10) consists of 1D anionic $[\{Zn(BH_4)_3\}_n]^{n-}$ chains with tetrahedrally coordinated Zn atoms, which are connected in three dimensions through the Na$^+$ ions. NaZn$_2$(BH$_4$)$_5$ meanwhile, is unstable and slowly decomposes to NaZn(BH$_4$)$_3$ at room temperature. The structure of NaZn$_2$(BH$_4$)$_5$ (also monoclinic space group $P2_1/c$) consists of Na$^+$ cations and isolated complex dimeric $[Zn_2(BH_4)_5]^-$ anions in which trigonal planar centers of Zn are each coordinated to one bridging and two terminal BH$_4$ groups.

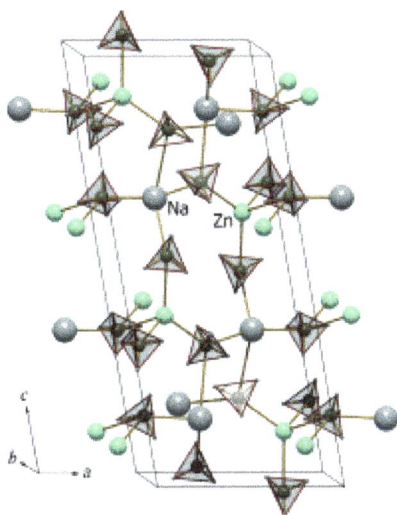

Figure 10. Crystal structure of NaZn(BH$_4$)$_3$; Zn atoms in blue, B in brown, Na in dark grey, and H in light grey [77]. Reprinted with permission from [77], copyright 2009 John Wiley & Sons.

Pure NaZn(BH$_4$)$_3$, synthesized by a solution route, releases hydrogen coupled with borane and diborane BH$_3$, and B$_2$H$_6$, giving a total weight loss of 29 wt% between 80 °C and 200 °C [78]. Nanoconfinement of NaZn(BH$_4$)$_3$ in SBA-15 however leads to borane-free hydrogen evolution across a temperature range of 50–150 °C from onset to completion. The activation energy for dehydrogenation was reduced to 38.9 kJ·mol^{-1} in the nanoconfined solid; a reduction of 5.3 kJ·mol^{-1} compared to that of bulk NaZn(BH$_4$)$_3$.

The novel mixed-cation mixed-anion borohydride chloride, NaY(BH$_4$)$_2$Cl$_2$ was prepared by mechanochemical synthesis from NaBH$_4$-YCl$_3$ mixtures followed by annealing (with Na$_3$YCl$_6$ and Na(BH$_4$)$_{1-x}$Cl$_x$ as impurity phases) [79]. The structure of $NaY(BH_4)_2Cl_2$ is pseudo-orthorhombic (monoclinic space group $P2/c$) and isotypic with the high temperature polymorph of $NaYCl_4$. The borohydride chloride is comprised of edge- and corner-sharing yttrium-centered and sodium-centered octahedra (Y is coordinated by four Cl and two BH_4 ligands, whereas Na is coordinated by two Cl and four BH_4 ligands; Figure 11).

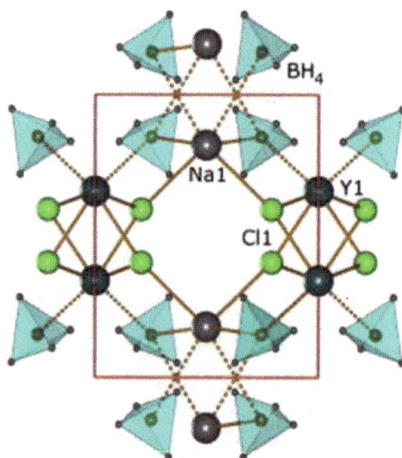

Figure 11. Crystal structure of $NaY(BH_4)_2Cl_2$ at $T \sim 230\,°C$ [79]. Reprinted from [79] with permission from the International Association of Hydrogen Energy.

In situ synchrotron PXD studies show that $NaY(BH_4)_2Cl_2$ decomposes to Na_3YCl_6 with amorphous yttrium borides the likely other products. The decomposition is an endothermic process that occurs at *ca.* 300 °C and the observed thermogravimetric analysis (TGA) mass losses suggest that no significant amount of diborane is released during the decomposition.

The thermal decomposition of ball milled mixtures of $NaBH_4$ with the chlorides of the 3d transition metals and cadmium (M) has been systematically investigated (Table 2) [75]. In contrast to predictions from theory, which in many cases have suggested the formation of mixed metal borohydrides [80], all the above reactions involve the substitution of BH_4^- by Cl^- and the formation of cubic NaCl-type $Na(BH_4)_{1-x}Cl_x$ solid solutions (with presumed amorphous transition metal borides as the other product in most cases). Samples containing Sc, Mn and Zn release <0.05 mol gas per mol of Na atoms during milling whereas the remainder of the 3d transition metal samples release *ca.* 1 mol of gas per mol of Na atoms with Fe- and Co-containing samples reaching maximum gas release most quickly (after 1 h).

Table 2. Structural and thermal decomposition parameters for ball milled $NaBH_4$ + MCl_n powders (mixed in in 3:1 (MCl_2) or 4:1 (MCl_3) ratios). Note: the parameters for M = Sc were not reported [75]. Reprinted from [75] with permission from Elsevier; copyright 2012.

Starting reagents	$Na(BH_4)_{1-x}Cl_x$ cell parameter, $a/\text{Å}$	Cl content, x	Decomposition $T/^\circ C$	Mass loss at $T \leqslant 600\,^\circ C/wt\%$
NaCl only	5.6400(5)	1	-	-
$NaBH_4/TiCl_2$	5.7685(3)	0.71	401	2.7
$NaBH_4/VCl_3$	5.7306(4)	0.79	391	3.0
$NaBH_4/CrCl_3$	5.7383(2)	0.77	397	-
$NaBH_4/MnCl_2$	5.7863(4)	0.68	146	5.7
$NaBH_4/FeCl_3$	5.7407(4)	0.77	397	0.2
$NaBH_4/CoCl_2$	5.8011(3)	0.65	413	3.6
$NaBH_4/NiCl_2$	5.7837(9)	0.68	391	4.0
$NaBH_4/CuCl_2$	5.7801(3)	0.69	343	3.7
$NaBH_4/ZnCl_2$	5.6576(2)	0.92	103	21.7
$NaBH_4/CdCl_2$	5.7572(8)	0.74	521	-
$NaBH_4$ only	6.13080(10)	0	-	-

In subsequent thermal desorption experiments, the maximum release temperature ranges from 103 °C (Zn) to 521 °C (Cd) and $NaBH_4$–NaCl samples demonstrate decomposition at *ca.* 500 °C, similar to pure $NaBH_4$. That Sc, Mn, and Zn form other stable compounds in addition to $Na(BH_4)_{1-x}Cl_x$ (e.g., $NaZn(BH_4)_3$ which decomposes between 92 °C and 112 °C) explains the low decomposition temperatures observed (*i.e.*, below 230 °C). In fact, the trends in decomposition temperatures can be related to these compounds and to kinetic effects rather than to a systematic destabilisation of the $NaBH_4$ cubic structure. V, Ni and Ti, for example, form borides, which can act as catalysts whereas for Cu and Cd the high desorption temperatures would indicate the presence of metallic Cd and Cu, which have no beneficial effect on decomposition.

4. Closing Remarks

$NaBH_4$ is undoubtedly a very interesting material for hydrogen storage due to its high hydrogen density (10.6 wt%), low cost and relative air stability. However, the high dehydrogenation temperature, slow kinetics and poor reversibility are challenges that have to be overcome before the borohydride could be considered for practical applications. Several approaches to tackle both thermodynamic and kinetic issues have been employed ranging from catalysis through nano-engineering and additive destabilization to chemical modification. From the discussions in the sections above, it is apparent that each input can make a successful impact in terms of modifying thermodynamic stability, reducing dehydrogenation temperatures, improving equilibrium pressures, lowering activation energies or optimizing reversible capacity. While sorption kinetics can be enhanced, catalysis alone is not capable of altering the thermodynamics of uptake and release and hence reversibility.

Various chemical destabilization, doping and "composite" approaches confront these issues via creating alternative reaction pathways for uptake and release. These strategies can either increase of decrease gravimetric capacity depending on the "activity" of the added components to making (and breaking) bonds with hydrogen. So-called nano-engineering methods such as nanoconfinement or the formation of core-shell nanostructures can combine the benefits of catalysis and chemical modification, but often to the detriment of gravimetric capacity. The benefits and drawbacks of each approach relative to the hydrogen storage performance of $NaBH_4$ itself are summarized in Table 3.

Table 3. Comparison of materials modification approaches relative to $NaBH_4$ itself (where "+" signifies an improvement and "−" signifies a decline).

Strategy	T_d [a]	Kinetics	wt% H_2 [b]	Cyclability	References
Catalytic doping	+	+	−	+	[32,33]
Nano-confinement	+	+	−	+	[29,36–39]
H− destabilisation	+	+	+	+	[27,41–45]
F− destabilisation	+	+	−	+	[40,59–65]
$H^{\delta+}$-$H^{\delta-}$ "composites"	+	+	−	−	[66–71]
Bimetallic Na borohydrides	+	+	−	−	[74–80]

[a] Dehydrogenation onset temperature; and [b] practically realizable gravimetric capacity.

Clearly, all these strategies demonstrably help overcome one or more of the limitations of pristine $NaBH_4$ as a thermally-driven hydrogen store, but thus far, no modified materials can simultaneously meet all the major performance criteria required for mobile applications. The screening of more suitable catalysts and additives, developing new techniques to fabricate nanomaterials, discovering suitable lightweight mesoporous hosts, restricting loss of molten Na and greater understanding of the mechanisms involved in hydrogen release and uptake are some of the principal objectives towards making $NaBH_4$-based systems viable. However, it may be a radically different approach in which $NaBH_4$ is a component part that finally delivers a practical solution.

Acknowledgments: The research post for Jianfeng Mao has received funding from the European Union's Seventh Framework Programme (FP7/2007–2013) for the Fuel Cells and Hydrogen Joint Technology Initiative under Grant Agreement Number 303447.

Conflicts of Interest: The authors declare no conflict of interest.

References

1. Schlapbach, L.; Züttel, A. Hydrogen-storage materials for mobile applications. *Nature* **2001**, *414*, 353–358.
2. Eberle, U.; Felderhoff, M.; Schüth, F. Chemical and physical solutions for hydrogen storage. *Angew. Chem. Int. Ed.* **2009**, *48*, 6608–6630.

3.   Yang, J.; Sudik, A.; Wolverton, C.; Siegel, D.J. High capacity hydrogen storage materials: Attributes for automotive applications and techniques for materials discovery. *Chem. Soc. Rev.* **2010**, *39*, 656–675.

4.   Reardon, H.; Hanlon, J.M.; Hughes, R.W.; Godula-Jopek, A.; Mandal, T.K.; Gregory, D.H. Emerging concepts in solid-state hydrogen storage the role of nanomaterials design. *Energy Environ. Sci.* **2012**, *5*, 5951–5979.

5.   U.S. Department of Energy, Office of Energy Efficiency and Renewable Energy and the FreedomCAR and Fuel Partnership. Technical System Targets: Onboard Hydrogen Storage for Light-Duty Fuel Cell Vehicles. Available online: http://energy. gov/sites/prod/files/2014/03/f12/targets_onboard_hydro_storage.pdf (accessed on 23 December 2014).

6.   Urgnani, J.; Torres, F.J.; Palumbo, M.; Baricco, M. Hydrogen release from solid state $NaBH_4$. *Int. J. Hydrog. Energy* **2008**, *33*, 3111–3115.

7.   Martelli, P.; Caputo, R.; Remhof, A.; Mauron, P.; Borgschulte, A.; Züttel, A. Stability and decomposition of $NaBH_4$. *J. Phys. Chem. C* **2010**, *114*, 7173–7177.

8.   Milanese, C.; Garroni, S.; Girella, A.; Mulas, G.; Berbenni, V.; Bruni, G.; Suriach, S.; Baró, M.D.; Marini, A. Thermodynamic and kinetic investigations on pure and doped $NaBH_4$-MgH_2 system. *J. Phys. Chem. C* **2011**, *115*, 3151–3162.

9.   Züttel, A.; Rentsch, S.; Fischer, P.; Wenger, P.; Sudan, P.; Mauron, P.; Emmenegger, C. Hydrogen storage properties of $LiBH_4$. *J. Alloys Compd.* **2003**, *356–357*, 515–520.

10. Mao, J.F.; Guo, Z.P.; Liu, H.K.; Yu, X.B. Reversible hydrogen storage in titanium-catalyzed $LiAlH_4$–$LiBH_4$ system. *J. Alloys Compd.* **2009**, *487*, 434–438.

11. Mauron, P.; Buchter, F.; Friedrichs, O.; Remhof, A.; Bielmann, M.; Christoph, N.Z.; Züttel, A. Stability and reversibility of $LiBH_4$. *J. Phys. Chem. B* **2008**, *112*, 906–910.

12. Yu, X.B.; Grant, D.M.; Walker, G.S. Dehydrogenation of $LiBH_4$ destabilized with various oxides. *J. Phys. Chem. C* **2009**, *113*, 17945–17949.

13. Chłopek, K.; Frommen, C.; Léon, A.; Zabara, O.; Fichtner, M. Synthesis and properties of magnesium tetrahydroborate, $Mg(BH_4)_2$. *J. Mater. Chem.* **2007**, *17*, 3496–3503.

14. Li, H.W.; Kikuchi, K.; Nakamori, Y.; Miwa, K.; Towata, S.; Orimo, S. Effects of ball milling and additives on dehydriding behaviors of well-crystallized $Mg(BH_4)_2$. *Scr. Mater.* **2007**, *57*, 679–682.

15. Newhouse, R.J.; Stavila, V.; Hwang, S.-J.; Klebanoff, L.E.; Zhang, J.Z. Reversibility and improved hydrogen release of magnesium borohydride. *J. Phys. Chem. C* **2010**, *114*, 5224–5232.

16. Mao, J.F.; Guo, Z.P.; Poh, C.K.; Ranjbar, A.; Guo, Y.H.; Yu, X.B.; Liu, H.K. Study on the dehydrogenation kinetics and thermodynamics of $Ca(BH_4)_2$. *J. Alloys Compd.* **2010**, *500*, 200–205.

17. Kim, Y.; Hwang, S.-J.; Lee, Y.S.; Suh, J.Y.; Han, H.N.; Cho, Y.W. Hydrogen back-pressure effects on the dehydrogenation reactions of $Ca(BH_4)_2$. *J. Phys. Chem. C* **2012**, *116*, 25715–25720.

18. Riktor, M.D.; Sørby, M.H.; Chopek, K.; Fichtner, M.; Haubac, B.C. The identification of a hitherto unknown intermediate phase CaB_2H_X from decomposition of $Ca(BH_4)_2$. *J. Mater. Chem.* **2009**, *19*, 2754–2759.

19. Orimo, S.; Nakamori, Y.; Eliseo, J.; Züttel, A.; Jensen, C.M. Complex hydrides for hydrogen storage. *Chem. Rev.* **2007**, *107*, 4111–4132.

20. Li, H.W.; Yan, Y.G.; Orimo, S.; Züttel, A.; Jensen, C.M. Recent progress in metal borohydrides for hydrogen storage. *Energies* **2011**, *4*, 185–214.

21. Muir, S.S.; Yao, X.D. Progress in sodium borohydride as a hydrogen storage material: Development of hydrolysis catalysts and reaction systems. *Int. J. Hydrog. Energy* **2011**, *36*, 5983–5997.

22. U.S. Department of Energy Hydrogen Program. *Go/No-Go Recommendation for Sodium Borohydride for On-Board Vehicular Hydrogen Storage*; NREL/MP-150-42220; National Renewable Energy Laboratory (NREL): Golden, CO, USA, 2007.

23. Sigma-Aldrich. Available online: http://www.sigmaaldrich.com/united-states (accessed on 19 October 2014).

24. Santos, D.M.F.; Sequeira, C.A.C. Sodium borohydride as a fuel for the future. *Renew. Sustain. Energy Rev.* **2011**, *15*, 3980–4001.

25. Cakır, D.; Wijs, G.A.D.; Brocks, G. Native defects and the dehydrogenation of $NaBH_4$. *J. Phys. Chem. C* **2011**, *115*, 24429–24434.

26. Grochala, W.; Edwards, P. Thermal decomposition of the non-interstitial hydrides for the storage and production of hydrogen. *Chem. Rev.* **2004**, *104*, 1283–1315.

27. Mao, J.; Guo, Z.; Yu, X.; Liu, H. Improved hydrogen storage properties of $NaBH_4$ destabilized by CaH_2 and $Ca(BH_4)_2$. *J. Phys. Chem. C* **2011**, *115*, 9283–9290.

28. Garroni, S.; Milanese, C.; Pottmaier, D.; Mulas, G.; Nolis, P.; Girella, A.; Caputo, R.; Olid, D.; Teixdor, F.; Baricco, M.; *et al.* Experimental evidence of $Na_2[B_{12}H_{12}]$ and Na formation in the desorption pathway of the $2NaBH_4 + MgH_2$ System. *J. Phys. Chem. C* **2011**, *115*, 16664–16671.

29. Ngene, P.; van den Berg, R.; Verkuijlen, M.H.W.; de Jong, K.P.; de Jongh, P.E. Reversibility of the hydrogen desorption from $NaBH_4$ by confinement in nanoporous carbon. *Energy Environ. Sci.* **2011**, *4*, 4108–4115.

30. Friedrichs, O.; Remhof, A.; Hwang, S.-J.; Züttel, A. Role of $Li_2B_{12}H_{12}$ for the formation and decomposition of $LiBH_4$. *Chem. Mater.* **2009**, *22*, 3265–3268.

31. Caputo, R.; Garroni, S.; Olid, D.; Teixidor, F.; Suriñach, S.; Dolors Baró, M. Can $Na_2[B_{12}H_{12}]$ be a decomposition product of $NaBH_4$? *Phys. Chem. Chem. Phys.* **2010**, *12*, 15093–15100.

32. Mao, J.F.; Guo, Z.P.; Nevirkovets, I.P.; Liu, H.K.; Dou, S.X. Hydrogen de-/absorption improvement of $NaBH_4$ catalyzed by titanium-based additives. *J. Phys. Chem. C* **2012**, *116*, 1596–1604.

33. Humphries, T.D.; Kalantzopoulos, G.N.; Llamas-Jansa, I.; Olsen, J.E.; Hauback, B.C. Reversible hydrogenation studies of $NaBH_4$ milled with Ni-containing additives. *J. Phys. Chem. C* **2013**, *117*, 6060–6065.

34. De Jongh, P.E.; Adelhelm, P. Nanosizing and nanoconfinement: New strategies towards meeting hydrogen storage goals. *ChemSusChem* **2010**, *3*, 1332–1348.

35. Varin, R.A.; Chiu, C. Structural stability of sodium borohydride (NaBH$_4$) during controlled mechanical milling. *J. Alloys Compd.* **2005**, *397*, 276–281.

36. Ampoumogli, A.; Steriotis, T.; Trikalitis, P.; Giasafaki, D.; Bardaji, E.G.; Fichtner, M.; Charalambopoulou, G. Nanostructured composites of mesoporous carbons and boranates as hydrogen storage materials. *J. Alloys Compd.* **2011**, *509*, S705–S708.

37. Peru, F.; Garroni, S.; Campesi, R.; Milanese, C.; Marini, A.; Pellicer, E.; Baró, M.D.; Mulas, G. Ammonia-free infiltration of NaBH$_4$ into highly-ordered mesoporous silica and carbon matrices for hydrogen storage. *J. Alloys Compd.* **2013**, *580*, S309–S312.

38. Christian, M.L.; Aguey-Zinsou, K.F. Core–shell strategy leading to high reversible hydrogen storage capacity for NaBH$_4$. *ACS Nano* **2012**, *6*, 7739–7751.

39. Christian, M.L.; Aguey-Zinsou, K.F. Synthesis of core–shell NaBH$_4$@M (M = Co, Cu, Fe, Ni, Sn) nanoparticles leading to various morphologies and hydrogen storage properties. *Chem. Commun.* **2013**, *49*, 6794–6796.

40. Zhang, L.T.; Xiao, X.Z.; Fan, X.L.; Li, S.Q.; Ge, H.W.; Wang, Q.D.; Chen, L.X. Fast hydrogen release under moderate conditions from NaBH$_4$ destabilized by fluorographite. *RSC Adv.* **2014**, *4*, 2550–2556.

41. Mao, J.F.; Yu, X.B.; Guo, Z.P.; Liu, H.K.; Wu, Z.; Ni, J. Enhanced hydrogen storage performances of NaBH$_4$–MgH$_2$ system. *J. Alloys Compd.* **2009**, *479*, 619–623.

42. Garroni, S.; Pistidda, C.; Brunelli, M.; Vaughan, G.B.M.; Suriñach, S.; Baró, M.D. Hydrogen desorption mechanism of 2NaBH$_4$ + MgH$_2$ composite prepared by high-energy ball milling. *Scr. Mater.* **2009**, *60*, 1129–1132.

43. Dornheim, M. Tailoring Reaction Enthalpies of Hydrides. In *Handbook of Hydrogen Storage*; Hirscher, M., Ed.; Wiley-VCH: New York, NY, USA, 2010; pp. 187–214.

44. Mao, J.F.; Yu, X.B.; Guo, Z.P.; Poh, C.K.; Liu, H.K.; Wu, Z.; Ni, J. Improvement of the LiAlH$_4$-NaBH$_4$ system for reversible hydrogen storage. *J. Phys. Chem. C* **2009**, *113*, 10813–10818.

45. Afonso, G.; Bonakdarpour, A.; Wilkinson, D.P. Hydrogen storage properties of the destabilized 4NaBH$_4$/5Mg$_2$NiH$_4$ composite system. *J. Phys. Chem. C* **2013**, *117*, 21105–21111.

46. Li, G.Q.; Matsuo, M.; Deledda, S.; Hauback, B.C.; Orimo, S. Dehydriding property of NaBH$_4$ combined with Mg$_2$FeH$_6$. *Mater. Trans.* **2014**, *55*, 1141–1143.

47. Garroni, S.; Milanese, C.; Girell, A.; Marini, A.; Mulas, G.; Meneñdez, E.; Pistidda, C.; Dornheim, M.; Suriñach, S.; Baró, M.D. Sorption properties of NaBH$_4$/MH$_2$ (M = Mg, Ti) powder systems. *Int. J. Hydrog. Energy* **2010**, *35*, 5434–5441.

48. Pottmaier, D.; Pistidda, C.; Groppo, E.; Bordiga, S.; Spoto, G.; Dornheim, M.; Baricco, M. Dehydrogenation reactions of 2NaBH$_4$ + MgH$_2$ system. *Int. J. Hydrog. Energy* **2011**, *36*, 7891–7896.

49. Nwakwuo, C.C.; Pistidda, C.; Dornheim, M.; Hutchison, J.L.; Sykes, J.M. Microstructural analysis of hydrogen absorption in 2NaH + MgB$_2$. *Scr. Mater.* **2011**, *64*, 351–354.

50. Pistidda, C.; Garroni, S.; Minella, C.; Dolci, F.; Jensen, T.R.; Nolis, P.; Bösenberg, U.; Cerenius, Y.; Lohstroh, W.; Fichtner, M.; *et al.* Pressure effect on the 2NaH + MgB$_2$ hydrogen absorption reaction. *J. Phys. Chem. C* **2010**, *114*, 21816–21823.

51. Pistidda, C.; Pottmaier, D.; Karimi, F.; Garroni, S.; Rzeszutek, A.; Tolkiehn, M.; Fichtner, M.; Lohstroh, W.; Baricco, M.; Klassen, T.; *et al.* Effect of NaH/MgB$_2$ ratio on the hydrogen absorption kinetics of the system NaH + MgB$_2$. *Int. J. Hydrog. Energy* **2014**, *39*, 5030–5036.

52. Garroni, S.; Minella, C.B.; Pottmaier, D.; Pistidda, C.; Milanese, C.; Marini, A.; Enzo, S.; Mulas, G.; Dornheim, M.; Baricco, M.; *et al.* Mechanochemical synthesis of NaBH$_4$ starting from NaH-MgB$_2$ reactive hydride composite system. *Int. J. Hydrog. Energy* **2013**, *38*, 2363–2369.

53. Pistidda, C.; Barkhordarian, G.; Rzeszutek, A.; Garroni, S.; Bonatto Minella, C.; Baró, M.D.; Nolis, P.; Bormann, R.; Klassen, T.; Dornheim, M. Activation of the reactive hydride composite 2NaBH$_4$ + MgH$_2$. *Scr. Mater.* **2011**, *64*, 1035–1038.

54. Mulas, G.; Campesi, R.; Garroni, S.; Napolitano, E.; Milanese, C.; Dolci, F.; Pellicer, E.; Baró, D.; Marini, A. Hydrogen storage in 2NaBH$_4$ + MgH$_2$ mixtures: Destabilization by additives and nanoconfinement. *J. Alloys Compd.* **2012**, *536*, S236–S240.

55. Lemke, C.H. Sodium and sodium alloys. In *Kirk-Othmer Encyclopedia of Chemical Technology*; John Wiley & Sons: New York, NY, USA, 1983; pp. 181–204.

56. Young, J.A. Sodium fluoride. *J. Chem. Educ.* **2002**, *79*.

57. Yin, L.C.; Wang, P.; Kang, X.D.; Sun, C.H.; Cheng, H.M. Functional anion concept: Effect of fluorine anion on hydrogen storage of sodium alanate. *Phys. Chem. Chem. Phys.* **2007**, *9*, 1499–1502.

58. Rude, L.H.; Filsø, U.; D'Anna, V.; Spyratou, A.; Richter, B.; Hino, S.; Zavorotynska, O.; Baricco, M.; Sørby, M.H.; Hauback, B.C.; *et al.* Hydrogen–fluorine exchange in NaBH$_4$–NaBF$_4$. *Phys. Chem. Chem. Phys.* **2013**, *15*, 18185–18194.

59. Mao, J.F.; Guo, Z.P.; Liu, H.K.; Dou, S.X. Reversible storage of hydrogen in NaF–MB$_2$ (M = Mg, Al) composites. *J. Mater. Chem. A* **2013**, *1*, 2806–2811.

60. Zhang, Z.G.; Wang, H.; Zhu, M. Hydrogen release from sodium borohydrides at low temperature by the addition of zinc fluoride. *Int. J. Hydrog. Energy* **2011**, *36*, 8203–8208.

61. Kalantzopoulos, G.N.; Guzik, M.N.; Deledda, S.; Heyn, R.H.; Mullera, J.; Hauback, B.C. Destabilization effect of transition metal fluorides on sodium borohydride. *Phys. Chem. Chem. Phys.* **2014**, *16*, 20483–20491.

62. Chong, L.N.; Zou, J.X.; Zeng, X.Q.; Ding, W.J. Mechanisms of reversible hydrogen storage in NaBH$_4$ through NdF$_3$ addition. *J. Mater. Chem. A* **2013**, *1*, 3983–3991.

63. Zou, J.X.; Li, L.J.; Zeng, X.Q.; Ding, W.J. Reversible hydrogen storage in a 3NaBH$_4$/YF$_3$ composite. *Int. J. Hydrog. Energy* **2012**, *37*, 17118–17125.

64. Chong, L.N.; Zou, J.X.; Zeng, X.Q.; Ding, W.J. Effects of La fluoride and La hydride on the reversible hydrogen sorption behaviors of NaBH$_4$: A comparative study. *J. Mater. Chem. A* **2014**, *2*, 8557–8570.

65. Chong, L.N.; Zou, J.X.; Zeng, X.Q.; Ding, W.J. Study on reversible hydrogen sorption behaviors of a 3NaBH$_4$/HoF$_3$ composite. *Int. J. Hydrog. Energy* **2014**, *39*, 14275–14281.

66. Lu, J.; Fang, Z.Z.; Sohn, H.Y. A dehydrogenation mechanism of metal hydrides based on interactions between $H^{\delta+}$ and H^-. *Inorg. Chem.* **2006**, *45*, 8749–8754.

67. Chater, P.A.; Anderson, P.A.; Prendergast, J.W.; Walton, A.; Mann, V.S.J.; Book, D.; David, W.I.F.; Johnson, S.R.; Edwards, P.P. Synthesis and characterization of amide–borohydrides: New complex light hydrides for potential hydrogen storage. *J. Alloys Compd.* **2007**, *446–447*, 350–354.

68. Somer, M.; Acar, S.; Koz, C.; Kokal, I.; Hohn, P.; Cardoso-Gil, R.; Aydemir, U.; Akselrud, L. α- and β-$Na_2[BH_4][NH_2]$: Two modifications of a complex hydride in the system $NaNH_2$–$NaBH_4$; syntheses, crystal structures, thermal analyses, mass and vibrational spectra. *J. Alloys Compd.* **2010**, *491*, 98–105.

69. Wu, C.; Bai, Y.; Yang, J.H.; Wu, F.; Long, F. Characterizations of composite $NaNH_2$–$NaBH_4$ hydrogen storage materials synthesized via ball milling. *Int. J. Hydrog. Energy* **2012**, *37*, 889–893.

70. Drozd, V.; Saxena, S.; Garimella, S.V.; Durygin, A. Hydrogen release from a mixture of $NaBH_4$ and $Mg(OH)_2$. *Int. J. Hydrog. Energy* **2007**, *32*, 3370–3375.

71. Varin, R.A.; Parviz, R. Hydrogen generation from the ball milled composites of sodium and lithium borohydride ($NaBH_4$/$LiBH_4$) and magnesium hydroxide ($Mg(OH)_2$) without and with the nanometric nickel (Ni) additive. *Int. J. Hydrog. Energy* **2012**, *37*, 1584–1293.

72. Nakamori, Y.; Miwa, K.; Ninoyiya, A.; Li, H.; Ohba, N.; Towata, S.; Züttel, A.; Orimo, S. Correlation between thermodynamical stabilities of metal borohydrides and cation electronegativities: First-principles calculations and experiments. *Phys. Rev. B* **2006**, *74*.

73. Nakamori, Y.; Li, H.W.; Kikuchi, K.; Aoki, M.; Miwa, K.; Towata, S.; Orimo, S. Thermodynamical stabilities of metal-borohydrides. *J. Alloys Compd.* **2007**, *446–447*, 296–300.

74. Seballos, L.; Zhang, J.Z.; Rönnebro, E.; Herbergd, J.L.; Majzoub, E.H. Metastability and crystal structure of the bialkali complex metal borohydride $NaK(BH_4)_2$. *J. Alloys Compd.* **2009**, *476*, 446–450.

75. Llamas-Jansa, I.; Aliouane, N.; Deledda, S.; Fonneløp, J.E.; Frommen, C.; Humphries, T.; Lieutenant, K.; Sartori, S.; Sørby, M.H.; Hauback, B.C. Chloride substitution induced by mechano-chemical reactions between $NaBH_4$ and transition metal chlorides. *J. Alloys Compd.* **2012**, *530*, 186–192.

76. Černý, R.; Severa, G.; Ravnsbæk, D.B.; Filinchuk, Y.; D'Anna, V.; Hagemann, H.; Haase, D.; Jensen, C.M.; Jensen, T.R. $NaSc(BH_4)_4$: A novel scandium-based borohydride. *J. Phys. Chem. C* **2010**, *114*, 1357–1364.

77. Ravnsbæk, D.; Filinchuk, Y.; Cerenius, Y.; Jakobsen, H.J.; Besenbacher, F.; Skibsted, J.; Jensen, T.R. A series of mixed-metal borohydrides. *Angew. Chem. Int. Ed.* **2009**, *48*, 6659–6663.

78. Xia, G.L.; Li, L.; Guo, Z.P.; Gu, Q.F.; Guo, Y.H.; Yu, X.B.; Liu, H.K.; Liu, Z.W. Stabilization of $NaZn(BH_4)_3$ via nanoconfinement in SBA-15 towards enhanced hydrogen release. *J. Mater. Chem. A* **2013**, *1*, 250–257.

79. Ravnsbæk, D.B.; Ley, M.B.; Lee, Y.S.; Hagemann, H.; D'Anna, V.; Cho, Y.W.; Filinchuk, Y.; Jensen, T.R. A mixed-cation mixed-anion borohydride $NaY(BH_4)_2Cl_2$. *Int. J. Hydrog. Energy* **2012**, *37*, 8428–8438.

80. Hummelshøj, J.S.; Landis, D.D.; Voss, J.; Jiang, T.; Tekin, A.; Bork, N.; Dułak, M.; Mortensen, J.J.; Adamska, L.; Andersin, J.; *et al.* Density functional theory based screening of ternary alkali-transition metal borohydrides: A computational material design project. *J. Chem. Phys.* **2009**, *131*.

Hydrazine Borane and Hydrazinidoboranes as Chemical Hydrogen Storage Materials

Romain Moury and Umit B. Demirci

Abstract: Hydrazine borane $N_2H_4BH_3$ and alkali derivatives (*i.e.*, lithium, sodium and potassium hydrazinidoboranes $MN_2H_3BH_3$ with M = Li, Na and K) have been considered as potential chemical hydrogen storage materials. They belong to the family of boron- and nitrogen-based materials and the present article aims at providing a timely review while focusing on fundamentals so that their effective potential in the field could be appreciated. It stands out that, on the one hand, hydrazine borane, in aqueous solution, would be suitable for full dehydrogenation in hydrolytic conditions; the most attractive feature is the possibility to dehydrogenate, in addition to the BH_3 group, the N_2H_4 moiety in the presence of an active and selective metal-based catalyst but for which further improvements are still necessary. However, the thermolytic dehydrogenation of hydrazine borane should be avoided because of the evolution of significant amounts of hydrazine and the formation of a shock-sensitive solid residue upon heating at >300 °C. On the other hand, the alkali hydrazinidoboranes, obtained by reaction of hydrazine borane with alkali hydrides, would be more suitable to thermolytic dehydrogenation, with improved properties in comparison to the parent borane. All of these aspects are surveyed herein and put into perspective.

Reprinted from *Energies*. Cite as: Moury, R.; Demirci, U.B. Hydrazine Borane and Hydrazinidoboranes as Chemical Hydrogen Storage Materials. *Energies* **2015**, *8*, 3118–3141.

1. Introduction

Access to energy has been one of the most important events in recent human history. The world entered into a new era with technological development related to widespread use of coal in the 19th century. With oil emerging in the 20th century, the world has entered into another era, characterized by faster technological progress, which has, unfortunately, negatively impacted the environment. Nowadays, and in light of past experience, a new era must begin. The 21st century could be that of the hydrogen century. Hydrogen is attractive owing to abundance via various sources, high mass energy density (120 MJ/kg) and oxidation into water. However, the transition C_xH_y ($\leqslant 25$ $w_t\%$ H) → H_2 (100 $w_t\%$ H), *i.e.*, the development of a near-future energy economy, is very challenging. Important technical/scientific issues touching production, storage and end-use have to be addressed [1–3].

Storage of hydrogen is particularly critical and problematic, mainly because molecular hydrogen is a gas, even the lightest one. Accordingly, it has a low volumetric energy density (10.7 kJ$\cdot$L^{-1} at $27\,^\circ$C and 1 bar). Solutions have been investigated in order to make safe and efficient technologies emerge. First, the conventional storage methods (*i.e.*, compressed gas up to 700 bars and cryogenic liquid at $-253\,^\circ$C) were considered while the efforts have been concentrated on storage system (*i.e.*, tank, pipes, and so on) in terms of safety and performance. Then, alternative methods, involving materials, which are generally called hydrogen storage materials, emerged [4–6].

Hydrogen storage materials enable a safer storage than the compressed and cryogenic technologies, and naturally carry 7–20 $w_t\%$ H. Depending on their nature, there is distinction between physical storage (*i.e.*, cryo-adsorption) and chemical storage. With the former, porous materials store molecular hydrogen in conditions ($-196\,^\circ$C and 10–120 bars of H_2) that are milder than those for cryogenic liquid [4–7]. With respect to chemical hydrogen storage materials, atomic hydrogen is chemically bonded to a heteroatom and molecular hydrogen is released by solvolysis or thermolysis [4–6]. Borohydrides [8] and nitrogen-containing boranes (also called B–N–H compounds or boron- and nitrogen-based materials) [9] are typical examples.

Hydrazine borane $N_2H_4BH_3$ is one of the most recent boron- and nitrogen-based materials in the field of hydrogen storage. Though discovered and known since more than fifty years [10], the current energy context has been an opportunity to revive scientific interest on it, especially in view of the high gravimetric hydrogen density (15.4 $w_t\%$ H). Hence, since 2009, hydrazine borane and new derivative compounds, the alkali hydrazinidoboranes $MN_2H_3BH_3$, have positioned themselves as being potential candidates for chemical hydrogen storage, then focusing more and more attention. This is the core topic of the present review, which for the first time aims at specifically focusing on these materials, giving a timely and detailed overview about fundamentals, and tentatively discussing application prospects on the basis of the recent achievements.

2. Brief Historical View of Hydrazine Borane

Hydrazine borane was first reported by Goubeau and Ricker in 1961, in an original paper written in German [10]. The article provides experimental details about the synthesis as well as useful data about the molecular and crystal structures. Interestingly, it is mentioned the formation of a shock-sensitive solid residue upon the release of 2 equivalents of H_2. Later, in 1967, Gunderloy stressed on the shock-sensitivity and flammability of hydrazine borane but no further detail can be found in the report [11]. Yet, one year later, the same author wrote in a patent that hydrazine borane "is highly stable at room temperature ($25\,^\circ$C) and is neither impact nor friction sensitive" [12]. With hindsight, the contradiction does not appear to be so critical, since the borane-hydrazine compounds were demonstrated to be potential solid-state monopropellants for rocket devices [12–14] and fast hydrogen generating systems [15,16].

From an academic point of view, little research was carried out on hydrazine borane from 1961 to 2009. In 1971, the standard enthalpy was determined [17]. In 1997, hydrazine borane was used as precursor of porous boron nitride obtained by self-propagating high-temperature synthesis (Equation (1)) [18]. In 1999, the structure of hydrazine borane and that of its protonated analogue were calculated by the density functional theory method [19].

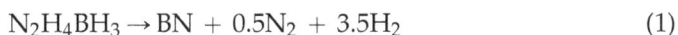

$$N_2H_4BH_3 \rightarrow BN + 0.5N_2 + 3.5H_2 \qquad (1)$$

In the 2000s, ammonia borane NH_3BH_3 was the only boron- and nitrogen-based material under intense research for chemical hydrogen storage [20]. One of the strategies was to destabilize it by chemical modification (synthesis of derivatives) [21]. This is in this context that hydrazine borane, which can be seen as a derivative of ammonia borane, emerged in 2009. The same year, Hamilton $et\ al.$ [22] dedicated very few lines to pristine hydrazine borane in a review paper about the boron- and nitrogen-based materials and, on the basis of the aforementioned 1960s' literature, suggested unsuitability for chemical hydrogen storage.

3. Hydrazine Borane

3.1. Synthesis

The original synthesis procedure of hydrazine borane (Equation (2)) is based on the reaction of sodium borohydride $NaBH_4$ with hydrazine sulfate $(N_2H_4)_2SO_4$ in dioxane at around 30 °C for 5–15 h [10]. It may be qualified as the classical procedure, reused by Hügle $et\ al.$ in 2009 [23] and then, revisited and improved in terms of yield, purity and overall cost by Moury $et\ al.$ in 2012 [24]. This is today the main procedure for the preparation of hydrazine borane at lab-scale.

Hydrazine borane can also be synthesized by reaction of sodium borohydride with magnesium chloride $MgCl_2$ either in hexahydrated form $MgCl_2 \cdot 6H_2O$ implying then the use of iced hydrazine N_2H_4 (Equation (3)) or in the form of a tetrahydrazinate $MgCl_2 \cdot 4N_2H_4$ (Equation (4)) with tetrahydrofuran C_4H_8O as solvent [12,25]. Instead of the chloride salt, a hydrazine salt $N_2H_4 \cdot HX$ with X = Cl or CH_3COO can be used (Equation (5)), the reaction taking place in tetrahydrofuran at temperatures between 50 and 100 °C [11,12,26]. The BH_3 source can be changed also. Trimethylamine borane $N(CH_3)_3BH_3$ can be reacted with hydrazine (Equation (6)) in benzene C_6H_6 at 50 °C for several hours [27].

$$NaBH_4 + 0.5(N_2H_4)_2SO_4 \rightarrow N_2H_4BH_3 + 0.5Na_2SO_4 + 0.5H_2 \qquad (2)$$

$$2NaBH_4 + MgCl_2 \cdot 6H_2O + 2N_2H_4 \rightarrow 2N_2H_4BH_3 + 2NaCl + 2H_2 + Mg(OH)_2 + 4H_2O \qquad (3)$$

$$2NaBH_4 + MgCl_2 \cdot 4N_2H_4 \rightarrow 2N_2H_4BH_3 + 2NaCl + Mg(N_2H_3)_2 + 2H_2 \qquad (4)$$

$$NaBH_4 + N_2H_4 \cdot HCl \rightarrow N_2H_4BH_3 + NaCl + H_2 \qquad (5)$$

$$N(CH_3)_3BH_3 + N_2H_4 \rightarrow N_2H_4BH_3 + N(CH_3)_3 \qquad (6)$$

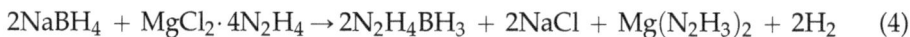

In chemistry, synthesis generally makes two or more reactants react in order to get the targeted molecule. This was the classical strategy for the reactions mentioned above. Often, the first attempts fail. Sometimes, a surprising result stands out, like the formation of hydrazine borane while trying to get ammonia borane. Sutton *et al.* [28,29] were widely involved in finding an efficient chemical route to form ammonia borane from one of its solid residue, polyborazylene. Hydrazine was tentatively used as reducing agent of polyborazylene in tetrahydrofuran. After 12 h of reaction under stirring in room conditions, hydrazine borane was found to form. This is somehow an alternative route for synthesizing hydrazine borane. This would be also a way of regeneration, provided the thermolysis of hydrazine borane mostly leads to polyborazylene.

The heat of formation of solid-state hydrazine borane was determined by pyrolysis in a bomb calorimeter under 29.6 bars of argon. Hydrazine borane decomposed into boron nitride BN (Equation (1)). The heat of formation of solid-state hydrazine borane was found to be 42.7 ± 0.4 kJ·mol^{-1} [17].

3.2. Molecular and Structural Analyses

The FTIR spectrum of hydrazine borane (Figure 1a) is typical of a boron- and nitrogen-based material, with numerous vibration bands, especially those ascribed to the N–H and B–H stretching regions (2600–3500 and 2100–2600 cm^{-1}). Compared to the spectrum of ammonia borane, it is roughly comparable, but shows several additional bands of different intensity [10]. Particularly, there are two small bands at 1915 and 2015 cm^{-1} (B–H stretching region), suggesting strong interactions between H of BH_3 and other elements. Another example is the band at 910 cm^{-1} in the BN–N asymmetric and N–N symmetric stretching region [24].

The solution-state ^{11}B NMR spectrum of hydrazine borane (Figure 1b) shows a signal at δ between -20 and -17.1 ppm, and the ^{11}B{^{1}H} spectrum a quartet (1:3:3:1) characteristic of the BH_3 group ($^1J_{BH}$ 94 $\pm$ 1 Hz) [23,24,30]. The presence of the N_2H_4 moiety can be verified in the ^{1}H NMR spectrum (Figure 1c) via two singlets at δ 3.44 ppm (NH_2–N) and δ 5.45 ppm (NH_2–B). The BH_3 group is also confirmed by a quartet (1:1.1:1.1:1) centered at δ 1.41 ppm due to the heteronuclear coupling between ^{11}B and ^1H and some small signals (three visible over the δ range 1.12–1.72 ppm and four overlapped) attributed to the heteronuclear coupling between ^{10}B and ^1H. The solid-state ^{11}B NMR spectrum (Figure 1d) shows two signals (due to quadrupolar coupling) centered at about δ −24 ppm and whose sharpness indicates high crystallinity.

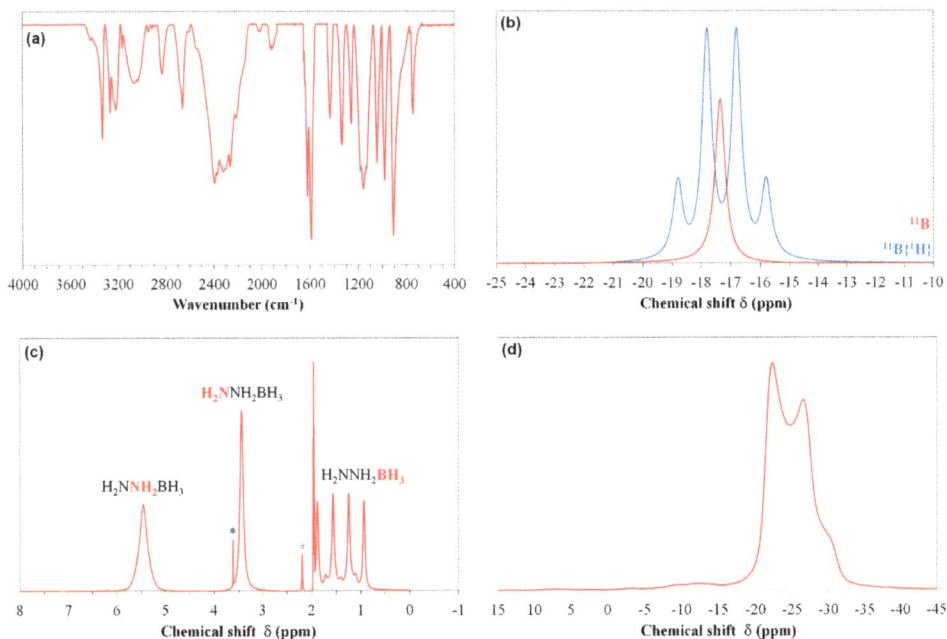

Figure 1. Molecular identification of hydrazine borane $N_2H_4BH_3$. (**a**) FTIR spectrum; (**b**) solution-state ^{11}B and $^{11}B\{^1H\}$ NMR spectra; (**c**) solution-state ^1H NMR spectrum; and (**d**) solid-state ^{11}B NMR spectrum. Adapted from [24]— Reproduced by permission of the Physical Chemistry Chemical Physics (PCCP) Owner Societies.

Hydrazine borane is a white crystalline solid and a Lewis acid-base adduct. By XRD of a single crystal, Goubeau and Ricker reported an orthorhombic *Pccn* (56) space group [10]. More recently, the structure was solved using an orthorhombic *Pbcn* (60) space group with all of the B, N and H atoms belonging to the 8 d sites; Further, the cell parameters were refined [24,31,32]. As shown in Table 1, the cell parameters *a*, *b* and *c* are in good agreement. Neutron diffraction experiment permitted to obtain correct coordinates of the H atoms [31]. The N–B, N–N, N–H and B–H bonds as well as the N–H···H–B and N–H···N interactions were described. The nature of the N–B coordinative (or dative) bond (1.596 Å) exhibits an electrostatic feature mainly, but with substantial contribution of covalence with a large electron population of

~2.1 electrons and a small donation of ~0.05 electron from the Lewis base (N) to the Lewis acid (B). With respect to the N–H$\cdots$H–B intermolecular weak interaction (2.01(1)–2.41(1) Å) [31], it allows a head-to-tail network of the hydrazine borane molecules (Figure 2), which rationalizes the solid state of the material [24]. The N–H$\cdots$N intermolecular interactions (2.114 Å) occur with the head-to-tail network, according to planes parallel to the a-axis. Of note is a dipole moment of 4.18 D determined by Goubeau and Ricker for hydrazine borane [10].

Table 1. Crystallographic data of hydrazine borane (HB) from various works (with ref. as reference and No. as number).

Feature	HB in ref. [10]	HB in ref. [31]	HB in ref. [24]	HB in ref. [32]
Analyzed sample	Single crystal	Single crystal	Single crystal	Powder
Crystal size (mm^3)	2.5 × 0.5 × 0.5	0.3 × 0.3 × 0.2	0.45 × 0.5 × 0.5	–
Temperature (K)	not given	95	173	Room
Crystal system	Orthorhombic	Orthorhombic	Orthorhombic	Orthorhombic
Space group (No.)	*Pccn* (56)	*Pbcn* (60)	*Pbcn* (60)	*Pbcn* (60)
Z	8	8	8	8
a (Å)	13.05	12.974(2)	12.9788(5)	13.1227(11)
b (Å)	5.12	5.070(1)	5.0616(2)	5.1000(5)
c (Å)	9.55	9.507(1)	9.5087(4)	9.5807(9)
B–N bond (Å)	–	1.596	1.587	1.592
N–N bond (Å)	–	1.452	1.452	1.458

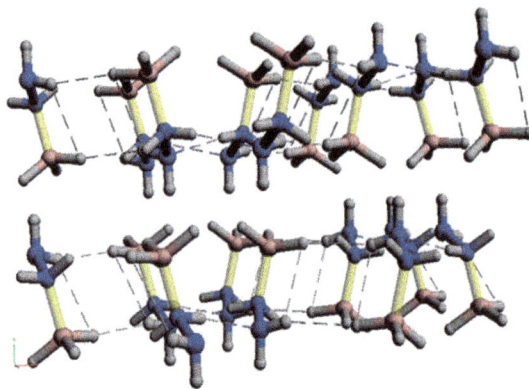

Figure 2. Head-to-tail network of the hydrazine borane molecules $N_2H_4BH_3$ determined from XRD data [24]—Reproduced by permission by the PCCP Owner Societies. The red arrow indicates the x axis and the green one the y axis. The intermolecular interactions are shown by the grey dashed lines.

3.3. Stability and Solubility

Moury *et al.* [24] analyzed hydrazine borane in solid state after storage for one month in an argon-filled glove box and room conditions. The XRD, ^{11}B NMR and FTIR spectroscopy techniques were used. No difference was observed between the results obtained after synthesis and those collected one month later. The authors emphasized the stability of hydrazine borane under inert and dry atmosphere. Data about long-term stability (e.g., up to one year) and stability under air are nevertheless missing.

Goubeau and Ricker [10] considered a large number of solvents to evaluate qualitatively the solubility of hydrazine borane. The solvents were classified into five categories such as: extremely soluble for e.g., water, methanol and pyridine; very soluble for e.g., ethanol, dioxane and tetrahydrofuran; soluble for dimethylaniline and acetyl acetate; moderately soluble for diethyl ether and *n*-butyl acetate; and insoluble for e.g., petroleum ether, benzole and chloroform. Elsewhere, the solubility of hydrazine borane in water was found to be low, *i.e.*, 6 g $N_2H_4BH_3$ in 100 g H_2O [30].By ^{11}B NMR analyses, hydrazine borane solved in dioxane was found to be stable after one month in room conditions and under argon atmosphere [24]. Hydrazine borane could thus be kept solved in dioxane after the filtration following the synthesis. Moury *et al.* [24] also focused on the stability in water because hydrazine borane was intended to be dehydrogenated by hydrolysis in the presence of a catalyst. With deionized water at an initial pH of 6.8 and the same storage conditions than those used for dioxane, hydrazine borane hydrolyzed such that 99%, 98% and 93% of it remained unchanged after two days, one week, and one month, respectively. Karahan *et al.* [30] reported hydrolysis of <5% of hydrazine borane over four days of storage under air at room temperature. They also reported great stability in methanol but without further details. Moury *et al.* [24] envisaged stabilization of the aqueous solution by increasing the pH to 8 with the help of sodium hydroxide NaOH. Similar action has been efficient for hindering the spontaneous hydrolysis of sodium borohydride [9]. After one month, improved stability was noticed: 97% of hydrazine borane remained unchanged. The formation of a borate by-product by hydrolysis was evidenced by large signals at δ higher than 10 ppm. One can thus reasonably conclude that hydrogen evolution by spontaneous hydrolysis of hydrazine borane is negligible at the time scale of one catalytic hydrolysis experiment (<10 min), and long-term storage of aqueous alkaline solution of hydrazine borane may be satisfactorily.

4. Liquid-State Chemical Hydrogen Storage

4.1. Introductive Remark

Like sodium borohydride and ammonia borane [33], hydrazine borane in aqueous solution is a potential liquid-state chemical hydrogen storage material, which means that the main challenge is the catalytic dehydrogenation. Since 2011, mainly three groups, who were already much involved in hydrolyses of sodium borohydride and ammonia borane, have focused on catalytic dehydrogenation of hydrazine borane. This is discussed hereafter.

4.2. Hydrolysis of the BH$_3$ Group of Hydrazine Borane

The Özkar's group (Middle East Technical University, Ankara, Turkey) have focused their efforts on the development of metal-based catalysts of different forms for hydrolysis of the BH$_3$ group of hydrazine borane (Equation (7)). Systematic works were performed where, in addition to the characterization of the catalyst (in fresh and/or used states), the hydrolysis reaction was analyzed in terms of turnovers, turnover frequency, power law and thermodynamic data (apparent activation energy, activation enthalpy and activation entropy). The lifetime and isolability/reusability of the catalysts were also systematically considered.

$$N_2H_4BH_3 + 2H_2O \rightarrow N_2H_5^+ + BO_2^- + 3H_2 \qquad (7)$$

Some catalysts were *ex-situ* prepared: e.g., Rh/Al$_2$O$_3$, Ru/Al$_2$O$_3$, Rh(0) or Ru(0) nanoparticles supported on the zeolite Y [30]; Rh(0) supported on hydroxyapatite Ca$_{10}$(OH)$_2$(PO$_4$)$_6$ [34]. Catalytic precursors were used to get some other catalysts by *in-situ* reduction owing to the reducing properties of hydrazine borane: e.g., RhCl$_3$ and RuCl$_3$ [30]; poly(4-styrenesulfonic acid-co-maleic-acid) stabilized Ni(II) [35]. With such catalytic materials, different results were obtained in terms of hydrogen generation rate and catalytic lifetime. For the reason that will be evoked in the next sub-section and because focusing on the catalytic activity is off topic here, the reader is referred to the following review articles for more information about the catalysts, their performance and the kinetic parameters [32,36–38].

The hydrolysis reaction was reported to take place according to the reaction shown by Equation (6) However, the solution-state ^{11}B NMR spectrum reported by the authors shows a signal centered at 12.5 ppm, preceded by a shoulder at around 10 ppm. This is typical of the presence, in equilibrium, of the base tetrahydroxyborate anion B(OH)$_4^-$ and the acid counterpart boric acid B(OH)$_3$ [39]. Furthermore, in the hydrolysis conditions, the anhydrous borate anion BO$_2^-$ cannot exist, the hydrated form B(OH)$_4^-$ being the thermodynamically stable phase [40]. Consequently, the hydrolysis more likely takes place according to the reactions described by Equation (8) ($\Delta_r H = -244.5$ kJ·mol^{-1}) and Equation (9) ($\Delta_r H = -208.7$ kJ·mol^{-1}). The theoretical gravimetric hydrogen storage capacity will thus be impacted (Figure 3): 7.3 w_t% for N$_2$H$_4$BH$_3$-2H$_2$O in Equation (7); 6 w_t% for N$_2$H$_4$BH$_3$-3H$_2$O in Equation (8); and 5.1 w_t% for N$_2$H$_4$BH$_3$-4H$_2$O in Equation (9).

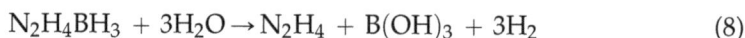

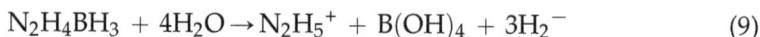

$$N_2H_4BH_3 + 3H_2O \rightarrow N_2H_4 + B(OH)_3 + 3H_2 \qquad (8)$$

$$N_2H_4BH_3 + 4H_2O \rightarrow N_2H_5^+ + B(OH)_4 + 3H_2^- \qquad (9)$$

Karahan *et al.* [30] optimized the effective gravimetric hydrogen storage capacities. For a $H_2O/N_2H_4BH_3$ mole ratio of 6.6, 2.8 moles of H_2 per mole of $N_2H_4BH_3$ were released within 4 min at 25 °C. Taking into account the weight of the *in-situ* formed Rh(0) catalyst (from $RhCl_3$), this means an effective capacity of <3.4 w_t%. In harsher conditions, namely for a $H_2O/N_2H_4BH_3$ mole ratio of 2, complete hydrolysis was achieved within less than 6 h.

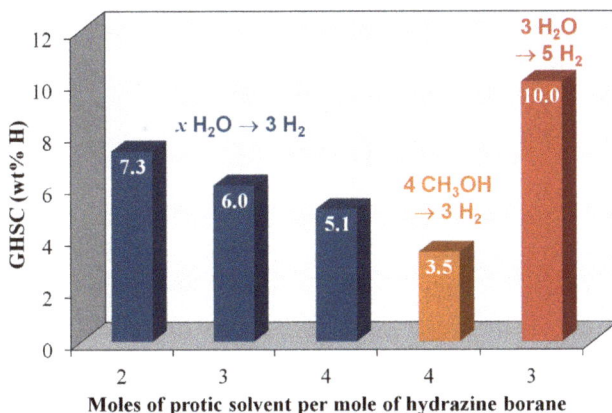

Figure 3. Theoretical gravimetric hydrogen storage capacity (denoted GHSC, in w_t% H) for the couples $N_2H_4BH_3$-xH_2O (with x equal to 2, 3 and 4) and $N_2H_4BH_3$-$4CH_3OH$. In blue and orange, the BH_3 group hydrolyzes only into 3 moles of H_2 (Equations (7)–(10)). In red, the BH_3 group and the N_2H_4 moiety generates 5 moles of H_2 (Equation (12)).

The absence of ammonia NH_3 as gaseous by-product was controlled with acid/base indicator or a trap of aqueous hydrochloric acid HCl placed upstream from the hydrolysis reactor [30,35]. In other words, this result suggests that either the N_2H_4 moiety is decomposed into NH_3 that remains solved in the aqueous medium kept at 25 °C or the catalyst is inactive towards the dehydrogenation of N_2H_4. If the latter hypothesis is correct, the attractiveness of the catalyst would then be negatively affected; This is discussed hereafter.

It is worth mentioning that, like for sodium borohydride and ammonia borane, catalytic methanolysis of hydrazine borane (Equation (10)) was also investigated [41,42]. With the help of solution-state ^{11}B NMR and FTIR spectroscopic methods, the by-product hydrazinium tetramethoxyborate $N_2H_5B(OCH_3)_4$ was found to form. The advantage of this reaction over hydrolysis has not been demonstrated. Further, when water and methanol are put together, the hydrolysis reaction predominates [41]. The theoretical gravimetric hydrogen storage capacity of the couple $N_2H_4BH_3$-$4CH_3OH$ is low (3.5 w_t%; Figure 3).

$$N_2H_4BH_3 + 4\,CH_3OH \rightarrow N_2H_5{}^+ + B(OCH_3)_4 + 3\,H_2{}^- \qquad (10)$$

In the light of this overview, considering that ammonia borane is richer in hydrogen than hydrazine borane (19.5 w_t% vs. 15.4 w_t%), and taking into account that the couple NH_3BH_3-4H$_2$O (Equation (11)) is richer in hydrogen than the couple $N_2H_4BH_3$-4H$_2$O (Equation (9); $\Delta_rH = -155.8$ kJ·mol^{-1}) with 5.8 w_t% vs. 5.1 w_t%, that raises a question. What is the advantage of hydrazine borane over ammonia borane?

$$NH_3BH_3 + 4\,H_2O \rightarrow NH_4{}^+ + B(OH)_4 + 3\,H_2{}^- \qquad (11)$$

4.3. Hydrolysis of the BH$_3$ Group and Dehydrogenation of the N$_2$H$_4$ Moiety of Hydrazine Borane

The Demirci's group (University of Montpellier, France) in collaboration with the Xu's group (AIST, Osaka, Japan) joined their efforts to answer the question asked above. They demonstrated that the great advantage of hydrazine borane is that the N_2H_4 moiety can be dehydrogenated. In the case of the NH_3 group of ammonia borane, dehydrogenation is thermodynamically impossible in room conditions, limiting thus the theoretical gravimetric hydrogen storage capacity of NH_3BH_3-3H$_2$O to 7.1 w_t%. With $N_2H_4BH_3$-3H$_2$O (Equation (12); $\Delta_rH = -193.9$ kJ·mol^{-1}), the gravimetric hydrogen storage capacity that could be ideally obtained is 10 w_t% (Figure 3).

$$N_2H_4BH_3 + 3H_2O \rightarrow B(OH)_3 + 5H_2 + N_2 \qquad (12)$$

Every catalyst made of, e.g., nickel, cobalt, rhodium, ruthenium and platinum, is especially active in hydrolysis of the BH$_3$ group (and also of the BH$_4{}^-$ anion), it is just a matter of rate [43,44]. The challenge is thus to find the metal-based catalyst that is active towards the dehydrogenation of the N_2H_4 moiety. Xu's group had developed active nickel-based alloyed nanoparticles for selective dehydrogenation of hydrous hydrazine N_2H_4·H$_2$O at temperatures over the range 20–70 °C [45]. Indeed, the dehydrogenation of N_2H_4·H$_2$O (Equation (13)) competes with the decomposition (Equation (14)).

$$N_2H_4\cdot H_2O \rightarrow 2\,H_2 + N_2 + H_2O \qquad (13)$$

$$N_2H_4\cdot H_2O \rightarrow 4/3\,NH_3 + 1/3\,N_2 + H_2O \qquad (14)$$

A first and successful attempt envisaged the use of nickel-platinum nanoparticles [46]. While the monometallic catalysts were found to be active in the hydrolysis of the BH$_3$ group only, the bimetallic alloy $Ni_{0.89}Pt_{0.11}$ showed to be active and 93% selective in the dehydrogenation of the N_2H_4 moiety, owing to geometric and electronic effects. However, the hydrogen evolution showed two regime of kinetics (Figure 4), with a fast first step attributed to the hydrolysis of BH$_3$

and a second one with slower kinetics due to the dehydrogenation of N_2H_4. The former reaction is at least 40 times faster. In fact, during the first step, when the BH_3 groups are hydrolyzed, some of the N_2H_4 moieties interact with the catalytic surface and decompose [47,48]. The second regime of kinetics (*i.e.*, dehydrogenation of N_2H_4) determines the total time that the dehydrogenation of hydrazine borane really needs.

Other catalytic solutions were investigated: e.g., metal salts to form *in-situ* the catalytically active phase [49]; supported nickel [50]; Ni@NiPt core shell nanoparticles [51]. With respect to the nickel-based bimetallic alloy nanoparticles, several metals were tested: Pt, Rh, Ru, Ir, Fe, Co and Pd [52,53]. Though very active and almost-100% hydrogen selective, the Pt-, Rh- and Ir-containing systems were found not to be stable after the first cycle because of nickel surface enrichment, explained by borate-induced segregation [47,52]. Nickel, like cobalt, strongly adsorbs borates via M–O–B bonds, leading to the catalyst deactivation [54–56]. However, the presence of both Ni and the noble metal is essential and would be the optimum conditions to activate the bonds of the N_2H_4 moiety of hydrazine borane [52]. Recently, another group has contributed to the development of field. Li *et al.* [57] reported the synthesis and use of alumina supported Ni@(RhNi-alloy) nanocomposites that showed a H_2 selectivity of <95%.

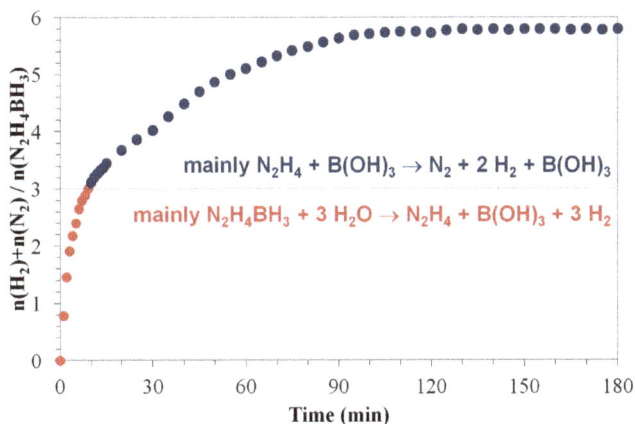

Figure 4. Stepwise catalytic dehydrogenation of aqueous hydrazine borane in near-room conditions and in the presence of $Ni_{0.89}Pt_{0.11}$ nanoparticles. The gas evolution is composed of a first step, *i.e.*, fast hydrolysis of the BH_3 group, followed by a second one, *i.e.*, slow dehydrogenation of the N_2H_4 moiety. Adapted from Ref. [46] with permission from The Royal Society of Chemistry.

Among the catalytic solutions investigated so far, the best performance was achieved with nanoporous carbon-supported $Ni_{0.6}Pt_{0.4}$ nanoalloys [58]. It is considered the best for the following three reasons: (*i*) It is 100% selective, generating five moles of H_2 and one mole of N_2 per mole of $N_2H_4BH_3$; (*ii*) It greatly improves the rates of the second regime of kinetics, making the reaction complete within <5 min at 30 °C; and (*iii*) It is stable in terms of activity and selectivity over five cycles.

Up to now, the catalytic activity and selectivity have been the major concerns. The heterogeneous catalysts were evaluated in favorable conditions, that is, using diluted aqueous hydrazine borane solutions. The effective gravimetric hydrogen storage capacities were therefore low (<0.5 w_t%) and not realistic for practical applications. The highest capacity ever reported is 1.2 w_t% for a $H_2O/N_2H_4BH_3$ molar ratio of 42, but in these conditions the selectivity of the $Ni_{0.89}Pt_{0.11}$ catalyst was negatively affected [47]. This reveals another issue to overcome with the catalyst used in this reaction: it has to be selective whatever the concentration of the borane. This is the only way to get high effective gravimetric hydrogen storage capacities and to make aqueous hydrazine borane viable for liquid-state chemical hydrogen storage.

5. Solid-State Chemical Hydrogen Storage

5.1. Pristine Hydrazine Borane

The behavior of hydrazine borane under heating at constant temperature (70, 85, 100 and 200 °C) was first investigated by Goubeau and Ricker [10]. Melting at 61 °C was reported. Upon melting, the borane foamed because of its decomposition, which was then visually observed. Besides hydrogen, hydrazine in gaseous state was found to evolve whereas no gaseous boron-containing by-product (e.g., diborane B_2H_6) was detected. The solid residue forming upon the evolution of 1 mole of H_2 and 0.24–0.47 mole of N_2H_4 per mole of $N_2H_4BH_3$ (Equation (15)) was reported to be constituted by the structural unit $NHBH_2$. It was inert towards acidic and basic aqueous solutions. The evolution of the second mole of H_2 took place when the solid was kept at 200 °C for 10 h (Equation (16)). The as-obtained solid residue was found to be shock-sensitive and explosive.

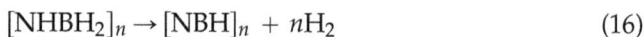

$$nN_2H_4BH_3 \rightarrow [NHBH_2]_n + n/2N_2H_4 + nH_2 \qquad (15)$$

$$[NHBH_2]_n \rightarrow [NBH]_n + nH_2 \qquad (16)$$

With the help of correlated molecular orbital theory, Vinh-Son *et al.* [59] concluded that the energy of the B–N dative bond of gas-phase hydrazine borane is larger than that of gas-phase ammonia borane (~130 *vs.* ~110 kJ· mol^{-1}) and that such higher thermodynamic stability could be an advantageous feature for chemical hydrogen storage. However, experimental works performed by Hügle *et al.* [23] did not confirm: at 65 and 100 °C, less than one mole of H_2 per mole of $N_2H_4BH_3$ was liberated in 43 and 30 h, respectively. At higher temperatures, the hydrogen release was faster, with about 1.4 moles of H_2 per mole of $N_2H_4BH_3$ in less than 1 h at 140 °C. Slightly different results were reported by Moury et al. [24]. They found 1.8 moles of H_2 evolving at 140 °C, but with similar kinetics. Hence, both groups concluded that pristine hydrazine borane is not suitable for chemical hydrogen storage.

Moury et al. [24] analyzed the thermolytic decomposition of hydrazine in dynamic conditions (5 °C· min^{-1}), by TGA and DSC, and over the range 25–400 °C. It was confirmed that the melting occurs at around 60 °C with an enthalpy of 15 kJ· mol^{-1}. Together with melting, hydrazine borane dehydrogenated in small extent (1.2 $w_t\%$ H_2 at 95 °C). The main decomposition (mass loss of 28.7 $w_t\%$) took place at 105–160 °C and had a heat of 20 kJ· mol^{-1}. The analysis of the evolving products revealed the presence of hydrogen and of the unwanted hydrazine, confirming the observations made by Goubeau and Ricker [10]. There was a third mass loss (4.3 $w_t\%$), due to some dehydrogenation (14 kJ· mol^{-1}). Most importantly, it was reported that the solid residue forming upon this decomposition was shock-sensitive and made the authors strongly stress on the unsuitability of hydrazine borane for chemical hydrogen storage. In fact, hydrazine borane was recently proposed as being a possible hypergolic fuel for propellant systems [60,61].

Goubeau and Ricker [10] and, later, Moury et al. [24] proposed some likely structural units of the solid residue forming upon the release of the first equivalent of hydrogen (Figure 5). Any identification was found to be very difficult because of insolubility of the residue in organic solvents, reactivity in water and alcohols (solvolysis), and amorphous state [23]. Furthermore, like for thermolysis of ammonia borane, the IR and solid-state NMR spectroscopy methods do not enable gaining insight about the exact nature of the solid residue [20–24]. For the solid residue forming upon the release of the second equivalent, none of the groups were able to recover it in safe conditions. Accordingly, there has been no mechanism proposed for the decomposition of hydrazine borane.

Like pristine hydrazine borane, neat ammonia borane is ineffective for chemical hydrogen storage. Hence, different strategies were envisaged to destabilize it: (i) Chemical doping; (ii) Dispersion in solvent (organic or ionic liquid) with/without the presence of a homogeneous metal-based catalyst; and (iii) Nanoconfinement into a porous host [22]. Strategies (i) and (ii) were also envisaged in the case of hydrazine borane. They are discussed hereafter. A fourth strategy (iv) was also considered: it consists in elaborating derivatives of the borane [22]. This was also applied to hydrazine borane. This is reported below.

165

Structural units formed from $N_2H_4BH_3$	Structural units formed from $N_2H_4BH_3$ and BH_3
$NH - NH_2 - BH_2$	$BH_2 - NH - NH_2 - BH_2$
(cyclic structures)	(cyclic structures)
And many other possible structural units involving several $N_2H_4BH_3$ molecules and/or BH_3 moieties	(cyclic structures)

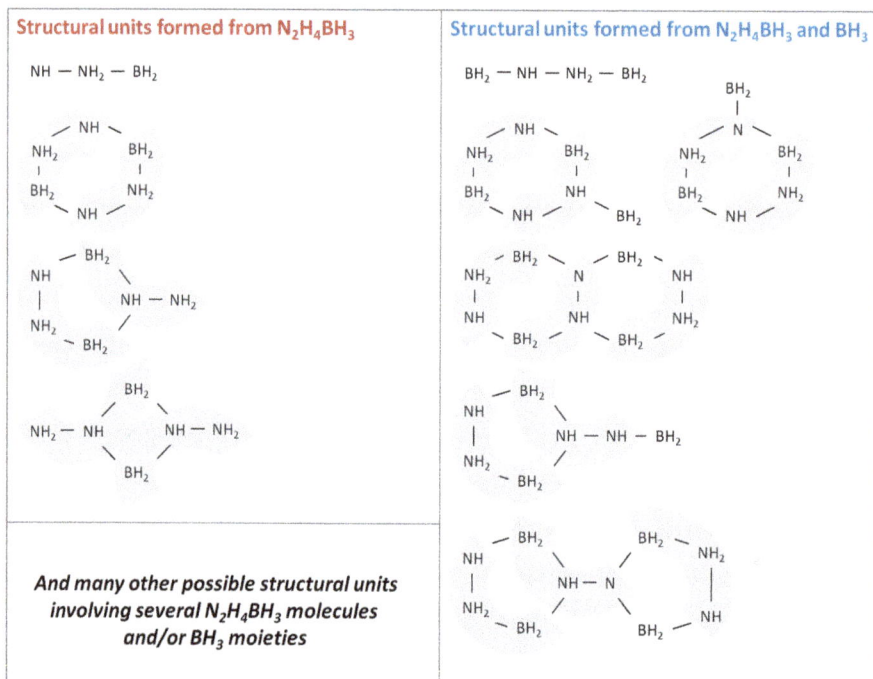

Figure 5. Examples of likely structural units of the solid by-products formed by the decomposition of hydrazine borane $N_2H_4BH_3$ over the range 25–200 °C [24]—reproduced by permission of the PCCP Owner Societies.

5.2. Chemical Doping of Hydrazine Borane

Hügle *et al.* [23] chose lithium hydride LiH to destabilize hydrazine borane. The 1:1 mixture (14.8 w_t% H) was constituted of 4 $H^{\delta-}$ (from BH_3 of hydrazine borane and LiH) and 4 $H^{\delta+}$ (from N_2H_4 of hydrazine borane). In comparison to pristine hydrazine borane, better dehydrogenation kinetics was reported. For example, at 150 °C more than three moles of H_2 per mole of the mixture were released within 4.5 h. The presence of lithium hydride improved not only the dehydrogenation extent, but also the dehydrogenation kinetics and the purity of the liberated hydrogen. Only traces of ammonia were detected.

Chemical doping was also studied by Toche *et al.* [62]. Borohydrides (LiBH$_4$ or NaBH$_4$) were added to hydrazine borane. The mixtures borane-borohydride had a molar ratio 4:1, namely with one $H^{\delta+}$ for one $H^{\delta-}$. They showed improved dehydrogenation properties in comparison to not only hydrazine borane but also to the borohydride. Indeed, the borane destabilized the borohydrides, making them dehydrogenate at temperatures much lower than the temperature 400 °C reported for the pristine borohydrides. The mixtures were capable of liberating hydrogen as soon as 50 °C. For example, the four moles of $N_2H_4BH_3$ and one mole of $LiBH_4$ released 12.1 moles of H_2 upon heating up to 300 °C (5 °C·min^{-1}), but 1.1 moles of N_2H_4 also evolved. It was proposed the formation of a solid residue of empirical formulae $LiB_5N_{5.8}H_{3.4}$ (Equation (17)), which consisted likely of polyborazylene- and/or boron nitride-like compounds. Interestingly, the presence of boron nitride was reported. It would have formed at <500 °C whereas the formation of such ceramic material generally occurs at >1000 °C [63].

$$4N_2H_4BH_3 + LiBH_4 \rightarrow LiB_5N_{5.8}H_{3.4} + 1.1N_2H_4 + 12.1H_2 \qquad (17)$$

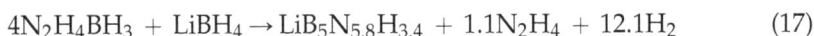

The formation of boron nitride from hydrazine borane at <300 °C was also reported in another contribution. Petit *et al.* [64] focused on the destabilization of hydrazine borane by ammonia borane, and *vice versa*, in equimolar amounts. The destabilization took place at 30 °C. The binary solid melted because one of the boranes disrupted the intermolecular $H^{\delta+}\cdots H^{\delta-}$ network of the other borane. At temperatures higher than 75 °C, the binary mixture explosively decomposed, liberating hydrogen and hydrazine. Alternatively, the sample was treated at 90 °C for 2 h so that a stable solid, *i.e.*, a mixture of oligomers from both boranes, was obtained. The stable solid was then investigated for chemical hydrogen storage. It was able to liberate *ca.* 11.4 w_t% of almost pure H_2 from 75 to 300 °C according to a two-step process. Traces of ammonia were detected during the first decomposition step (75–150 °C). Very fast dehydrogenation was observed in the second step, which occurred at around 200 °C. The solid residue was found to be insoluble and stable in organic solvents as well as in water, and was analyzed by FTIR and XRD evidencing the formation of orthorhombic boron nitride. The presence of an amorphous phase of boron nitride could not be excluded.

The findings reported by Toche *et al.* [62] and Petit *et al.* [64] are very interesting from the point of view of ceramics chemistry. However, the formation of boron nitride is a drawback from the point of view of chemical hydrogen storage, since boron nitride cannot be recycled to close the hydrogen cycle with the aforementioned boron-based materials.

5.3. Dispersion and Catalysis of Hydrazine Borane

To date, just one paper reported the destabilization of hydrazine borane by dispersion in an organic solvent (tetrahydrofuran) and in the presence of a homogeneous catalyst (group 4 metallocene alkyne complexes of the type $Cp'_2M(L)(\eta^2-(CH_3)_3SiC_2Si(CH_3)_3)$ with Cp' as substituted or unsubstituted η^5-cyclopentadienyl and M as Ti (no L) or Zr (L = pyridine)) [65]. Up to four moles of H_2 and N_2 per mole of hydrazine borane evolved at 25 or 50 °C, in the best case within 32 h. The solid residue was analyzed but the task was tough because of the reasons mentioned above. By elemental analysis, it was found a B/N ratio of 1:1.03, suggesting that 3.5 moles of H_2 and 0.5 mole of N_2 evolved. XRD and IR analyses were in line with the results reported by Moury *et al.* [24]. They were indicative of the formation of a mixture of cyclic [H_2N–NH–BH_2] structures (Figure 5) and boron nitride species. Inspired from the process of Sutton *et al.* [28,29], the spent fuel was tentatively reduced by hydrazine in tetrahydrofuran at 50 °C. The attempt was successful in some extent. Further optimizations would be in progress (not reported yet in the open literature).

5.4. Chemical Modification of Hydrazine Borane

Chemical modification of hydrazine borane by reaction with an alkaline hydride was first introduced by Hügle *et al.* in 2009 [23]. However, the authors preferred to investigate the 1:1 mixture of lithium hydride and hydrazine borane (both in solid states; 14.8 w_t% H) to avoid losing one equivalent of hydrogen (Equation (17)). In 2012, Wu *et al.* [32] reported the preparation of the first sample of lithium hydrazinidoborane $LiN_2H_3BH_3$ (11.6 w_t% H) by ball-milling (200 rpm, 1 h) lithium hydride and hydrazine borane under inert atmosphere. The structure was solved using a monoclinic $P2_1/c$ (14) space group with all atoms in 4e sites. The adduct $LiN_2H_3BH_3 \cdot 2N_2H_4BH_3$ (13.9 w_t% H) was obtained in a similar way by increasing the amount of hydrazine borane.

$$N_2H_4BH_3 + LiH \rightarrow LiN_2H_3BH_3 + H_2 \qquad (18)$$

The formation of lithium hydrazinidoborane $LiN_2H_3BH_3$ can be explained by the reaction of the strong Lewis base H^- of lithium hydride with one of the protic hydrogen $H^{\delta+}$ on the middle NH_2 group of hydrazine borane and its replacement by the lithium cation Li^+ (Figure 6) [66]. Qian *et al.* [67] found by first-principles calculations that the most stable structure would be the one where the lithium cation Li^+ substitutes one of the hydrides $H^{\delta-}$ of the BH_3 group. No detail is provided on fate of the strong Lewis base H^- of lithium hydride and the substituted $H^{\delta-}$. This result is surprising, as even for amidoboranes, such an unexpected substitution has never been observed [21].

Moury *et al.* [66] recently argued that they were also working on lithium hydrazinidoborane when Wu *et al.* [32] published their experimental report. However, the former researchers were not able to compare favorably the structure of lithium hydrazinidoborane they mechano-synthesized to the compound proposed by later authors. Moury *et al.* [66] demonstrated that lithium hydrazinidoborane is a polymorphic material with a stable low-temperature phase with orthorhombic *Pbca* (61) space group, atoms standing in the 8c sites, and a metastable high-temperature phase as described by Wu *et al.* [32]. The former was called the β phase because discovered after the metastable phase, then called α. The crystallographic data for both phases are reported in Table 2. In their report, Moury *et al.* highlighted a phase transition from the β phase to the α phase at about 95 °C. The phase transition is of first order, the volume of the β phase (648.8 Å^3) being almost twice that of the α one (328.3 Å^3). The Li$\cdots$Li distance decreases from 3.49 Å for the β phase to 3.31 Å for the α one, explaining the better stability of the former phase.

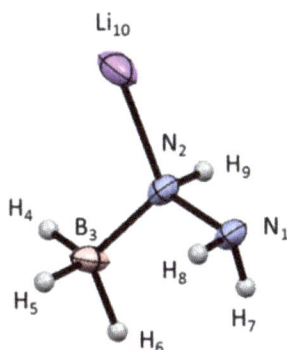

Figure 6. Asymmetric unit with labels of lithium hydrazinidoborane LiN$_2$H$_3$BH$_3$. Reprinted with permission from [66]. Copyright 2014 American Chemical Society.

Table 2. Crystallographic data of alkali hydrazinodoboranes at room temperature (No. as number).

MN$_2$H$_3$BH$_3$	α-LiN$_2$H$_3$BH$_3$	β-LiN$_2$H$_3$BH$_3$	NaN$_2$H$_3$BH$_3$	KN$_2$H$_3$BH$_3$
Reference	[32,66]	[66]	[68]	[69]
Crystal system	Monoclinic	Orthorhombic	Monoclinic	Monoclinic
Space group (No.)	$P2_1/c$ (14)	$Pbca$ (61)	$P2_1/n$ (14)	$P2_1$ (4)
a (Å)	5.8503(11)	10.25182(11)	4.97437(11)	6.72102(23)
b (Å)	7.4676(11)	8.47851(10)	7.95806(15)	5.89299(20)
c (Å)	8.8937(15)	7.46891(8)	9.29232(19)	5.77795(17)
β (°)	122.329(6)	–	93.8137(11)	108.2595(13)
B–N bond (Å)	1.539	1.549(2)	1.537(6)	1.541
N–N bond (Å)	1.469	1.495(2)	1.453(5)	1.463

The solid state of lithium hydrazinidoborane is explained by the presence of an intermolecular head-to-tail H$^{\delta+}\cdots$H$^{\delta-}$ interaction that form chains (*i.e.*, parallel plans on which the H$^{\delta+}\cdots$H$^{\delta-}$ network extends). In the β phase, the network was identified according to the definition of Klooster *et al.* [70], with a B–H$\cdots$H angle slightly bent (106.8°), a N–H$\cdots$H angle almost linear (171.2°) and a H$\cdots$H distance of 2.25 Å [66]. For both phases, the lithium is in tetrahedral environment (Figure 7). Two corners of the tetrahedron are occupied by the BH$_3$ moiety and interaction occurs through the BH$_2$ edges of the –BH$_3$ tetrahedron. Of the two remaining corners, one is occupied by the central N of the N_2H_3 moiety and the other by the terminal N through the lone electron pair. Thereby, these interactions induce a strong electronic modification in the $[N_2H_3BH_3]^-$ entities. Such coordination leads to modified B–H bond polarization, decreased electron density around the boron element, and, thus, increased reactivity of the hydridic $H^{\delta-}$ hydrogen. Further, compared to hydrazine borane, the B–N bond is shortened owing to strong electron donation from N to Li^+ and the N–N bond is stretched due to the interactions between Li^+ and the lone electrons pair of the terminal N (Tables 1 and 2). These results may rationalize the improved thermal dehydrogenation properties described hereafter [32].

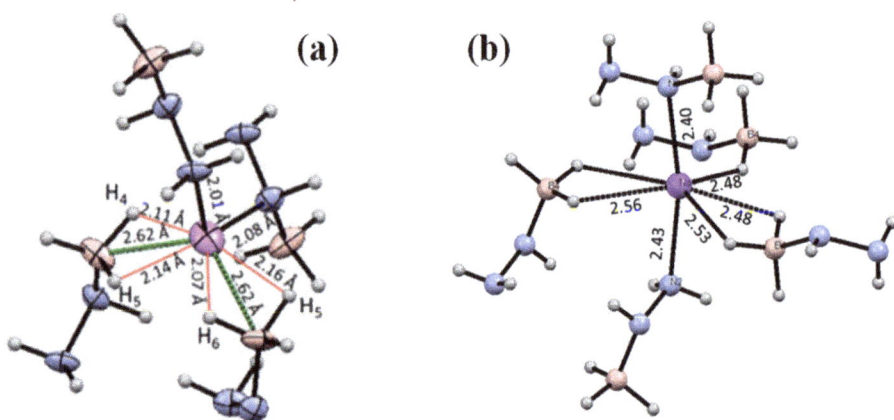

Figure 7. (a) Coordination of Li^+ in lithium hydrazinidoborane $LiN_2H_3BH_3$ (reprinted with permission from [66]; copyright 2014 American Chemical Society); (b) Coordination of Na^+ in sodium hydrazinidoborane $NaN_2H_3BH_3$ (reprinted with permission from [71]; copyright 2013 Wiley).

The thermal dehydrogenation of lithium hydrazinidoborane is indeed more attractive than that of hydrazine borane. By TGA, the α phase starts to generate hydrogen below 70 °C but most is liberated over the range 100–200 °C [32]. It was measured the release of 9.5 $w_t\%$ of H_2 at 200 °C. It was also detected the release of 0.7 $w_t\%$ of N_2 and 0.1 $w_t\%$ of NH_3. With respect to the β phase, the decomposition starts at 40 °C and follows a five-step process over the range 40–400 °C [66]. Up to 144 °C, a mass loss of 7.8 $w_t\%$ was observed due to the liberation of H_2 and a small amount of N_2 (<1 $w_t\%$). The liberation of ammonia (along with H_2 and N_2) takes place at >144 °C. No traces of hydrazine, diborane or borazine were detected. In isothermal conditions, the β phase liberated 2.6 equivalents of H_2 in 1 h at 150 °C (vs. 1.4 equivalents for hydrazine borane) and without traces of ammonia. Apparent activation energy of 58 kJ· mol^{-1} was calculated [66]. With respect to the α phase, 2.4 equivalents of H_2 were released in 1 h at 130 °C [32].

Wu *et al.* [32] and then Moury *et al.* [66] suggested that the dehydrogenation of lithium hydrazinidoborane can be explained by the combination of protic $H^{\delta+}$ and hydridic $H^{\delta-}$ hydrogens that are in intermolecular interactions. The latter authors proposed a series of reaction mechanisms (Figure 8) and suggested the appearance of *bis*(lithium hydrazide) of diborane $[(LiN_2H_3)_2BH_2]^+[BH_4]^-$ as reaction intermediate. Nonetheless, a recent work on the isotopomer α-$LiN_2H_3BD_3$ gave evidence of the occurrence of homopolar pathways [72]. The initial dehydrogenation would be due to the reaction between two protic $H^{\delta+}$ hydrogens (homopolar N–H$\cdots$H–N pathway) and it would be followed by the N–H$\cdots$H–B and B–H$\cdots$H–B pathways. It is generally admitted that the dehydrogenation of boranes is complex [71], and the reference [72] is further evidence of such complexity.

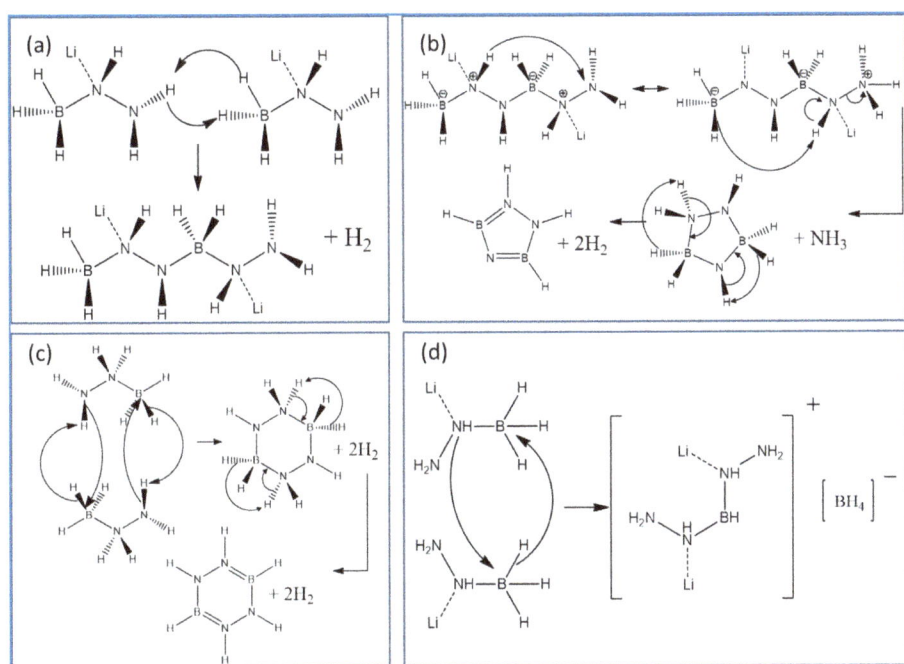

Figure 8. Proposition of mechanisms for the formation of monomeric units and *bis*(lithium hydrazide) of diborane $[(LiN_2H_3)_2BH_2]^+[BH_4]^-$. (a) Linear dimer with boron in sp^3 hybridation; (b) 5-Ring center monomer formed by proton exchange, then cyclization with NH_3 release and finally dehydrocyclization, where boron is in sp^2 hybridation; (c) 6-Ring center monomer formed by cyclization of two hydrazine borane monomers, followed by dehydrocyclization; and (d) Formation of *bis*(lithium hydrazide) of diborane $[(LiN_2H_3)_2BH_2]^+[BH_4]^-$. Reprinted with permission from [66]. Copyright 2014 American Chemical Society.

Moury *et al.* [68] investigated the sodium derivative. Sodium hydride NaH showed high reactivity towards hydrazine borane. An explosive reaction took place when the reactants were put into contact under inert atmosphere. The authors circumvented the problem by working in cold conditions ($\leqslant 30\ ^\circ$C). Sodium hydrazinidoborane $NaN_2H_3BH_3$ (8.8 w_t% H) was then successfully synthesized. The molecular structure was verified by the NMR and FTIR spectroscopy methods. The crystal structure was determined by powder XRD. A monoclinic structure with a space group $P2_1/n$ (14) with all the atoms in the 4e sites (Table 2) was found. Like for lithium hydrazinidoborane, the sodium cation Na^+ replaces one of the protic $H^{\delta+}$ hydrogens of the middle NH_2, but unlike Li in $LiN_2H_3BH_3$, Na in $NaN_2H_3BH_3$ is surrounded by five hydrazinidoborane entities to fulfill the coordination sphere of Na (Figure 7). The coordination is obtained via the BH_2 edges, which activate the B–H bonds, as well as with the central and terminal nitrogen atoms in $[N_2H_3BH_3]^-$. Compared to the parent hydrazine borane, sodium hydrazinidoborane is destabilized. Like for lithium hydrazinidoborane, the DSC results suggested a complex reaction, involving at least four successive exothermic processes. Sodium hydrazinidoborane starts its dehydrogenation at around 60 $^\circ$C. Below 100 $^\circ$C, it is able to release 6 w_t% of gases, mainly hydrogen. Nitrogen and traces of both ammonia and hydrazine were detected. At 150 $^\circ$C, the overall mass loss is 7.6 w_t%. In a further work, Moury *et al.* [73] reported that the addition of an excess of 5 w_t% of sodium hydride leads to the formation of a sample of sodium hydrazinidoborane that is able to liberate *ca.* 8.8 w_t% of pure hydrogen at 160 $^\circ$C.

The last hydrazinidoborane reported so far is the potassium one. In our laboratory, we found that the solid-state reaction between potassium hydride and hydrazine borane is extremely reactive in inert and room conditions, even more reactive than sodium hydride [74]. Such an issue was recently addressed by Chua *et al.* [69]. The preparations of sodium and potassium hydrazinidoboranes were performed in an autoclave while using tetrahydrofuran as dispersion medium. The synthesis of the potassium derivative $KN_2H_3BH_3$ (7.2 w_t% H) was confirmed by spectroscopy. The crystal structure was defined as being monoclinic with a space group $P2_1$ (4) (Table 2). Like the other hydrazinidoboranes, the substitution of one of the protic $H^{\delta+}$ hydrogens of the middle NH_2 by the potassium cation K^+ leads to a compound with changed bonding chemistry in comparison to hydrazine borane. Accordingly, potassium hydrazinidoborane showed improved dehydrogenation properties, with e.g., hydrogen evolution from ~50 $^\circ$C and mass loss of 7.3 w_t% at 180 $^\circ$C due to hydrogen and a small amount of ammonia. In isothermal conditions, at 88 $^\circ$C, potassium hydrazinidoborane was able to liberate 1.8 equivalents of H_2 (60% of its theoretical H) within 1 h, whereas hydrazine borane released only 0.5 equivalent of H_2 (15% of its theoretical H). Chua *et al.* [69] emphasized the complexity of the dehydrogenation mechanisms. In particular, they suggested that the initial step would be characterized by the formation of dimers like $NH_2NH(M)BH_2–NHNH(M)BH_3$ and $NH_2NH(M)BH=NNH(M)BH_3$ (with M as Li, Na or K) obtained by intermolecular reactions between $H^{\delta+}$ in one molecule and $H^{\delta-}$ in another one. This is in agreement with the hypotheses reported by Moury *et al.* [66,68]. Also, Chua *et al.* [74] highlighted a clear correlation between the cation size and dehydrogenation properties of the metal hydrazinidoborane. In fact, the bigger the cation is, the lower the onset temperature of dehydrogenation.

The formation of an alkali derivative from hydrazine borane leads thus to a material more suitable for chemical hydrogen storage. Indeed, the dehydrogenation properties are improved in terms of onset temperature, kinetics and purity of hydrogen (*i.e.*, inhibition of the formation of the unwanted hydrazine and ammonia). Further, unlike for hydrazine borane, there is no mention of the formation of any shock-sensitive solid residue.

6. Conclusions and Outlook

Hydrazine borane $N_2H_4BH_3$ is under investigation for chemical hydrogen storage since the late 2010s when it was suggested as having a good potential in the field, especially as an alternative to ammonia borane NH_3BH_3. It was then studied for solid- and liquid-state chemical hydrogen storage, as was the case for ammonia borane. In fact, hydrazine borane faces the same challenges as ammonia borane.

Hydrazine borane is quite stable in water at neutral and basic pH, which is a first attractive feature for catalytic dehydrogenation in room conditions. Like for ammonia borane, the BH_3 group of hydrazine borane is easily hydrolyzed in the presence of a metal-based catalyst (homo- or heterogeneous), with almost three moles of hydrogen liberated with fast kinetics. Yet, in such a context, ammonia borane is more attractive in terms of gravimetric hydrogen storage capacities. In fact, interest on hydrazine borane does not arise unless the N_2H_4 moiety can be dehydrogenated. Nickel-based bimetallic nanoalloys showed to be efficient catalysts for both hydrolysis of BH_3 and selective dehydrogenation of N_2H_4. Important achievements were reported in the recent years but more needs to be done for improving the kinetics of the latter reaction, which is still too slow in comparison to that of the former reaction. Also, more needs to be done to improve the catalysts stability in successive uses. The suitability of hydrazine borane for liquid-state chemical hydrogen storage has also to be evidenced by improving and optimizing the effective gravimetric hydrogen storage capacity, with the ideal target of 10 wt%. This is in fact a critical issue as, beyond ammonia borane, sodium borohydride is also a candidate with high potential for liquid-state chemical hydrogen storage.

It could be undoubtedly stated that pristine hydrazine borane is not suitable for solid-state chemical hydrogen storage. Like ammonia borane, it decomposes under heating, mainly at >100 °C, and liberates substantial amounts of unwanted gaseous by-products along with hydrogen. The release of hydrazine is particularly problematic. Unlike ammonia borane, the extensive decomposition of hydrazine borane leads to the formation of a shock-sensitive and explosive solid residue. Such features serve hydrazine borane negatively. However, hydrazine borane should not be definitely discarded. Like ammonia borane, it may be destabilized with the help of a chemical additive. Also, it is an attractive reactant to get derivatives by reaction with alkaline hydrides. The preparation of the lithium, sodium and potassium hydrazinidoboranes $MN_2H_3BH_3$ (M as Li, Na and K) were reported to be successful by either ball-milling or solvent approach in autoclave. The as-obtained materials were fully characterized and then assessed for chemical hydrogen storage. By comparison to hydrazine borane, the derivatives showed improved dehydrogenation properties, in terms of release of almost pure hydrogen at low temperatures. However, the reaction mechanisms are still unknown and the solid residues not well identified. Like for ammonia borane, progresses on these aspects are likely to be limited because of the difficult characterization of the spent fuel. There is another grey area in relation to the solid residues, and one has to ensure that, unlike the solid residue of hydrazine borane, there is no risk with them. To date, there is no clear report about the stability, reactivity and sensitivity of the spent fuels stemming from the thermal dehydrogenation of the hydrazinidoboranes.

Last but not least, closing the hydrogen cycle with hydrazine borane is an important issue to address if future technological application is foreseen. Like in hydrolyses of sodium borohydride and ammonia borane, the by-products are boric acid $B(OH)_3$ and tetrahydroxyborate $B(OH)_4{}^-$. Their recyclability, via chemical recycling, has been considered over the past decade and mutual efforts could permit to develop efficient and cost-effective processes. It is important to note that the B–O bond is as strong as the C–O bond of carbon dioxide and a breakthrough should be expected to address the related cost issue. Recyclability into sodium borohydride would be a first important achievement as this compound is the main reactant for the synthesis of hydrazine borane. Otherwise, the development of aqueous hydrazine borane as hydrogen carrier could not be realized. Like in thermolysis of ammonia borane and derivatives, the recyclability of the solid residues forming upon thermal dehydrogenation of the hydrazinidoborane derivatives is much dependent on the identification of these solid residues. The difficult characterization of the solid residues is today the limiting factor. That is when the recyclability could be achievable by reduction, as that has been done for polyborazylene, a model residue of thermolyzed ammonia borane. In any case, the viability of hydrazine borane for chemical hydrogen storage will be closely related to closing the hydrogen cycle. In addition, even if the hydrogen cycle is closed, boranes could only be envisaged in off-board refueling systems due to the lack of direct rehydrogenation, but this is not essentially a drawback.

Up to here, hydrazine borane has been much compared to ammonia borane. So the question remains as to its interest compared to the latter. It would be premature to answer the question, mainly because of the discrepancy in the efforts dedicated to each borane. Ammonia borane has been under investigation for about ten years, whereas hydrazine borane has had focused attention since 2009. One may, however, state that hydrazine borane could be more attractive for liquid-state chemical hydrogen storage provided the aforementioned issues are addressed and that the derivatives of these boranes have a potential for solid-state chemical hydrogen storage. There is still room for improvement and the advances of the next years should help in shedding light on the real potential of hydrazine borane and derivatives.

Acknowledgments: UBD acknowledges ANR (JCJC BoraHCx), DGA (PhD thesis of RM), CNRS (PhD thesis of RM) and TUBITAK (post-doc of ç. Çakanyildirim) for financial supports over the period 2009–2014.

Author Contributions: Umit B. Demirci wrote the paper; Romain Moury contributed in writing the paper while specifically focusing on parts about crystallography and the hydrazinidoborane compounds.

Conflicts of Interest: The authors declare no conflict of interest.

References and Notes

1. Conte, M.; di Mario, F.; Iacobazzi, A.; Mattucci, A.; Moreno, A.; Ronchetti, M. Hydrogen as future energy carrier: The ENEA point of view on technology and application prospects. *Energies* **2009**, *2*, 150–179.

2. Armaroli, N.; Balzani, V. The hydrogen issue. *ChemSusChem* **2011**, *4*, 21–36.

3. Mazloomi, K.; Gomes, C. Hydrogen as an energy carrier: Prospects and challenges. *Renew. Sustain. Energy Rev.* **2012**, *16*, 3024–3033.

4. Eberle, U.; Felderhoff, M.; Schüth, F. Chemical and physical solutions for hydrogen storage. *Angew. Chem. Int. Ed.* **2009**, *48*, 6608–6630.

5. Dalebrook, A.F.; Gan, W.; Grasemann, M.; Moret, S.; Laurenczy, G. Hydrogen storage: Beyond conventional methods. *Chem. Commun.* **2013**, *49*, 8735–8751.

6. Chamoun, R.; Demirci, U.B.; Miele, P. Cyclic dehydrogenation-(re)hydrogenation with hydrogen storage materials: An overview. *Energy Technol.* **2015**.

7. Chang, F.; Zhou, J.; Chen, P.; Chen, Y.; Jia, H.; Saad, S.M.I.; Gao, Y.; Cao, X.; Zheng, T. Microporous and mesoporous materials for gas storage and separation: A review. *Asia Pac. J. Chem. Eng.* **2013**, *8*, 618–626.

8. Li, H.W.; Yan, Y.; Orimo, S.I.; Züttel, A.; Jensen, C.M. Recent progress in metal borohydrides for hydrogen storage. *Energies* **2011**, *4*, 185–214.

9. Moussa, G.; Moury, R.; Demirci, U.B.; Şener, T.; Miele, P. Boron-based hydrides for chemical hydrogen storage. *Int. J. Energy Res.* **2013**, *37*, 825–842.

10. Goubeau, V.J.; Ricker, E. Borinhydrazin und seine pyrolyseprodukte. *Z. Anorg. Allg. Chem.* **1961**, *310*, 123–142.

11. Gunderloy, F.C., Jr. Hydrazine–Mono- and—Bisborane. *Inorg. Synth.* **1967**, *9*, 13–16.

12. Gunderloy, F.C., Jr. Process for Preparing Hydrazine Monoborane. U.S. Patent 3375087, 26 March 1968.

13. Hough, W.V.; Hashman, J.S. Borane-Hydrazine Compounds. U.S. Patent 3298799, 1967.

14. Artz, G.D.; Grant, L.R. Solid Propellant Hydrogen Generator. U.S. Patent 4468263, 1984.

15. Bratton, F.H.; Reynolds, H.I. Hydrogen Generating System. U.S. Patent 3419361, 1968.

16. Edwards, L.J. Hydrogen Generating Composition. U.S. Patent 3450638, 1969.

17. Kirpiche, E.P.; Rubtsov, Y.I.; Manelis, G.B. Standard enthalpies for hydrazine borane and hydrazine-*bis*-borane. *Zhurnal Neorg. Khim.* **1971**, *16*, 2064–2064.

18. Borovinskaya, I.P.; Bunin, V.A.; Merzhanov, A.G. Self-propagating high-temperature synthesis of high-porous boron nitride. *Mendeleev Commun.* **1997**, *7*, 47–48.

19. Rasul, G.; Prakash, G.K.S.; Olah, G.A. B–H bond protonation in mono- and diprotonated borane complexes H_3BX (X = N_2H_4, NH_2OH, and H_2O_2) involving hypercoordinate boron. *Inorg. Chem.* **1999**, *38*, 5876–5878.

20. Jepsen, L.H.; Ley, M.B.; Lee, Y.S.; Cho, Y.W.; Dornheim, M.; Jensen, J.O.; Filinchuk, Y.; Jørgensen, J.E.; Besenbacher, F.; Jensen, T.R. Boron–nitrogen based hydrides and reactive composites for hydrogen storage. *Mater. Today* **2014**, *17*, 129–135.

21. Chua, Y.S.; Chen, P.; Wu, G.; Xiong, Z. Development of amidoboranes for hydrogen storage. *Chem. Commun.* **2011**, *47*, 5116–5129.

22. Hamilton, C.W.; Baker, R.T.; Staubitz, A.; Manners, I. B–N compounds for chemical hydrogen storage. *Chem. Soc. Rev.* **2009**, *38*, 279–293.

23. Hügle, T.; Kühnel, M.F.; Lentz, D. Hydrazine borane: A promising hydrogen storage material. *J. Am. Chem. Soc.* **2009**, *131*, 7444–7446.

24. Moury, R.; Moussa, G.; Demirci, U.B.; Hannauer, J.; Bernard, S.; Petit, E.; van der Lee, A.; Miele, P. Hydrazine borane: Synthesis, characterization, and application prospects in chemical hydrogen storage. *Phys. Chem. Chem. Phys.* **2012**, *14*, 1768–1777.

25. Gunderloy, F.C., Jr. Reactions of the borohydride group with the proton donors hydroxylammonium, methoxyammonium, and hydrazinium-magnesium ions. *Inorg. Chem.* **1963**, *2*, 221–222.

26. Gunderloy, F.C., Jr. Preparation of Solid Boron Compounds. U.S. Patent 3159451, 1 December 1964.

27. Uchida, H.S.; Hefferan, G.T. Production of Hydrazine Boranes. U.S. Patent 3119652, 28 January 1964.

28. Sutton, A.; Gordon, J.C.; Ott, K.C.; Burrell, A.K. Regeneration of Ammonia Borane from Polyborazylene. U.S. Patent 20100272622, 28 October 2010.

29. Sutton, A.D.; Burrell, A.K.; Dixon, D.A.; Garner, E.B., III; Gordon, J.C.; Nakagawa, T.; Ott, K.C.; Robinson, J.P.; Vasiliu, M. Regeneration of ammonia borane spent fuel by direct reaction with hydrazine and liquid ammonia. *Science* **2011**, *331*, 1426–1420.

30. Karahan, S.; Zahmakiran, M.; Özkar, S. Catalytic hydrolysis of hydrazine borane for chemical hydrogen storage: Highly efficient and fast hydrogen generation system at room temperature. *Int. J. Hydrog. Energy* **2011**, *36*, 4958–4966.

31. Mebs, S.; Grabowsky, S.; Förster, D.; Kickbusch, R.; Hartl, M.; Daemen, L.L.; Morgenroth, W.; Luger, P.; Paulus, B.; Lentz, D. Charge transfer via the dative N–B bond and dihydrogen contacts. Experimental and theoretical electron density studies of small Lewis acid-base adducts. *J. Phys. Chem. A* **2010**, *114*, 10185–10196.

32. Wu, H.; Zhou, W.; Pinkerton, F.E.; Udovic, T.J.; Yildirim, T.; Rush, J.J. Metal hydrazinoborane $LiN_2H_3BH_3$ and $LiN_2H_3BH_3 \cdot 2N_2H_4BH_3$: Crystal structures and high-extent dehydrogenation. *Energy Environ. Sci.* **2012**, *5*, 7531–7535.

33. Lu, Z.H.; Xu, Q. Recent progress in boron- and nitrogen-based chemical hydrogen storage. *Funct. Mater. Lett.* **2012**, *5*, 1230001.

34. çelik, D.; Karahan, S.; Zahmakiran, M.; Özkar, S. Hydrogen generation from the hydrolysis of hydrazine-borane catalyzed by rhodium(0) nanoparticles supported on hydroxyapatite. *Int. J. Hydrog. Energy* **2011**, *37*, 5143–5151.

35. Şencanli, S.; Karahan, S.; Özkar, S. Poly(4-styrene acid-co-maleic acid) stabilized nickel(0) nanoparticles: Highly active and cost effective in hydrogen generation from the hydrolysis of hydrazine borane. *Int. J. Hydrog. Energy* **2013**, *38*, 1493–14700.

36. Yadav, M.; Xu, Q. Liquid-phase chemical hydrogen storage materials. *Energy Environ. Sci.* **2012**, *5*, 9698–9725.

37. Li, P.Z.; Xu, Q. Metal-nanoparticles catalyzed hydrogen generation from liquid-phase chemical hydrogen storage materials. *J. Chin. Chem. Soc.* **2012**, *59*, 1181–1189.

38. Lu, Z.H.; Yao, Q.; Zhang, Z.; Yang, Y.; Chen, X. Nanocatalysts for hydrogen generation from ammonia borane and hydrazine borane. *J. Nanomater.* **2014**, *2014*, 729029.

39. Moussa, G.; Moury, R.; Demirci, U.B.; Miele, P. Borates in hydrolysis of ammonia borane. *Int. J. Hydrog. Energy* **2013**, *38*, 7888–7895.

40. Marrero-Alfonso, E.Y.; Beaird, A.M.; Davis, T.A.; Matthews, M.A. Hydrogen generation from chemical hydrides. *Ind. Eng. Chem. Res.* **2009**, *48*, 3703–3712.

41. Karahan, S.; Zahmakiran, M.; Özkar, S. Catalytic methanolysis of hydrazine borane: A new and efficient hydrogen generation system under mild conditions. *Dalton Trans.* **2012**, *41*, 4918–4918.

42. Özhava, D.; Kiliçaslan, N.Z.; Özkar, S. PVP-stabilized nickel(0) nanoparticles as catalyst in hydrogen generation from the methanolysis of hydrazine borane or ammonia borane. *Appl. Catal. B* **2014**, *162*, 573–582.

43. Demirci, U.B. The hydrogen cycle with the hydrolysis of sodium borohydride: A statistical approach for highlighting the scientific/technical issues to prioritize in the field. *Int. J. Hydrog. Energy* **2015**.

44. Jiang, H.L.; Xu, Q. Catalytic hydrolysis of ammonia borane for chemical hydrogen storage. *Catal. Today* **2011**, *170*, 56–63.

45. Singh, S.K.; Xu, Q. Nanocatalysts for hydrogen generation from hydrazine. *Catal. Sci. Technol.* **2013**, *3*, 1889–1900.

46. Hannauer, J.; Akdim, O.; Demirci, U.B.; Geantet, C.; Herrmann, J.M.; Miele, P.; Xu, Q. High-extent dehydrogenation of hydrazine borane $N_2H_4BH_3$ by hydrolysis of BH_3 and decomposition of N_2H_4. *Energy Environ. Sci.* **2011**, *4*, 3355–3358.

177

47. Çakanyildirim, Ç.; Petit, E.; Demirci, U.B.; Moury, R.; Petit, J.F.; Xu, Q.; Miele, P. Gaining insight into the catalytic dehydrogenation of hydrazine borane in water. *Int. J. Hydrog. Energy* **2012**, *37*, 15983–15991.
48. Zhong, D.C.; Aranishi, K.; Singh, A.K.; Demirci, U.B.; Xu, Q. The synergistic effect of Rh-Ni catalysts on the highly-efficient dehydrogenation of aqueous hydrazine borane for chemical hydrogen storage. *Chem. Commun.* **2012**, *48*, 11945–11947.
49. Hannauer, J.; Demirci, U.B.; Geantet, C.; Herrmann, J.M.; Miele, P. Transition metal-catalyzed dehydrogenation of hydrazine borane $N_2H_4BH_3$ via the hydrolysis of BH_3 and the decomposition of N_2H_4. *Int. J. Hydrog. Energy* **2012**, *37*, 10758–10767.
50. Çakanyıldırım, Ç.; Demirci, U.B.; Xu, Q.; Miele, P. Supported nickel catalysts for the decomposition of hydrazine borane $N_2H_4BH_3$. *Adv. Energy Res.* **2013**, *1*, 1–12.
51. Clémençon, D.; Petit, J.F.; Demirci, U.B.; Xu, Q.; Miele, P. Nickel- and platinum-containing core@shell catalysts for hydrogen generation of aqueous hydrazine borane. *J. Power Sourc.* **2014**, *260*, 77–81.
52. Çakanyıldırım, Ç.; Demirci, U.B.; Şener, T.; Xu, Q.; Miele, P. Nickel-based bimetallic nanocatalysts in high-extent dehydrogenation of hydrazine borane. *Int. J. Hydrog. Energy* **2012**, *37*, 9722–9729.
53. Ben Aziza, W.; Demirci, U.B.; Xu, Q.; Miele, P. Bimetallic nickel-based nanocatalysts for hydrogen generation from aqueous hydrazine borane: Investigation of iron, cobalt and palladium as the second metal. *Int. J. Hydrog. Energy* **2014**, *39*, 16919–16926.
54. Kim, J.H.; Kim, K.T.; Kang, Y.M.; Kim, H.S.; Song, M.S.; Lee, Y.J.; Lee, P.S.; Lee, J.Y. Study on degradation of filamentary Ni catalyst on hydrolysis of sodium borohydride. *J. Alloys Compd.* **2004**, *379*, 222–227.
55. Demirci, U.B.; Miele, P. Cobalt in $NaBH_4$ hydrolysis. *Phys. Chem. Chem. Phys.* **2010**, *12*, 14651–14665.
56. Demirci, U.B.; Miele, P. Cobalt-based catalysts in hydrolysis of $NaBH_4$ and NH_3BH_3. *Phys. Chem. Chem. Phys.* **2014**, *16*, 6872–6885.
57. Li, C.; Dou, Y.; Liu, J.; Chen, Y.; He, S.; Wei, M.; Evans, D.G.; Duan, X. Synthesis of supported Ni@(RhNi-alloy) nanocomposites as an efficient catalyst towards hydrogen generation from $N_2H_4BH_3$. *Chem. Commun.* **2013**, *49*, 9992–9994.
58. Zhu, Q.L.; Zhong, D.C.; Demirci, U.B.; Xu, Q. Controlled synthesis of ultrafine surfactant-free NiPt nanocatalysts towards efficient and complete hydrogen generation from hydrazine borane at room temperature. *ACS Catal.* **2014**, *4*, 4261–4268.
59. Vinh-Son, N.; Swinnen, S.; Matus, M.H.; Nguyen, M.T.; Dixon, D.A. The effect of the NH_2 substituent on NH_3: Hydrazine as an alternative for ammonia in hydrogen release in the presence of boranes and alanes. *Phys. Chem. Chem. Phys.* **2009**, *11*, 6339–6344.
60. Gao, H.; Shreeve, J.M. Ionic liquid solubilized boranes as hypergolic fluids. *J. Mater. Chem.* **2012**, *22*, 11022–11024.
61. Zhang, Q.; Shreeve, J.M. Ionic liquid propellants: Future fuels for space propulsion. *Chem. Eur. J.* **2013**, *19*, 15446–15451.
62. Toche, F.; Chiriac, R.; Demirci, U.B.; Miele, P. Borohydride-induced destabilization of hydrazine borane. *Int. J. Hydrog. Energy* **2014**, *39*, 9321–9329.

63. Frueh, S.; Kellett, R.; Mallery, C.; Molter, T.; Willis, W.S.; King'ondu, C.; Suib, S.L. Pyrolytic decomposition of ammonia borane to boron nitride. *Inorg. Chem.* **2011**, *50*, 783–792.

64. Petit, J.F.; Moussa, G.; Demirci, U.B.; Toche, F.; Chiriac, R.; Miele, P. Hydrazine borane-induced destabilization of ammonia borane, and *vice versa*. *J. Hazard. Mater.* **2014**, *278*, 158–162.

65. Thomas, J.; Klahn, M.; Spannenberg, A.; Beweries, T. Group 4 metallocene catalysed full dehydrogenation of hydrazine borane. *Dalton Trans.* **2013**, *42*, 14668–14672.

66. Moury, R.; Demirci, U.B.; Ban, V.; Filinchuk, Y.; Ichikawa, T.; Zeng, L.; Goshome, K.; Miele, P. Lithium hydrazinidoborane: A polymorphic material with potential for chemical hydrogen storage. *Chem. Mater.* **2014**, *26*, 3249–3255.

67. Qian, Z.; Pathak, B.; Ahuja, R. Energetic and structural analysis of $N_2H_4BH_3$ inorganic solid and its modified material for hydrogen storage. *Int. J. Hydrog. Energy* **2013**, *38*, 6718–6725.

68. Moury, R.; Demirci, U.B.; Ichikawa, T.; Filinchuk, Y.; Chiriac, R.; van der Lee, A.; Miele, P. Sodium hydrazinidoborane: A chemical hydrogen-storage material. *ChemSusChem* **2013**, *6*, 667–673.

69. Chua, Y.S.; Pei, Q.; Ju, X.; Zhou, W.; Udovic, T.J.; Wu, G.; Xiong, Z.; Chen, P.; Wu, H. Alkali metal hydride modification on hydrazine borane for improved dehydrogenation. *J. Phys. Chem. C* **2014**, *118*, 11244–11251.

70. Klooster, W.T.; Koetzle, T.F.; Siegbahn, P.E.M.; Richardson, T.B.; Crabtree, R.H. Study of the N–H$\cdots$H–B dihydrogen bond including the crystal structure of BH_3NH_3 by neutron diffraction. *J. Am. Chem. Soc.* **1999**, *121*, 6337–6343.

71. Al-Kukhun, A.; Hwang, H.T.; Varma, A. Mechanistic studies of ammonia borane dehydrogenation. *Int. J. Hydrog. Energy* **2013**, *38*, 169–179.

72. Tan, Y.; Chen, X.; Chen, J.; Gu, Q.; Yu, X. The decomposition of α-LiN$_2$H$_3$BH$_3$: An unexpected hydrogen release from a homopolar proton-proton pathway. *J. Mater. Chem. A* **2014**, *2*, 15627–15632.

73. Moury, R.; Petit, J.F.; Demirci, U.B.; Ichikawa, T.; Miele, P. Pure hydrogen-generating "doped" sodium hydrazinidoborane. *Int. J. Hydrog. Energy* **2015**.

74. Unpublished results. Typically, 10 mg of hydrazine borane were weighted and transferred in an agate mortar in an argon-filled glove box where the water and oxygen concentrations were kept below 0.1 ppm. Then, KH was added onto hydrazine borane with the help of a spatula. The addition was made carefully, almost grain by grain, and at each addition the reactivity was immediate, with an explosive formation of a brown gas and dispersion of products around the mortar.

LaNi$_5$-Assisted Hydrogenation of MgNi$_2$ in the Hybrid Structures of La$_{1.09}$Mg$_{1.91}$Ni$_9$D$_{9.5}$ and La$_{0.91}$Mg$_{2.09}$Ni$_9$D$_{9.4}$

Roman V. Denys, Volodymyr A. Yartys, Evan MacA. Gray and Colin J. Webb

Abstract: This work focused on the high pressure PCT and *in situ* neutron powder diffraction studies of the LaMg$_2$Ni$_9$-H$_2$ (D$_2$) system at pressures up to 1,000 bar. LaMg$_2$Ni$_9$ alloy was prepared by a powder metallurgy route from the LaNi$_9$ alloy precursor and Mg powder. Two La$_{3-x}$Mg$_x$Ni$_9$ samples with slightly different La/Mg ratios were studied, La$_{1.1}$Mg$_{1.9}$Ni$_9$ (sample 1) and La$_{0.9}$Mg$_{2.1}$Ni$_9$ (sample 2). *In situ* neutron powder diffraction studies of the La$_{1.09}$Mg$_{1.91}$Ni$_9$D$_{9.5}$ (1) and La$_{0.91}$Mg$_{2.09}$Ni$_9$D$_{9.4}$ (2) deuterides were performed at 25 bar D$_2$ (1) and 918 bar D$_2$ (2). The hydrogenation properties of the (1) and (2) are dramatically different from those for LaNi$_3$. The Mg-containing intermetallics reversibly form hydrides with ΔH_{des} = 24.0 kJ/mol$_{H2}$ and an equilibrium pressure of H$_2$ desorption of 18 bar at 20 °C (La$_{1.09}$Mg$_{1.91}$Ni$_9$). A pronounced hysteresis of H$_2$ absorption and desorption, ~100 bar, is observed. The studies showed that LaNi$_5$-assisted hydrogenation of MgNi$_2$ in the LaMg$_2$Ni$_9$ hybrid structure takes place. In the La$_{1.09}$Mg$_{1.91}$Ni$_9$D$_{9.5}$ (1) and La$_{0.91}$Mg$_{2.09}$Ni$_9$D$_{9.4}$ (2) (a = 5.263/5.212; c = 25.803/25.71 Å) D atoms are accommodated in both Laves and CaCu$_5$-type slabs. In the LaNi$_5$ CaCu$_5$-type layer, D atoms fill three types of interstices; a deformed octahedron [La$_2$Ni$_4$], and [La(Mg)$_2$Ni$_2$] and [Ni$_4$] tetrahedra. The overall chemical compositions can be presented as LaNi$_5$H$_{5.6/5.0}$ + 2*MgNi$_2$H$_{1.95/2.2}$ showing that the hydrogenation of the MgNi$_2$ slab proceeds at mild H$_2$/D$_2$ pressure of just 20 bar. A partial filling by D of the four types of the tetrahedral interstices in the MgNi$_2$ slab takes place, including [MgNi$_3$] and [Mg$_2$Ni$_2$] tetrahedra.

Reprinted from *Energies*. Cite as: Denys, R.V.; Yartys, V.A.; Gray, E.M.; Webb, C.J. LaNi$_5$-Assisted Hydrogenation of MgNi$_2$ in the Hybrid Structures of La$_{1.09}$Mg$_{1.91}$Ni$_9$D$_{9.5}$ and La$_{0.91}$Mg$_{2.09}$Ni$_9$D$_{9.4}$. *Energies* **2015**, *8*, 3198–3211.

1. Introduction

Despite significant differences in chemistry between La and Mg, magnesium forms a very extensive solid solution in the LaNi$_3$ intermetallic alloy, crystallizing with a PuNi$_3$ type trigonal structure. Up to 67% of La atoms can be replaced by Mg to form a LaMg$_2$Ni$_9$ intermetallic compound. The LaNi$_3$ crystal structure is formed by a stacking of the LaNi$_5$ (*Haucke* CaCu$_5$ type) and MgNi$_2$ (*Laves* type) slabs along the trigonal 00z axis (LaNi$_5$ + 2MgNi$_2$ = LaMg$_2$Ni$_9$). Studies of hydrogen

180

absorption–desorption properties of the $LaMg_2Ni_9$ [1,2] have shown that it forms a hydride containing up to 1.2 wt% H ($\sim$0.8 H/M; $LaMg_2Ni_9H_{9.6}$).

The building blocks of $LaMg_2Ni_9$—$LaNi_5$ and $MgNi_2$—are well characterized individually as hydride-forming intermetallic compounds. The thermodynamics and structural features of their interaction with hydrogen are quite different. At room temperature, $LaNi_5$ forms a saturated $LaNi_5H_{6.7}$ hydride and shows a reversible interaction with hydrogen at hydrogen pressures slightly exceeding atmospheric pressure. Hydrogen atoms fill tetrahedral La_2Ni_2, $LaNi_3$ and Ni_4 sites in the hydride crystal structure [3].

In contrast, hydrogenation of the Laves phase $MgNi_2$ compound is possible only at hydrogen pressures close to 30 kbar, while maintaining an interaction temperature of 300 °C. Formation of $MgNi_2H_3$ results in a complete rebuilding of the metal sublattice. Hydrogen atoms in the orthorhombic structure of trihydride fill two different sites, the Mg_4Ni_2 octahedra and the positions within the buckled Ni nets, consequently forming directional Ni-H bonds [4].

A gradual increase of Mg content in $La_{3-x}Mg_xNi_9$ is accompanied by a linear decrease of the volumes of the unit cells. Interestingly, a substantial contraction takes place not only for the $(La,Mg)_2Ni_4$ slabs, but also for Mg-free $CaCu_5$-type $LaNi_5$ slabs. Hydrogen interaction with the $La_{3-x}Mg_xNi_9$ alloys has been investigated by *in situ* synchrotron X-ray, neutron powder diffraction, theoretical modeling, electrochemical studies as metal hydride battery anode materials, rapid solidification and pressure–composition–temperature studies [1,2,5–10]. In the whole substitution range, $La_{3-x}Mg_xNi_9$ alloys form intermetallic hydrides with H/M ratios ranging from 0.77 to 1.16. Magnesium influences structural features of the hydrogenation process and determines various aspects of the hydrogen interaction with intermetallics causing: (a) more than a 1,000-fold increase in the equilibrium pressures of hydrogen absorption and desorption for the Mg-rich $LaMg_2Ni_9$ as compared to the Mg-poor $La_{2.3}Mg_{0.7}Ni_9$ and a substantial modification of the thermodynamics of the formation–decomposition of the hydrides; (b) an increase of the reversible hydrogen storage capacities following increase of Mg content in the $La_{3-x}Mg_xNi_9$ to $\sim$1.5 wt% H for La_2MgNi_9; (c) improvement of the resistance against hydrogen-induced amorphisation and disproportionation and (d) change of the mechanism of the hydrogenation from anisotropic to isotropic. Thus, optimisation of the magnesium content provides different possibilities for improving properties of the studied alloys as hydrogen storage and battery electrode materials. Studies of the thermodynamics and crystal chemistry of the $RE_2MgNi_9H_{12-13}$ (RE = La and Nd) hydrides showed that La substitution by Pr or Nd causes destabilization of the formed hydrides without affecting their hydrogen storage capacities and leaves unchanged the most important features of their crystal structures [11].

Observed values of H capacities in the $LaMg_2Ni_9$-based hydride of 9.6 atoms H/f.u. cannot be explained by exclusive hydrogen insertion into the $LaNi_5$ slabs, and requires H incorporation into the $MgNi_2$ blocks of the structure to reach the experimentally observed H/M ratios. Thus, studies of the thermodynamics and crystal chemistry of $La_{3-x}Mg_xNi_9$-H_2 systems are very interesting and important from the point of view of the effect of magnesium on the behaviours of the metal-hydrogen systems. The goal of the present study was to study two alloy compositions formed close to the limiting value of the magnesium solubility in $LaNi_3$, $LaMg_2Ni_9$, by performing *in situ* neutron powder diffraction studies of the deuterated $La_{0.91}Mg_{2.09}Ni_9$ and $La_{1.09}Mg_{1.91}Ni_9$ and by studying the thermodynamics of the metal-hydrogen interactions by measurements of the PCT diagrams.

2. Experimental

$La_{1.09}Mg_{1.91}Ni_9$ and $La_{0.91}Mg_{2.09}Ni_9$ alloys were prepared by a powder metallurgy route from $LaNi_5$ alloy precursor, Mg and Ni. Initial metals La, Mg and Ni with high purity exceeding 99.9% were used in the synthesis. $LaNi_5$ precursor was prepared by arc melting of a stoichiometric 1:5 mixture of La and Ni.

The powder mixture $LaNi_5$ + Mg + Ni was ball milled under protective atmosphere of argon gas in a SPEX 8000D mill for 8 h. After the milling process, the mixture was placed into a tantalum crucible and then annealed in Ar atmosphere in the sealed stainless steel containers at 600–1000 °C. Two samples with a slightly different stoichiometry were prepared. Their stoichiometric compositions were: sample 1: $La_{1.09(1)}Mg_{1.91(1)}Ni_9$; sample 2: $La_{0.91(1)}Mg_{2.09(1)}Ni_9$.

The first sample was annealed at 800 °C for 8 h and then at 600 °C for 8 h. The second sample was annealed at 1000 °C for 2 h and, later, at 800 °C for 12 h. The samples were quenched into a mixture of water and ice after the annealing. A small excess of Mg (5 wt%) was introduced into the initial mixtures to compensate for its sublimation at high temperatures.

The homogeneity of the prepared samples was characterized by XRD. Laboratory powder X-ray diffraction data were collected with a Siemens D5000 diffractometer (Oslo, Norway) equipped with a Ge primary monochromator giving Cu $K\alpha_1$ radiation. Initial phase-structural analysis was performed by X-ray powder diffraction using a Bruker D8 Advance diffractometer (Kjeller, Norway) with Cu-Kα radiation. High-resolution SR XRD data were collected at the Swiss-Norwegian Beamlines (SNBL, BM01B) at ESRF, Grenoble, France. A monochromatic beam with $\lambda = 0.5009(1)$ Å was provided by a double Si monochromator. A 2θ angular range of $1°$–$50.5°$ was scanned with a detector bank consisting of six scintillation detectors mounted in series with $1.1°$ separation. The data were binned to the step size $\Delta 2\theta = 0.003°$. The instrumental contribution to the line broadening was evaluated by refining the profile parameters for a standard Si sample.

In situ neutron powder diffraction studies were performed at HRPT diffractometer, SINQ, PSI, Switzerland using a wavelength of $\lambda = 1.494$ Å. The deuteride of sample 1 was synthesized at 25 bar D_2 and $-30\,°C$ ($P_{eq.}$ for absorption ~20 bar); it was synthesized and studied by NPD using a thin walled stainless steel sample cell (6 mm OD). The deuteride of sample 2 was synthesized at 950 bar D_2 and measured at 912 bar D_2 at room temperature. The experimental setup for the *in situ* NPD study consisted of a high-pressure Sieverts' manometric hydrogenator connected to a high-pressure sample cell made of a null matrix coherent scattering alloy (Zr–Ti) with a thin stainless steel inner liner.

Powder diffraction data were analysed by the Rietveld whole-profile refinement method using the General Structure Analysis System (GSAS) [12] and FULLPROF [13] software packages. Pressure-composition-temperature isotherms were measured at $-40, -20, 0$ and $20\,°C$.

3. Results and Discussion

3.1. XRD Characterization of the Initial Intermetallic Alloys $La_{0.91}Mg_{2.09}Ni_9$ and $La_{1.09}Mg_{1.91}Ni_9$

XRD characterization of two studied alloys $La_{0.91}Mg_{2.09}Ni_9$ and $La_{1.09}Mg_{1.91}Ni_9$ showed that they both contain $PuNi_3$ trigonal $La_{3-x}Mg_xNi_9$ as the main phase constituents (80% for sample 1 and 75% for sample 2). The common secondary constituent was identified as a $LaNi_5$ binary intermetallic. Furthermore, sample 1 contained an admixture of the $MgNi_2$ Laves-type intermetallic phase, while sample 2 contained a cubic $MgNi_3$ intermetallic compound recently also observed during the studies of the $MgNi_2$-H_2 system [4]. $MgNi_3$ compound (sp.gr. $Pm\bar{3}m$; $a = 3.7185(5)$ Å) has an $AlCu_3$-type structure and earlier it was synthesized by high-energy ball milling of a mixture of Mg and Ni metals [14]. We assume that in present study $MgNi_3$ was synthesised already during the reactive ball milling and remained stable during the consecutive annealing at 1,000 and 800 °C. As an example, Figure 1 shows an excellent fit of the experimental X-ray powder diffraction pattern collected for the sample 2, $La_{0.91}Mg_{2.09}Ni_9$.

Crystallographic data for the studied intermetallic samples obtained from the refinements of the XRD pattern are listed in Table 1.

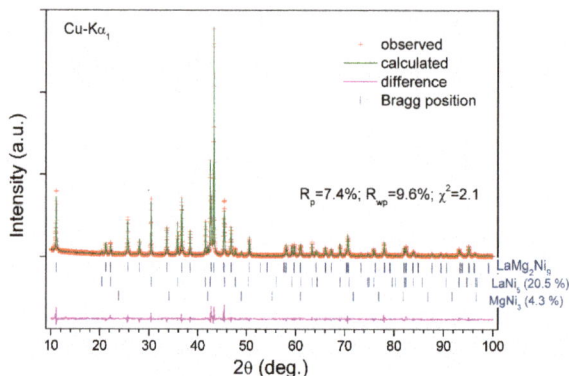

Figure 1. XRD pattern of $La_{0.91}Mg_{2.09}Ni_9$ (sample 2) (Cu-Kα_1 radiation).

Table 1. Crystal structure data for the $La_{0.91}Mg_{2.09}Ni_9$ and $La_{1.09}Mg_{1.91}Ni_9$ alloys from Rietveld refinements of the X-ray diffraction data. *PuNi$_3$* type of structure, space group $R\bar{3}m$.

Alloy	Sample 1	Sample 2
Source of experimental data	SR XRD collected at BM01B, SNBL using a wavelength $\lambda = 0.5009(1)$ Å	Siemens D5000 diffractometer, Cu Kα_1 radiation
Composition of AB$_3$ phase	$La_{1.09(1)}Mg_{1.91(1)}Ni_9$	$La_{0.91(1)}Mg_{2.09(1)}Ni_9$
Unit cell parameters:		
a (Å)	4.94024(8)	4.8986(1)
c (Å)	23.8188(4)	23.957(1)
V (Å^3)	503.44(1)	497.86(2)
Atomic parameters:		
La1/Mg1 in 3a (0, 0, 0)		
$U_{iso} \times 100$ (Å^2)	0.43(5)	2.1(2)
n_{Mg}, ($n_{La} = 1-n_{Mg}$)	0.0(−)	0.09(1)
La2/Mg2 in 6c (0, 0, z)		
z	0.1453(3)	0.1471(6)
$U_{iso} \times 100$ (Å^2)	1.2(3)	0.5(3)
n_{Mg}, ($n_{RE} = 1-n_{Mg}$)	0.954(5)	1.0(−)
Ni1 in 3b (0, 0, $\frac{1}{2}$)		
$U_{iso} \times 100$ (Å^2)	0.7(1)	0.8(3)
Ni2 in 6c (0, 0, z)		
z	0.3335(2)	0.3334(4)
$U_{iso} \times 100$ (Å^2)	0.13(8)	1.8(3)
Ni3 in 18h (x, $-x$, z)		
x	0.5009(3)	0.5014(6)
z	0.08529(8)	0.0854(2)
$U_{iso} \times 100$ (Å^2)	0.57(5)	1.4(2)
R-factors of refinements		
R_p	8.9	7.4
R_{wp}	11.9	9.6
χ^2	2.0	2.1
Impurity phases	LaNi$_5$ 7.8(2) wt% MgNi$_2$ 12.0(2) wt%	LaNi$_5$ 20.5(2) wt% MgNi$_3$ 4.2(3) wt%

The crystallographic characteristics of LaNi$_3$ change significantly on Mg $\rightarrow$ La substitution; a decrease in the unit cell parameters takes place from a = 5.0842(2); c = 25.106(1) Å (LaNi$_3$) to a = 4.8986(1) (sample 2)-4.94024(8) (sample 1); c = 23.8188(4) Å (sample 1)-23.957(1) (sample 2). Furthermore, comparison of the data shows that the studied intermetallic samples exhibit significant differences in the volumes of the unit cells and c/a ratios. A shrinkage along [001] appears to be more pronounced ($\Delta c/c$, -5.1%) as compared to $\Delta a/a$, -3.7%. The overall volume contraction is quite significant reaching 10.5%–11.5%. The measured dimensions of the unit cells well agree with the data reported for the stoichiometric LaMg$_2$Ni$_9$ alloy studied by single crystal XRD (a = 4.9241, c = 23.875 Å; V = 501.3 Å^3 [15]), which shows intermediate values of a, c and V being in between the values for the samples 1 and 2, as it could be expected from comparison of their chemical compositions.

Refined volumes of the unit cells correlate with their chemical compositions and Mg/La ratios. Indeed, sample 1, La$_{1.09(1)}$Mg$_{1.91(1)}$Ni$_9$ with a larger unit cell has a higher content of lanthanum, while for sample 2, La$_{0.91(1)}$Mg$_{2.09(1)}$Ni$_9$ with a smaller unit cell, the content of lanthanum becomes smaller than 1 atom/f.u., and the content of Mg reaches overstoichiometric compositions with more than 2 Mg atoms/f.u. (La,Mg)$_3$Ni$_9$.

Comparison of the data presented in Table 1 with crystallographic data for the (La,Mg)$_3$Ni$_9$ intermetallics studied in [1] shows a linear dependence between the decrease of the unit cell volumes and the content of Mg in the alloys.

We note a very interesting feature of the crystal structure of La$_{0.91(1)}$Mg$_{2.09(1)}$Ni$_9$ where a partial substitution of La by Mg takes place within the CaCu$_5$ type layer in the position 6c. This contrasts with the behaviour of the alloys in the La-Mg-Ni system with compositions close to LaNi$_5$. In the latter case studies of phase equilibria showed no dissolution of an appreciable amount of Mg in LaNi$_5$ [16]. Thus, the present study demonstrates that the situation with Mg solubility in the LaNi$_5$ slabs of the LaNi$_3$ structure becomes different in the sample 2 La$_{0.91(1)}$Mg$_{2.09(1)}$Ni$_9$. Here LaNi$_5$, when influenced by the MgNi$_2$ slabs of the hybrid structure, becomes capable of forming solid solutions of such a type with experimentally refined composition of La$_{0.95}$Mg$_{0.05}$Ni$_5$. Thus, La$_{0.91(1)}$Mg$_{2.09(1)}$Ni$_9$ should be considered as the first reported case where a CaCu$_5$ type layer accommodates Mg atoms allowing a Mg content of 2.09 at./f.u. (La,Mg)$_3$Ni$_9$. Consequently, the limits of Mg solubility in LaNi$_3$ are not confined to LaMg$_2$Ni$_9$ and extend to the composition La$_{0.91}$Mg$_{2.09}$Ni$_9$.

3.2. Thermodynamics of the (La,Mg)$_3$Ni$_9$—H$_2$ systems

The hydrogenation/deuteration properties of the prepared La$_{1\pm0.1}$Mg$_{2\pm0.1}$Ni$_9$ intermetallics appear to be dramatically different from those for LaNi$_3$. While LaNi$_3$ is prone to the hydrogen-induced disproportionation, the Mg-containing intermetallics reversibly form hydrides with ΔH_{des} = 24.0 kJ/mol$_{H2}$ and equilibrium pressure of H$_2$ desorption of 20 bar at room temperature for La$_{1.09}$Mg$_{1.91}$Ni$_9$ (see Figure 2). A pronounced hysteresis of H$_2$ absorption and desorption is evidenced by a high value of H$_2$ absorption pressure, more than 100 bar higher than that for desorption.

For La$_2$MgNi$_9$ [6] at room temperature the values of plateau pressures are 0.05 and 0.1 bar for hydrogen desorption and absorption, respectively, ΔH_{des} = 35.9 kJ/mol$_{H2}$. Equilibrium pressure of hydrogen desorption for La$_{0.91}$Mg$_{2.09}$Ni$_9$ is by more than 1000 times higher than that for La$_2$MgNi$_9$.

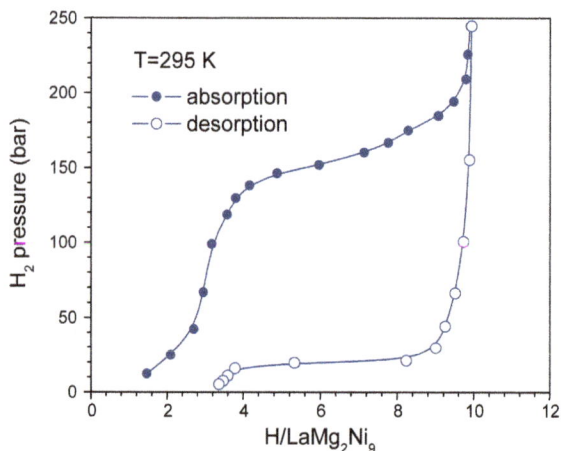

(a)

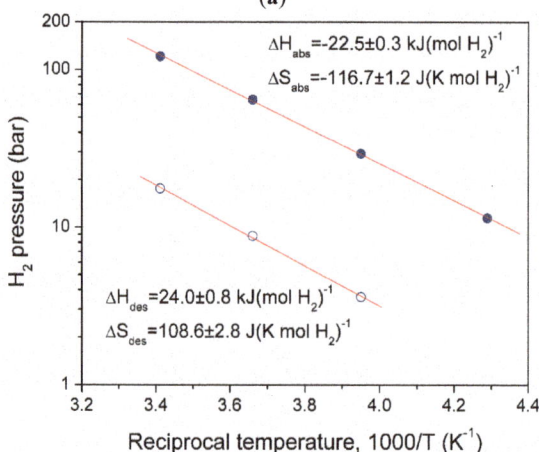

(b)

Figure 2. Room temperature isotherms of hydrogen absorption and desorption (a); and van't Hoff plots (b) for La$_{1.09}$Mg$_{1.91}$Ni$_9$-based hydride. At room temperature equilibrium pressure of hydrogen absorption is ~120 bar D$_2$, while for the desorption P$_{eq.}$ equals to ~20 bar D$_2$.

3.3. In situ NPD studies

In situ neutron powder diffraction studies of the $La_{1\pm0.1}Mg_{2\pm0.1}Ni_9D_{9.4-9.5}$ deuterides were performed at the Spallation Neutron Source SINQ accommodated at Paul Scherrer Institute (Villigen, Switzerland). Two samples, $La_{1.09}Mg_{1.91}Ni_9D_{9.5(3)}$ (sample 1) and $La_{0.9}Mg_{2.1}Ni_9D_{9.4(6)}$ (sample 2) were synthesised and studied under different conditions.

For the synthesis of $La_{1.09}Mg_{1.91}Ni_9D_{9.5}$, a 6 mm diameter stainless steel autoclave with a wall thickness of 0.2 mm was used. The synthesis was performed by saturating activated samples with deuterium gas (25 bar) at a sub–zero temperature of -30 °C. This was done in order to decrease the equilibrium pressure of hydrogen absorption-desorption in the $La_{1.09}Mg_{1.91}Ni_9$—D_2 system. The alloy absorbed deuterium to reach a composition $La_{1.09}Mg_{1.91}Ni_9D_{9.5}$ and was measured at 25 °C and deuterium pressure of 25 bar.

The second sample, $La_{0.9}Mg_{2.1}Ni_9D_{9.4}$, was synthesized at high pressure deuterium gas of 950 bar D_2. The studied sample was placed inside a TiZr sample cell with a stainless steel liner, which was used as a sample holder during the *in situ* NPD experiments (see Figure 3). The pressure during the NPD measurements performed at 20 °C was set to 912 bar D_2. No preliminary activation was applied prior to the synthesis.

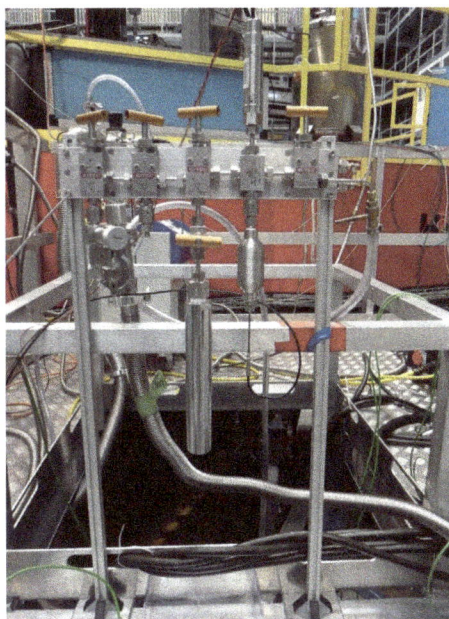

Figure 3. High pressure synthesis setup for the *in situ* NPD measurements at pressures up to 1000 bar D_2.

For the $La_{1.09}Mg_{1.91}Ni_9D_{9.5(5)}$ sample (No.1) at the highest applied deuterium pressure of 25 bar D_2, the deuteration resulted in the formation of a two-phase mixture of the α-solid solution of deuterium in the alloy and a corresponding β-deuteride. Such a mixture of the phase constituents was observed after allowing a deuteration time of ~20 h at interaction temperature of $-30°$ C. Since applied temperature-pressure conditions were rather close to the equilibrium ones (see Figure 2), the transformation was slow and was not completed on the time scale of the measurements performed. The second sample with a slightly higher content of magnesium, $La_{0.91}Mg_{2.09}Ni_9D_{9.3(7)}$ was saturated by deuterium at deuterium pressure of 950 bar and was equilibrated at 912 bar D_2 and 25 °C. Analysis of the diffraction pattern showed an excellent fit between the experimental data and calculated NPD profiles (Figure 4) and indicated a completeness of the transformation of the α-solid solution into the β-deuteride.

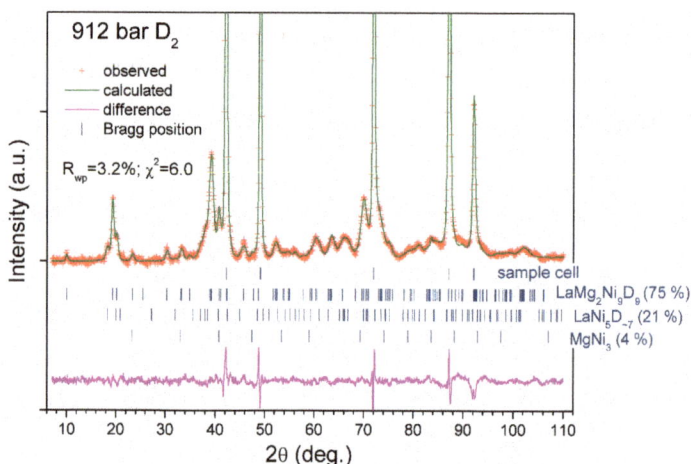

Figure 4. NPD pattern of $La_{0.9}Mg_{2.1}Ni_9D_{9.4(6)}$ (912 bar D_2, 298 K). Note that the most significant contributions to the difference intensities are coming from the sample cell. $R_p = 2.4\%$, $R_{wp} = 3.2$; $\chi^2 = 6.0$.

The results of the refinements of the NPD data for $La_{1.09}Mg_{1.91}Ni_9D_{9.5(5)}$ and for $La_{0.91}Mg_{2.09}Ni_9D_{9.4(6)}$ are summarized in Table 2. The data show a formation of very similar structures, with only minor differences in the occupancies of the specific D-sites of five various types. These sites are shown in Figure 5 and include four types of tetrahedral and one tetragonal bipyramid.

A partial filling by D atoms of the four types of the tetrahedral interstices takes place inside the $MgNi_2$ slab; these include two types of the $[MgNi_3]$ (18h and 6c) tetrahedra and two types of the $[Mg_2Ni_2]$ (36i and 18h) interstitial sites.

In addition, similar to the other studied $La_{3-x}Mg_xNi_9$-based deuterides, the remaining 5.0 or 5.6 at. D/f.u. form a standard hydrogen sublattice within the $LaNi_5$ slab which are statistically distributed within the four types of the interstices; hydrogen atoms partially occupy $[La_2Ni_4]$ octahedra, three types of $[Ni_4]$ tetrahedra, and two types of the $[LaMgNi_2]$ sites.

Table 2. Crystal structure data for the deuterated $La_{1\pm0.1}Mg_{2\pm0.1}Ni_9$ alloys (*PuNi$_3$* type, sp.gr. $R\bar{3}m$) from the Rietveld refinements of *in situ* neutron diffraction data.

Deuteride	$La_{1.09}Mg_{1.91}Ni_9D_{9.5(5)}$	$La_{0.91}Mg_{2.09}Ni_9D_{9.4(6)}$
Conditions	25 bar at 25 °C (prepared at −30 °C)	912 bar at 25 °C
Unit cell parameters:		
a (Å)	5.263(1)	5.212(1)
c (Å)	25.803(9)	25.71(1)
V (Å^3)	618.9(3)	604.8(3)
Unit cell parameters:		
$\Delta a/a$ (%)	6.5	6.4
$\Delta c/c$ (%)	8.3	7.3
$\Delta V/V$ (%)	23.0	21.6
$\Delta V/V[LaNi_5]$ (%)	20.4	20.7
$\Delta V/V[MgNi_2]$ (%)	25.4	22.2
Atomic parameters:		
La1/Mg1 in $3a$ (0, 0, 0) n_{Mg}, $(n_{La} = 1-n_{Mg})$	0.0(−)	0.09(−)
La2/Mg2 in $6c$ (0, 0, z) $\quad z$ $\quad U_{iso} \times 100$ (Å^2) $\quad n_{Mg}$, $(n_{RE} = 1-n_{Mg})$	1.0(−) 0.95(−)	1.0(−) 1.0(−)
Ni1 in $3b$ (0, 0, $\frac{1}{2}$) $\quad U_{iso} \times 100$ (Å^2)	1.0(−)	1.0(−)
Ni2 in $6c$ (0, 0, z) $\quad z$ $\quad U_{iso} \times 100$ (Å^2)	0.3279(7) 1.0(−)	0.3220(6) 1.0(−)
Ni3 in $18h$ (x, $-x$, z) $\quad x$ $\quad z$ $\quad U_{iso} \times 100$ (Å^2)	0.498(1) 0.0871(4) 1.0(−)	0.506(1) 0.0859(3) 1.0(−)
D1 in $18h$ (x, $-x$, z) $\quad x$ $\quad z$ $\quad n$	0.484(4) 0.023(1) 0.33(1)	0.496(3) 0.023(1) 0.31(2)
D2 in $6c$ (0, 0, z) $\quad z$ $\quad n$	0.390(1) 0.50(3)	0.385(1) 0.58(3)
D4' in $18h$ (x, $-x$, z) $\quad x$ $\quad z$ $\quad n$	0.814(3) 0.0626(9) 0.43(2)	0.792(2) 0.051(1) 0.33(3)
D5' in $18h$ (x, $-x$, z) $\quad x$ $\quad z$ $\quad n$	0.201(2) 0.120(1) 0.45(2)	0.192(3) 0.123(1) 0.35(2)

Table 2. *Cont.*

Deuteride	$La_{1.09}Mg_{1.91}Ni_9D_{9.5(5)}$	$La_{0.91}Mg_{2.09}Ni_9D_{9.4(6)}$
D6 in 18h (x, $-x$, z)		
x	0.819(4)	0.819(4)
z	0.117(1)	0.117(1)
n	0.20(2)	0.39(2)
$U_{iso} \times 100$ (Å²) for D1-D6	2.0(–)	2.0(–)
Atomic parameters:		
D distribution in the structure		
LaNi₅	5.6(3)	5.0(4)
2 MgNi₂	3.9(2)	4.4(2)
Shortest Metal—Hydrogen distances, Å		
La ... D	2.34(3)	2.29(2)
Mg ... D	1.97(3)	1.93(2)
Ni ... D	1.56(3)	1.53(2)
R-factors of refinements		
R_p	2.7	2.4
R_{wp}	3.4	3.2
χ^2	5.0	6.0
Secondary constituents	α-solid solution $La_{0.9}Mg_{2.1}Ni_9D_{0.9}$. Sp.gr. $R\bar{3}m$; a = 4.9459(2); c = 23.842(2) Å; V = 505.10(4). 0.3 D in D3 18h (0.15, 0.3, 0.085) and 0.6 D in D4 18h (0.3, 0.15, 0.085); 35.7(2) wt% LaNi₅D₇; Sp.gr. P6₃mc; a = 5.438(3), c= 8.598(5) Å; V = 220.3(2) Å³; 4.6(3) wt%. Atomic structure was taken from [3]. MgNi₂; MgNi₂ structure type; Sp.gr. P6₃/mmc; a = 4.8356(4), c = 15.850(3) Å; V = 320.97(5) Å³; 12.4(2) wt%. Atomic structure was taken from [4]. Sample holder: stainless steel; Sp.gr. $Fm\bar{3}m$; a = 3.598 Å.	LaNi₅D₇; Sp.gr. P6₃mc; a = 5.430(1), c = 8.606(4) Å; V = 219.8(2) Å³; 21.5(5) wt%. Atomic structure was taken from [3]. MgNi₃; AuCu₃ structure type; Sp.gr. $Pm\bar{3}m$; a = 3.7185 Å; 1 Mg in 1a: 0, 0, 0; 3 Ni in 3c: 1/2, 1/2, 0; 3.7(2) wt%. Sample holder: zero matrix TiZr alloy with Fe liner. The peaks from Fe liner are only observed. Sp.gr. $Fm\bar{3}m$; a = 3.5949(1) Å.

From the refinements of the NPD data we conclude that the overall chemical compositions $La_{1.09}Mg_{1.91}Ni_9D_{9.5}$/$La_{0.91}Mg_{2.09}Ni_9D_{9.4}$ can be presented as $LaNi_5H_{5.6}$/$LaNi_5H_{5.0}$ + 2*$MgNi_2H_{1.95}$/$MgNi_2H_{2.2}$. Thus, in the hybrid $La_{1\pm0.1}Mg_{2\pm0.1}Ni_9$ structure, a $LaNi_5$-assisted hydrogenation of the $MgNi_2$ slab proceeds at rather mild H_2/D_2 pressure conditions; the equilibrium D_2 desorption pressure is just 20 bar D_2. In contrast, the parent $MgNi_2$ intermetallic remains inert with respect to hydrogenation even at much higher hydrogen pressures as well as the conditions applied in the present study of 912 bar D_2 for sample 2.

The shortest Me–D distances in the studied deuterides are listed in Table 2 and are within the regular values for the La–H, Mg–H and Ni–H distances in the structures of the metal and intermetallic hydrides.

The data of the present study clearly shows an influence of the $LaNi_5$ and $MgNi_2$ layers in the hybrid $La_{1\pm0.1}Mg_{2\pm0.1}Ni_9$ structures on the hydrogenation of the other buildings blocks of the structure. $MgNi_2$ slabs accommodate hydrogen up to a composition $MgNi_2H_{2.2}$ at much lower pressures as compared to those required to form a hydride by the pure $MgNi_2$ intermetallic. In contrast, the $LaNi_5$ block absorbs 5.0–5.6 at.H/f.u., which is quite close to the maximum hydrogenation capacity of the title intermetallic alloy, $LaNi_5H_7$; however, hydrogen desorption from the $LaNi_5H_{5.0/5.6}$ block proceeds much easier, at significantly higher pressures of H_2/D_2 as compared to the individual $LaNi_5H_7$ hydride—as a result of influence of the $MgNi_2$ slab.

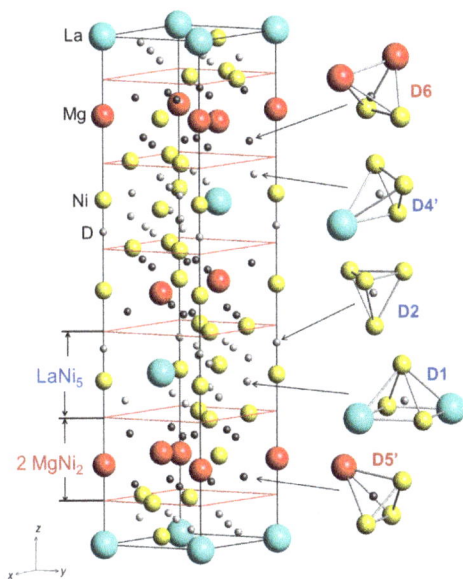

Figure 5. Crystal structure of $La_{1\pm0.1}Mg_{2\pm0.1}Ni_9D_{9.4-9.5}$ and types of the filled interstices.

4. Conclusions

$LaNi_5$-assisted hydrogenation of $MgNi_2$ is observed in the $LaMg_2Ni_9$ hybrid structure. Formation of $LaMg_2Ni_9D_{9.5}$ proceeds via an *isotropic* expansion of the trigonal unit cell. D atoms are accommodated in both Laves and $CaCu_5$-type slabs H atoms filling interstitial sites in both $LaNi_5$ and $MgNi_2$ structural fragments.

Limits of Mg solubility in $LaNi_3$ are not confined to $LaMg_2Ni_9$ and extend to the composition $La_{0.91}Mg_{2.09}Ni_9$ with a refined composition of the $CaCu_5$-type block of $La_{0.95}Mg_{0.05}Ni_5$.

Within the $LaNi_5$ $CaCu_5$-type layer, D atoms fill three types of interstices; a deformed octahedron $[La_2Ni_4]$, and two types of tetrahedra, $[LaNi_3]$ and $[Ni_4]$, to yield $LaNi_5D_{5-5.6}$ composition. D distribution is very similar to that in the individual β-$LaNi_5D_7$ deuteride.

In the $MgNi_2$ slab hydrogen atoms fill two types of tetrahedra, $[Mg_2Ni_2]$ and $[MgNi_3]$. The hydrogen sublattice formed is unique and is not formed in the studied structures of the Laves-type intermetallic hydrides.

A significant mutual influence of the $LaNi_5$ and $MgNi_2$ slabs causes a dramatic altering of their hydrogenation behaviours leading to:

(a)   significant decrease of the stability of the $LaNi_5$-type hydride;

(b)   much easier hydrogenation of the $MgNi_2$ slabs compared to the parent intermetallic compound;

(c)   increased hysteresis.

Acknowledgments: This work received support from the Research Council of Norway (project 223084 NOVELMAG "NOVEL MAGNESIUM BASED NANOMATERIALS FOR ADVANCED RECHARGEABLE BATTERIES") and is a part of the activities within the IEA Task32 "Hydrogen Based Energy Storage". The skillful assistance from the staff of the Swiss-Norwegian Beam Lines during the experimental studies at ESRF is gratefully acknowledged. Denis Sheptyakov (PSI, Switzerland) is sincerely thanked for the collaboration in the neutron powder diffraction experiments at HRPT, PSI (experiments 20090547, 20101348 and 20110612).

Author Contributions: All authors contributed extensively to the work presented in this paper. Volodymyr A. Yartys supervised the project. Evan MacA. Gray and Colin J. Webb designed and built the high pressure 1000 bar rig. Roman V. Denys, Volodymyr A. Yartys and Colin J. Webb jointly performed the in situ neutron powder diffraction experiments, while Roman V. Denys analyzed the NPD data. Roman V. Denys and Evan MacA. Gray measured the PCT diagrams. Volodymyr A. Yartys wrote the paper. All authors discussed the results and commented on the manuscript at all stages.

Conflicts of Interest: The authors declare no conflict of interest.

References

1.   Denys, R.V.; Yartys, V.A. Effect of magnesium on crystal structure and thermodynamics of the $La_{3-x}Mg_xNi_9$ hydrides. *J. Alloys Compd.* **2011,** *509* (Suppl. 2), S540–S548.

2.   Denys, R.V.; Yartys, V.A.; Webb, C.J. $LaNi_5$-assisted hydrogenation of $MgNi_2$ in the hybrid structure of $LaMg_2Ni_9D_{9.5}$. In Proceedings of the International Symposium on Metal-Hydrogen Systems MH2012, Fundamental and Applications, Kyoto, Japan, 21–26 October 2012. Poster Presentation. MoP38. Collected Abstracts. P.92.

3.   Lartigue, C.; Percheron Guégan, A.; Achard, J.C.; Soubeyroux, J.L. Hydrogen (deuterium) ordering in the β-$LaNi_5D_x$ phases: A neutron diffraction study. *J. Less-Common. Met.* **1985,** *113,* 127–148.

4.   Yartys, V.A.; Antonov, V.E.; Beskrovnyi, A.I.; Crivello, J.-C.; Denys, R.V.; Fedotov, V.K.; Gupta, M.; Kulakov, V.I.; Kuzovnikov, M.A.; Latroche, M.; *et al.* Hydrogen assisted phase transition in a trihydride $MgNi_2H_3$ synthesised at high H_2 pressures: thermodynamics, crystallographic and electronic structures. *Acta Mater.* **2015,** *82,* 316–327.

5.   Denys, R.V.; Yartys, V.A.; Webb, C.J. Hydrogen in $La_2MgNi_9D_{13}$. The role of magnesium. *Inorg. Chem.* **2012,** *51,* 4231–4238.

6. Nwakwuo, C.C.; Holm, T.H.; Denys, R.V.; Hu, W.; Maehlen, J.P.; Solberg, J.K.; Yartys, V.A. Effect of magnesium content and quenching rate on the phase structure and composition of rapidly solidified La_2MgNi_9 metal hydride battery electrode alloy. *J. Alloys Compd.* **2013**, *555*, 201–208.

7. Hu, W.-K.; Denys, R.V.; Nwakwuo, C.C.; Holm, T.H.; Maehlen, J.P.; Solberg, J.K.; Yartys, V.A. Annealing effect on phase composition and electrochemical properties of the Co-free La_2MgNi_9 anode for Ni-Metal Hydride batteries. *Electrochim. Acta* **2013**, *96*, 27–33.

8. Latroche, M.; Cuevas, F.; Hu, W.-K.; Sheptyakov, D.; Denys, R.V.; Yartys, V.A. Mechanistic and kinetic study of the electrochemical charge and discharge of La_2MgNi_9 by *in situ* powder neutron diffraction. *J. Phys. Chem. C* **2014**, *118*, 12162–12169.

9. Gabis, I.E.; Evard, E.A.; Voyt, A.P.; Kuznetsov, V.G.; Tarasov, B.P.; Crivello, J.-C.; Latroche, M.; Denys, R.V.; Hu, W.; Yartys, V.A. Modeling of metal hydride battery anodes at high discharge current densities. *Electrochim. Acta* **2014**, *147*, 73–81.

10. Yartys, V.A.; Denys, R.V. Structure-properties relationship in $RE_{3-x}Mg_xNi_9H_{10-13}$ (RE = La, Pr, Nd) hydrides for energy storage. *J. Alloys Compd.* **2015**.

11. Yartys, V.A.; Denys, R.V. Thermodynamics and crystal chemistry of the $RE_2MgNi_9H_{12-13}$ (RE=La and Nd) hydrides. *Chem. Met. Alloys* **2014**, *7*, 1–8.

12. Larson, A.C.; Dreele, R.B.V. General structure analysis system (GSAS). In *Los Alamos National Laboratory Report LAUR*; 2000; Alamos National Lab: Los Alamos, NM, USA, 2000; pp. 86–748.

13. Rodríguez-Carvajal, J. Recent advances in magnetic structure determination by neutron powder diffraction. *Phys. B: Condens. Matter* **1993**, *192*, 55–69.

14. Liu, G.; Xi, S.; Ran, G.; Zuo, K.; Li, P.; Zhou, J. A new phase $MgNi_3$ synthesized by mechanical alloying. *J. Alloys Compd.* **2008**, *448*, 206–209.

15. Kadir, K.; Yamamoto, H.; Sakai, T.; Uehara, I.; Kanehisa, N.; Kai, Y.; Eriksson, L. $LaMg_2Ni_9$, an example of the new AB_2C_9 structure type. *Acta Crystallogr.* **1999**, *C55*.

16. De Negri, S.; Giovannini, M.; Saccone, A. Phase relationships of the La–Ni–Mg system at 500 °C from 66.7 to 100 at% Ni. *J. Alloys Compd.* **2005**, *397*, 126–134.

Temperature Dependence of the Elastic Modulus of $(Ni_{0.6}Nb_{0.4})_{1-x}Zr_x$ Membranes: Effects of Thermal Treatments and Hydrogenation

Oriele Palumbo, Sergio Brutti, Francesco Trequattrini, Suchismita Sarker, Michael Dolan, Dhanesh Chandra and Annalisa Paolone

Abstract: Amorphous $(Ni_{0.6}Nb_{0.4})_{1-x}Zr_x$ membranes were investigated by means of X-ray diffraction, thermogravimetry, differential thermal analysis and tensile modulus measurements. Crystallization occurs only above 673 K, and even after hydrogenation the membranes retain their mainly amorphous nature. However, after exposure to gaseous hydrogen, the temperature dependence of the tensile modulus, M, displays large variations. The modulus of the hydrogen reacted membrane is higher with respect to the pristine samples in the temperature range between 298 K and 423 K. Moreover, a sharp drop in M is observed upon heating to approximately 473 K, well below the glass transition temperature of these glasses. We propose that the changes in the moduli as a function of temperature on the hydrogenated samples are due to the formation of nanocrystalline phases of Zr hydrides in $(Ni_{0.6}Nb_{0.4})_{1-x}Zr_x$-H membanes.

Reprinted from *Energies*. Cite as: Palumbo, O.; Brutti, S.; Trequattrini, F.; Sarker, S.; Dolan, M.; Chandra, D.; Paolone, A. Temperature Dependence of the Elastic Modulus of $(Ni_{0.6}Nb_{0.4})_{1-x}Zr_x$ Membranes: Effects of Thermal Treatments and Hydrogenation. *Energies* **2015**, *8*, 3944–3954.

1. Introduction

Alloy membranes with hydrogen selective properties have broad technological applications in the field of hydrogen production and purification [1–3]. Usually, thin, crystalline, Pd-based membranes are used for the purification of hydrogen due to their high permeability to hydrogen ($\sim$2.0 10^{-8} mol m^{-1} s^{-1} Pa$^{-0.5}$ for $Pd_{75}Ag_{25}$) and selectivity [4]. The operating temperatures of these membranes are higher than 573 K. Palladium, however, is an expensive and scarce element, therefore inexpensive and abundantly available alloys are being investigated as a replacement. Amorphous alloy membranes show promising properties [5], such as excellent mechanical strength, high thermal stability and soft magnetic properties [6–8].

A promising class of amorphous alloy membranes are the Ni-Zr or Ni-Nb-Zr ones. Hara *et al.* reported that Ni-Zr based alloys are stable, and do not become brittle in the temperature range between 473 K to 623 K while the hydrogen permeability of the $Ni_{64}Zr_{36}$ sample is of the order of 10^{-9} mol m^{-1} s^{-1} Pa$^{-0.5}$ [9–11]. Later studies showed that the hydrogen permeability of Ni-Nb-Zr amorphous samples is significantly enhanced by the addition of Zr to the alloys, reaching a permeability comparable to that of the Pd-based alloys, on the order of 10^{-9}–10^{-8} mol m^{-1} s^{-1} Pa$^{-0.5}$ [12–16].

The Ni-Nb-Zr membranes can be produced in the amorphous state by various techniques, such as planar flow cast or melt spinning. These amorphous alloys are stable at operating temperatures up to 623 K, but devitrify at temperatures higher than 780 K. In the Ni-Zr and Ni-Nb-Zr systems the crystallization temperature (T_x) decreases as the Zr content increases [15,17]; for example the T_x ~843 K for $Ni_{63.7}Zr_{36.3}$, that decreases to T_x ~653 K for $Ni_{30}Zr_{70}$. Another example is that of $Ni_{60}Nb_{40}$ with a T_x ~943 K which decreases to T_x ~843 K with the addition of 30 at.% Zr (*i.e.*, in the alloy $(Ni_{0.6}Nb_{0.4})_{70}Zr_{30}$). Permeation tests on Ni-Nb-Zr amorphous alloys are usually conducted in the temperature range of 673–723 K [12–16]; the permeability of Pd (1.8–2.0 10^{-8} mol m^{-1} s^{-1} Pa$^{-0.5}$) [14] is slightly higher than that of our alloys $(Ni_{0.6}Nb_{0.4})_{70}Zr_{30}$ (1.4 × 10^{-8} mol m^{-1} s^{-1} Pa$^{-0.5}$) at 723 K and of $(Ni_{0.6}Nb_{0.4})_{80}Zr_{20}$ (8 × 10^{-9} mol m^{-1} s^{-1} Pa$^{-0.5}$) at 673 K [14]. Typically, in a laboratory these experiments take one to two weeks during which the membranes are at elevated temperatures and are exposed to hydrogen. To terminate the experiment, it is important that all the hydrogen be desorbed from the membrane before cooling down to room temperature. If all the hydrogen is not removed, these membranes have a propensity to embrittle. The embrittlement of the samples with or without hydrogen is heavily influenced by their elastic properties, therefore, in this paper we investigate the changes of the elastic modulus of $(Ni_{0.6}Nb_{0.4})_{1-x}Zr_x$ membranes induced by thermal treatments and hydrogenation. One of the property we are interested is in the amorphous to amorphous phase transitions as a function of temperature.

2. Results

Two pristine membranes were investigated in the present study: (1) $(Ni_{0.6}Nb_{0.4})_{70}Zr_{30}$, thereafter referred to as Zr_{30}, and $(Ni_{0.6}Nb_{0.4})_{80}Zr_{20}$ as Zr_{20}. The effect of thermal treatments and/or hydrogenation on these two specimens will be determined. X-ray diffraction (XRD) data of the membranes are reported in Figure 1. The starting materials display a broad peak centered around 39.9° in Zr20 and 39.3° in Zr30, which confirms the manly amorphous nature of the pristine membranes.

Differential thermal analysis (DTA) measurements (in Figure 2) show the crystallization temperature. Two sharp peaks at 816 and 836 K, were observed for the Zr30 alloy, whereas Zr20 sample displayed one peak at ~868 K. We attribute these peaks to the crystallization process of the amorphous alloys, in agreement with previous reports [14,15,17]. The crystallization of Zr30 sample is a multistep process, as suggested by the weak structures (peaks) present in the DTA signal. A multistep crystallization is very common in amorphous alloys for hydrogen storage. It is worth noting that the temperatures at which our samples display the highest peaks of the DTA signals well agree with the known crystallization temperatures of the two alloys: T_x ~876 K for the Zr20 and 808 K for the Zr30 alloy [15].

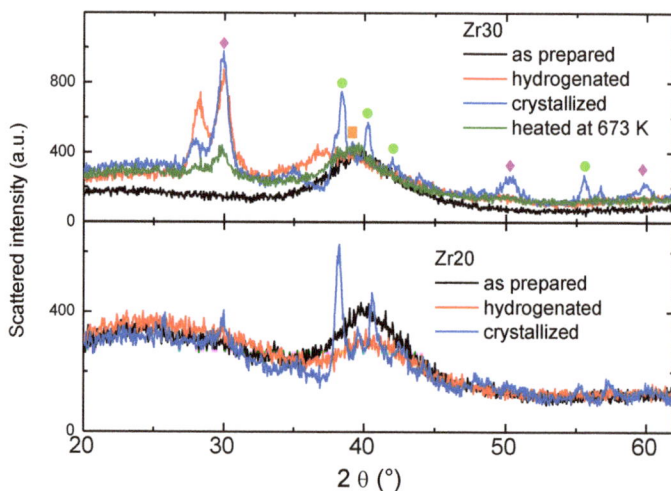

Figure 1. X-ray diffraction (XRD) spectra of all samples with attribution to crystalline phases (magenta diamonds: ZrO_2; green solid circles: $Ni_{10}Zr_7$; orange squares: $Ni_{48}Zr_{52}$) [18–22].

The occurrence of the crystallization process at high temperatures is confirmed by XRD data collected on samples Zr30 and Zr20 heated to 973 K in an argon environment (Figure 1). In Zr30 sample Bragg peaks are observed after such thermal treatments at ~38.4, 40.2, 42.1 and 55.5° 2θ that have been ascribed to $Ni_{48}Zr_{52}$ phase [21], and Bragg pea K located at 39.2° 2θ, that may possibly due to the presence of $Ni_{10}Zr_7$ phase [22]. Note that the Bragg peaks of ZrO_2 are due to the exposure of the sample to air for XRD experiments [18–20]. The Zr20 sample displays similar characteristics when heated to 973 K. For the sake of completeness we also heated a Zr30 sample only up to 673 K for 8 h in an argon atmosphere. In the case of the Zr30 sample heated to 673 K, (much lower than the T_x~816 K for Zr30), the XRD pattern shows the characteristic broad amorphous hump at ~40° 2θ, and a minor peak at 30.2°, ascribed to ZrO_2.

In order to check whether the hydrogenation process can induce crystallization, the membranes with the two different content of Zr were hydrogenated at 673 K by using a home-made Sieverts apparatus, starting with an initial hydrogen pressure of ~0.55 MPa. The hydrogenation process for Zr30 lasted 4 h, and 22 h for Zr20; with a final hydrogen content of ~0.4 wt%H for Zr20 and ~0.8 wt% H for Zr30, after which the experiments were terminated. In Figure 1 the high intensity Bragg peaks around $40°$ 2θ are retained in the hydrogenated samples, and therefore after hydrogenation the specimens remain mainly amorphous (see Figure 1). Only the hydrogenated Zr30 sample displays two Bragg peaks at 28.3 and $29.9°$ 2θ due to the formation of ZrO_2 [18–22].

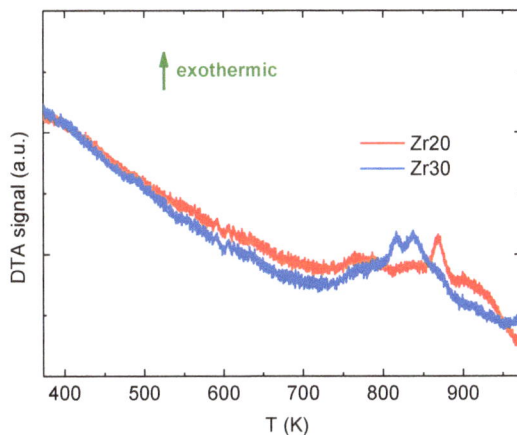

Figure 2. Differential thermal analysis (DTA) measurements on the two amorphous membranes conducted with a temperature rate of 4 K/min after subtraction of the background records with empty crucibles.

Dynamical Mechanical analysis (DMA) results yielded interesting results; the tensile modulus values, M, measured from the room temperature to 673 K are shown in Figure 3. At room temperature the tensile modulus of Zr20 (M = 1.0×10^{10} Pa) is higher than that of Zr30 (M = 6.5×10^9 Pa). The modulus decreases almost linearly up to 503 K for the Zr20 and up to 543 K for the Zr30 sample, showing a strong decrease of M by a factor of ~2 which is clearly visible. In another region of the spectrum above ~633 K one can observe, again, a much smoother linear decrease of the modulus. An almost linear decrease of the elastic modulus with increasing temperature is usually observed in many solids and can be explained by the Debye theory [23].

We could not measure the tensile modulus of samples in which a full crystallization was induced by a thermal treatment in argon at 923 K in the thermobalance, because the membranes become so fragile that they break as soon as they are mounted in the clamps of the Dynamic Mechanical Analyzer (DMA) used for the elastic modulus measurements. Therefore a direct comparison of the elastic properties of the crystallized samples could not be performed.

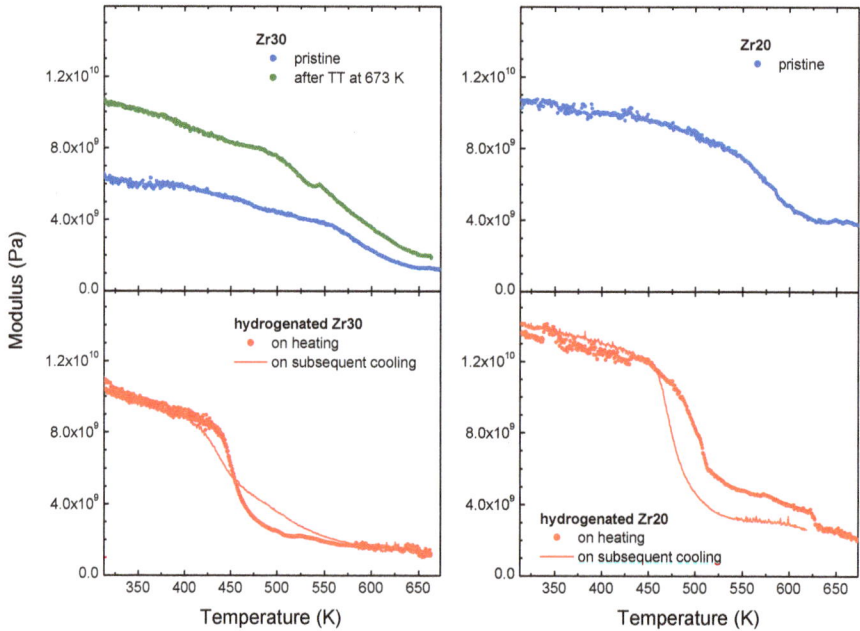

Figure 3. Dynamic Mechanical Analyzer (DMA) measurements of the membranes.

DMA measurements were repeated on the hydrogenated membranes. They were more fragile than the starting materials and in some cases small pieces of the hydrogenated samples broke up before measurements. In order to be sure that the hydrogenated membranes do not break during measurements we carefully measured the elastic moduli both on heating and cooling, comparing the values of M along the thermal cycle and checking that no sudden and irreversible change of the elastic modulus occurred. These DMA measurements of the hydrogenated membranes are reported in Figure 3. One can note that for both Zr concentrations the elastic modulus at room temperature and below 423 K is higher in hydrogenated samples than in the starting materials. Upon heating above the room temperature, M decreases almost linearly up to ~393 K for Zr30, and up to ~423 K for Zr20. Above such temperatures, M displays a strong and sharp decrease by a factor 3–4. For both samples such variation occurs in a very limited temperature range, 423 K to 673 K for the Zr30, and 443 K to 523 K for the Zr20 sample. For both specimens a smooth linear decrease of M is observed above 523 K.

The appearance of an abrupt decrease of M is strictly due to the presence of hydrogen in the hydrogenated samples and cannot be explained as an effect of the thermal cycle used during hydrogenation (in our case, heating at 673 K for at least 4 h). The proof of this comes from a measurement performed on a piece of the Zr30 membrane heated at 673 K for 8 h (sample Zr30_TT) in an argon atmosphere (a longer thermal treatment than that used for the hydrogenation process at the same temperature). Zr30_TT shows a smooth increase of the modulus, without the abrupt change presented by the hydrogenated membrane (see Figure 3). Indeed, the temperature dependence of M for sample Zr30_TT strictly resembles that of the pristine membrane, except that the absolute value is slightly higher. This increase of the absolute value of M in the thermally treated specimen can be possibly due the formation of nanometric crystallites, not detected by X-ray diffraction because of their small dimensions, whose presence increases the elastic stiffness of the sample.

The tensile modulus of the hydrogenated samples measured on subsequent cooling, after heating to 673 K, (in the Figure 3) again shows strong variations between 423 K and 523 K for the Zr30, and between 443 K to 523 K for the Zr20 sample. In the case of the Zr20 sample this feature stretches to a broader temperature range than the Zr30 sample with the modulus jump clearly visible.

To confirm if the hydrogenated sample are not completely desorbed, thermogravimetric analyses (TGA) measurements were performed on the hydrogenated Zr30 samples. Results show that hydrogen was not completely released by heating at 673 K as shown in Figure 4. The hydrogen gas starts to release above 473 K but there is incomplete dehydrogenation below ~873 K. One can note that above 873 K (Figure 4) the sample starts to oxidize, and the mass increases. For both the Zr20 and the Zr30 samples half of the initial hydrogen content is retained after thermal treatments up to ~673 K.

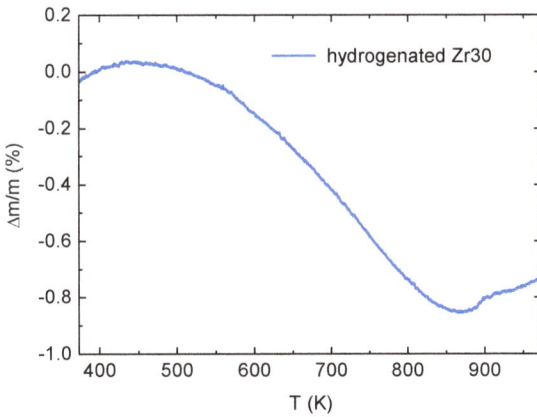

Figure 4. TGA scan of the hydrogenated Zr30 sample.

3. Discussion

A striking difference between the elastic moduli variation of the pristine (unhydrogenated) and the hydrogenated membranes is evident in Figure 3. In the hydrogenated Zr30 membrane we observe an abrupt decrease of the elastic modulus in an extremely limited range of 423 K to 473 K, and 443 K to 523 K for the Zr20 membrane sample. Such variations are completely absent in the pristine materials and they resemble the modulus changes occurring concomitantly with a phase transition. Indeed, no change in the shape and position of the modulus drop are visible when measurements are conducted with different frequencies (results not shown). Conversely, if the modulus drop was due to a thermally activated process, one would observe a shift of the modulus variations toward higher temperatures when M was measured at higher frequency [23,24].

Usually, these materials display a gradual decrease of the elastic modulus with increasing temperature [23]. However, materials undergoing phase transitions display strong deviations from such a trend and present sharp changes of M as a function of temperature around the phase transition temperature. Elastic moduli typically display remarkable variations in correspondence of phase transitions, independently of their order [24]. It should be noted that the DSC or DTA can easily detect first order phase transitions and measure latent heats of phase transitions, but are not well suited to detect phase transitions of the second or higher order; for these DMA measurements are critical.

The XRD measurements indicate that both the pristine and the hydrogenated $(Ni_{0.6}Nb_{0.4})_{1-x}Zr_x$ membranes investigated in the present study are mainly amorphous. However, some studies of similar samples indicate that at the microscopic level they are not completely amorphous: they actually possess some nano-crystalline inclusions [16,25,26]. Such nano-crystals are too small and diluted to cause a detectable diffraction signal in standard diffractometers [16,25,26]. Indeed, TEM analysis of $Ni_{64}Zr_{36}$ samples pointed out that after thermal treatments the surface shows the presence of fine precipitates, 500 nm in diameter, with a high concentration of Ni [16]. Moreover, some blistering possibly due to the precipitation of γ-ZrH was observed [16]. Also Jayala Kshmi et al. reported that nano-crystallization takes place in hydrogenated Zr- and Ni- based glasses, even though XRD is unable to detect crystalline precipitates [25,26]. Indeed, electron diffraction patterns show the formation of nanocrystals of γ-ZrH after hydrogenation the $Zr_{50}Ni_{27}Nb_{18}Co_5$ alloy, with dimensions of the order of ~2 nm and the formation of the Ni_2H phase in $Ni_{59}Zr_{16}Ti_{13}Nb_7Sn_3Si_2$ [26].

In view of this framework, we suggest that the modulus variation in the hydrogenated samples around 473 K is due to the formation on cooling of nanometric Zr hydride or its dissolution on heating. Indeed, Zr is an energy favorable site for the absorption of hydrogen, as pointed out for example by XAFS measurements and molecular dynamics simulations [27]. We suggest that in the Zr rich zones of the samples, hydrogen can form a solid solution with the Zr atoms. However, as temperature is decreases the solid solution undergoes a phase transition to a new stable hydride phase. On heating, the process is reversible. The Zr-H phase diagram [28,29] presents two regions with phase transitions in the temperature range around 473 K: the two lines defining the separation between the α and δ phases and the coexistence region between the δ hydride phase and α-Zr phase. We suggest that the modulus variations reported in Figure 3 correspond to the transition from the solid solution to the hydride on cooling and from the stable hydride to the solid solution on heating. Some modulus variations due to hydride formation and dissolution have also been previously observed in other Zr alloys, such as Zircaloy-4 or Zr-2.5Nb [30,31].

TGA measurements indicate that at least half of the hydrogen initially present in the samples is liberated after a thermal treatment at 673 K. On the other side DMA measurements find a reasonable recovery of the elastic modulus after the same thermal treatment. These results are not in contrast in view of the attribution of the modulus drop to the formation and dissolution of Zr hydride. Indeed Zr hydride is only a minority phase, whose presence is detected by DMA measurements. As Zr atoms are the energy preferred sites for absorption of hydrogen, Zr-H is the most stable of the hydride phases in the sample and therefore it will be the least affected by dehydrogenation. Therefore a partial release of hydrogen from the samples does not affect the presence of the Zr hydride.

In summary, we suggest that the present measurements of the elastic modulus of hydrogenated membranes strongly indicate that nanometer size Zr hydride can form or dissolve as a function of temperature and therefore the elastic properties of the amorphous materials display a dramatic variation around 473 K. In this way the modulus measurements become an easily way to detect the presence of nanocrystalline Zr hydrides.

4. Experimental Section

Crystalline $(Ni_{0.6}Nb_{0.4})_{100-x}Zr_x$ (x = 20 and 30) alloy ingots were prepared by arc melting in a purified Ar atmosphere. Amorphous ribbons ~100 μm thick and 30 mm wide have been prepared using the melt-spinning method [15].

Simultaneous TGA-DTA measurements were conducted by means of a Setaram Setsys Evolution 1200 TGA system (Caluire, France) [32]. The furnace was continuously flooded with high purity argon (60 ml/min). For each experiment a sample mass of ~10 mg was used.

XRD have been recorded at room temperature by using an X-Pert PANalytical theta-theta diffractometer (Almelo, The Netherlands) in the range 10–70° with a step of 0.025°, time/step = 3 s. Millimetric pieces of the membranes have been spread on the planar glass holder and then measured in the XRD chamber.

Dynamic Mechanical Analysis (DMA) was carried out on small membrane pieces 4–6 mm wide and 10–12 mm long by using a DMA 8000 (Perkin Elmer, Waltham, MA, USA) in the so-called "tension configuration" [33–35]. The storage modulus, M, was measured at frequencies of 1 and 10 Hz between room temperature and 673 K, that is the maximum temperature allowed by the system. All measurements were carried out in a flux of pure argon to prevent oxidation of samples. In order to directly compare DTA and DMA measurements, we conducted the two experiments at the same temperature rate (4 K/min).

Hydrogenation of the samples was obtained by means of a home-made Sieverts apparatus working up to a 20 MPa pressure and a 773 K temperature [36]. The amount of hydrogen exchanged between the gas atmosphere and the solid samples is measured by the pressure variation in calibrated cylinders connected by Swagelok tubes to the reaction chamber where the sample is placed. A homemade computer program developed in the Labview language is used to acquire the time evolution of the pressure detected by two micro-Baratron transducers working between 0 and 0.7 MPa and between 0 and 20 MPa, respectively, and of the temperature of the five cylinders and of the tubes connecting the cylinders and the reaction chamber. The real gas state equation is used to calculate the exchanged hydrogen moles. For all samples we chose to conduct hydrogenation processes at T = 673 K and p $\approx$ 0.55 MPa, because these are the temperature and pressure at which the membranes have practical application in gas reactors in order to separate H_2 from other gases. We extended the hydrogenation procedure until a steady state was obtained and no more hydrogen was absorbed by the membranes. The Zr30 sample, exposed to p $\approx$ 0.55 MPa, absorbs more hydrogen and in a shorter time than the Zr20 specimen. After only 4 h, Zr30 reaches a hydrogen concentration value of 0.8 wt%. On the contrary, in the same conditions, Zr20 absorbs only 0.4 wt% even after 22 h.

5. Conclusions

XRD measurements indicate that $(Ni_{0.6}Nb_{0.4})_{100-x}Zr_x$ (x = 20 and 30) membranes are mainly in an amorphous state both when prepared by means of melt spinning, and after hydrogenation at 673 K. However, the elastic modulus of the hydrogenated samples display a strong and sharp decrease that occurs in an extremely limited range between 423 K and 473 K for the Zr30 and 443 K to 523 K for the Zr20 sample. We suggest that these features are due to the dissolution on heating and to the formation on cooling of nanocrystalline Zr hydride. It is suggested that the DMA analysis is a method to detect the presence of nanocrystalline crystalline phases and possible local atomic order changes in the amorphous solids, such as the membranes.

Acknowledgments: This research has been partially supported by the Project FIRB 2010-Futuro in Ricerca "Hydrides as high capacity anodes for lithium ion batteries" funded by the Italian Ministry of Research. One of the authors (D.C.) would like to thank U.S. Department of Energy, Grant No. DE-NA-0002004 for the support of this project. Thanks are due to Prof. A. Colella and Prof. G. Mongelli for the diffraction experiments.

Author Contributions: Suchismita Saker, Michael Dolan and Dhanesh Chandra synthesized the samples. Sergio Brutti performed the XRD measurements. Oriele Palumbo, Francesco Trequattrini and Annalisa Paolone performed the DMA, TGA and DTA measurements. Annalisa Paolone and Dhanesh Chandra wrote and later edited, the manuscript.

Conflicts of Interest: The authors declare no conflict of interest.

References

1. Dolan, M.D. Non-Pd BCC alloy membranes for industrial hydrogen separation. *J. Membr. Sci.* **2010**, *362*, 12–28.

2. Hwang, K.; Lee, C.; Ryi, S.; Lee, S.; Park, J. A multi-membrane reformer for the direct production of hydrogen via a stream-reforming reaction of methane. *Int. J. Hydrog. Energy* **2011**, *37*, 6601–6607.

3. Paglieri, S.N.; Way, J.D. Innovations in palladium membrane research. *Sep. Purif. Rev.* **2002**, *31*, 1–169.

4. Serra, E.; Kemali, M.; Perujo, A.; Ross, D.K. Hydrogen and deuterium in Pd-25pct Ag alloy: Permeation, diffusion, solubilization, and surface reaction. *Metall. Mater. Trans. A* **1998**, 1023–1028.

5. Dolan, D.M.; Dave, N.C.; Ilyushech Kin, A.Y.; Morpeth, L.D.; McLennan, K.G. Comosition and operation of hydrogen-selective amorphous alloy membranes. *J. Membr. Sci.* **2006**, *285*, 30–55.

6. Zhang, T.; Inoue, A. Thermal and mechanical properties of Ti-Ni-Cu-Sn amorphous alloys with a wide supercooled liquid region before crystallization. *Mater. Trans. JIM* **1998**, *39*, 1001–1006.

7. Kimura, H.; Inoue, A.; Yamaura, S.I.; Sasamori, K.; Nishida, M.; Shinpo, Y.; Okouchi, H. Thermal stability and mechanical properties of glassy and amorphous Ni-Nb-Zr alloys produced by rapid solidification. *Mater. Trans. JIM* **2003**, *44*, 1167–1171.

8. Inoue, A.; Koshiba, H.; Zhang, T.; Makino, A. Wide supercooled liquid region and soft magnetic properties of $Fe_{56}Co_7Ni_7Zr_{0-10}Nb$ (or $Ta)_{0-10}B_{20}$ amourphous alloys. *J. Appl. Phys.* **1998**, *83*, 1957–1974.

9. Hara, S.; Hatakeyma, N.; Itoh, N.; Kimura, H.M.; Inoue, A. Hydrogen permeation through amorphous $Zr_{36-x}Hf_xNi_{64}$ alloy membranes. *J. Membr. Sci.* **2003**, *211*, 149–156.

10. Hara, S.; Sakaki, K.; Itoh, N.; Kimura, H.M.; Asami, K.; Inoue, A. An amorphous alloy membrane without noble metals for gaseous hydrogen separation. *J. Membr. Sci.* **2000**, *164*, 289–294.

11. Hara, S.; Hatakeyma, N.; Itoh, N.; Kimura, H.M.; Inoue, A. Hydrogen permeation through palladium-coated amorphous Zr-M-Ni (M = Ti, Hf) alloy membranes. *Desalination* **2002**, *144*, 115–120.

12. Yamaura, S.I.; Shimpo, Y.; Okouchi, H.; Nishida, M.; Kajita, O.; Kimura, H.; Inoue, A. Hydrogen permeation characteristics of melt-spun Ni-Nb-Zr amorphous alloy membranes. *Mater. Trans. JIM* **2003**, *44*, 1885–1890.

13. Yamaura, S.I.; Sakurai, M.; Hasegawa, M.; Wakoh, K.; Shimpo, Y.; Nishida, M.; Kimura, H.; Matsubara, E.; Inoue, A. Hydrogen permeation and structural features of melt-spun Ni-Nb-Zr amorphous alloys. *Acta Mater.* **2005**, *53*, 3703–3711.

14. Paglieri, S.N.; Pal, N.K.; Dolan, M.D.; Kim, S.M.; Chien, W.M.; Lamb, J.; Chandra, D.; Hubbard, K.M.; Moore, D.P. Hydrogen permeability, thermal stability and hydrogen embrittlement of Ni-Nb-Zr and Ni-Nb-Ta-Zr amorphous alloy membranes. *J. Membr. Sci.* **2011**, *378*, 42–50.

15. Kim, S.M.; Chandra, D.; Pal, N.K.; Dolan, M.D.; Chien, W.M.; Talekar, A.; Lamb, J.; Paglieri, S.N.; Flanagan, T.B. Hydrogen permeability and crystallization Kinetics in amorphous Ni-Nb-Zr alloys. *Int. J. Hydrog. Energy* **2012**, *37*, 3904–3913.

16. Adibhatla, A.; Dolan, M.D.; Chien, W.; Chandra, D. Enhancing the catalytic activity of Ni-based amorphous alloy membrane surfaces. *J. Membr. Sci.* **2014**, *463*, 190–195.

17. Kim, S.M.; Chien, W.M.; Chandra, D.; Pal, N.K.; Talekar, A.; Lamb, J.; Dolan, M.D.; Paglieri, S.N.; Flanagan, T.B. Phase transformation and crystallization Kinetics of melt-spun $Ni_{60}Nb_{20}Zr_{20}$ amorphous alloy. *J. Non-Cryst. Solids* **2012**, *358*, 1165–1170.

18. Adam, J.; Rogers, M.D. The crystal structure of ZrO_2 and HfO_2. *Acta Cryst.* **1959**, *12*, 951.

19. Smith, D.K.; Newkirk, H.W. The crystal structure of baddeleyite (monoclinic ZrO_2) and its relation to the polymorphism of ZrO_2. *Acta Cryst.* **1965**, *18*, 983–991.

20. Trufer, G. The crystal structure of tetragonal ZrO_2. *Acta Cryst.* **1962**, *15*, 1187.

21. Yan, Z.J.; Li, J.F.; He, S.R.; Zhou, Y.H. The relation between formation of compounds and glass forming ability for Zr-Al-Ni alloys. *Mater. Lett.* **2003**, *57*, 1840–1843.

22. Kirkpatrick, M.E.; Smith, J.F.; Larsen, W.L. Structure of the intermediate phases $Ni_{10}Zr_7$ and $Ni_{10}Hf_7$.. *Acta Cryst.* **1962**, *15*, 894–903.

23. Paolone, A.; Cordero, F.; Cantelli, R.; Costa, G.A.; Artini, C.; Vecchione, A.; Gombos, M. Magnetoelastic coupling in $RuSr_2GdCu_2O_8$. *J. Magn. Magn. Mater.* **2004**, *272–276*, 2106–2107.

24. Paolone, A.; Cantelli, R.; Scrosati, B.; Reale, P.; Ferretti, M.; Masquelier, C. Comparative study of the phase transition of $Li_{1+x}Mn_{2-x}O_4$ by anelastic spectroscopy and differential scanning calorimetry. *Electrochem. Commun.* **2006**, *8*, 113–117.

25. Jayalakshmi, S.; Fleury, E. High temperature mechanical properties of rapidly quenched $Zr_{50}Ni_{27}Nb_{18}Co_5$ amorphous alloy. *Met. Mater. Int.* **2009**, *15*, 701–711.

26. Jayalakshmi, S.; Park, S.O.; Kim, K.B.; Fleury, E.; Kim, D.H. Studies on hydrogen embrittlement in Zr- and Ni-based amorphous alloys. *Mater. Sci. Eng. A* **2007**, *449–451*, 920–923.

27. Fukuhara, M.; Fujima, N.; Oji, H.; Inoue, A.; Emura, S. Structures of the icosahedral clusters in Ni-Nb-Zr-H glassy alloys determined by first-principles molecular dynamics calculations and XAFS measurements. *J. Alloys Comps.* **2010**, *497*, 182–187.

28. Okamoto, H. H-Zr (Hydrogen-Zirconium). *J. Phase Equilib. Diff.* **2007**, *27*, 548–549.

29. Zuze, K.E.; Abriata, J.P. The H-Zr (Hydrogen-Zirconium) system. *Bull. Alloy Phase Diagram.* **1990**, *11*, 385–395.

30. Pan, Z.L.; Puls, M.P. Precipitation and dissolution pea Ks oh hydrides in Zr-2.5Nb during quasistatic thermal cycles. *J. Alloys Comps.* **2000**, *310*, 214–218.

31. Pan, Z.L.; Wang, N.; He, Z. Measurements of elastic modulus in Zr alloys For CANDU applications. In Proceedings of 11th International Conference on CANada Deuterium Uranium Fuel, Niagara Falls, ON, Canada, 17–20 October 2010.

32. Palumbo, O.; Paolone, A.; Rispoli, P.; Cantelli, R.; Autrey, T. Decomposition of NH_3BH_3 at sub-ambient pressures: A combined thermogravimetry–differential thermal analysis–mass spectrometry study. *J. Power Sources* **2010**, *195*, 1615–1618.

33. Teocoli, F.; Paolone, A.; Palumbo, O.; Navarra, M.A.; Casciola, M.; Donnadio, A. Effects of water freezing on the mechanical properties of Nafion membranes. *J. Polym. Sci. Part B Polym. Phys.* **2012**, *50*, 1421–1425.

34. Scipioni, R.; Gazzoli, D.; Teocoli, F.; Palumbo, O.; Paolone, A.; Ibris, N.; Brutti, S.; Navarra, M.A. Preparation and characterization of nanocomposite polymer membranes containing superacidic SnO_2 additives. *Membranes* **2014**, *4*, 123–142. PubMed]

35. Vitucci, F.M.; Manzo, D.; Navarra, M.A.; Palumbo, O.; Trequattrini, F.; Panero, S.; Bruni, P.; Croce, A.; Paolone, A. Low temperature phase transitions of 1-Butyl-1-methylpyrrolidinium bis(trifluoromethanesulfonyl)imide swelling a PVdF electrospun membrane. *J. Phys. Chem. C* **2014**, *118*, 5749–5755.

36. Palumbo, O.; Trequattrini, F.; Vitucci, F.M.; Bianchin, A.; Paolone, A. Study of the hydrogenation/dehydrogenation process in the Mg-Ni-C-Al system. *J. Alloys Compd.* 2015.

Enhanced Hydrogen Generation Properties of MgH$_2$-Based Hydrides by Breaking the Magnesium Hydroxide Passivation Layer

Liuzhang Ouyang, Miaolian Ma, Minghong Huang, Ruoming Duan, Hui Wang, Lixian Sun and Min Zhu

Abstract: Due to its relatively low cost, high hydrogen yield, and environmentally friendly hydrolysis byproducts, magnesium hydride (MgH$_2$) appears to be an attractive candidate for hydrogen generation. However, the hydrolysis reaction of MgH$_2$ is rapidly inhibited by the formation of a magnesium hydroxide passivation layer. To improve the hydrolysis properties of MgH$_2$-based hydrides we investigated three different approaches: ball milling, synthesis of MgH$_2$-based composites, and tuning of the solution composition. We demonstrate that the formation of a composite system, such as the MgH$_2$/LaH$_3$ composite, through ball milling and *in situ* synthesis, can improve the hydrolysis properties of MgH$_2$ in pure water. Furthermore, the addition of Ni to the MgH$_2$/LaH$_3$ composite resulted in the synthesis of LaH$_3$/MgH$_2$/Ni composites. The LaH$_3$/MgH$_2$/Ni composites exhibited a higher hydrolysis rate—120 mL/(g·min) of H$_2$ in the first 5 min—than the MgH$_2$/LaH$_3$ composite—95 mL/(g·min)—without the formation of the magnesium hydroxide passivation layer. Moreover, the yield rate was controlled by manipulation of the particle size via ball milling. The hydrolysis of MgH$_2$ was also improved by optimizing the solution. The MgH$_2$ produced 1711.2 mL/g of H$_2$ in 10 min at 298 K in the 27.1% ammonium chloride solution, and the hydrolytic conversion rate reached the value of 99.5%.

Reprinted from *Energies*. Cite as: Ouyang, L.; Ma, M.; Huang, M.; Duan, R.; Wang, H.; Sun, L.; Zhu, M. Enhanced Hydrogen Generation Properties of MgH$_2$-Based Hydrides by Breaking the Magnesium Hydroxide Passivation Layer. *Energies* **2015**, *8*, 4237–4252.

1. Introduction

The global energy crisis and our current ecological problems have stimulated the development of new clean energies [1–4]. Hydrogen has a high energy density of 142 MJ/kg, three times higher than that of petroleum, with 47 MJ/kg [5,6]. Upon combustion, the byproduct is water, which can subsequently be used to regenerate hydrogen [7–10]. Therefore, hydrogen is regarded as a potential candidate for clean and sustainable fuel to replace fossil fuels [11–13]. Currently, hydrogen is primarily produced through the processing of fossil fuels, biomass and water. The main methods include fossil fuel reforming [14], biological hydrogen production [15], generating hydrogen from water decomposition [16] and hydrogen production by metal or hydride hydrolysis [17]. Fuel processing of methane is the most common hydrogen production method in commercial use today [18,19]. Biohydrogen technologies include direct biophotolysis [20–22], indirect biophotolysis [23], photo-fermentation [24,25], and dark-fermentation [26–28]. Electrolysis technology is an effective method for transforming electrical energy into chemical energy and can be traced back to 1890 [29–31]. Photocatalytic decomposition of water was first used by Fujishima and Honda [32] and the catalytic efficiency reached 93% under the UV irradiation [33,34].

In recent years, more and more attention has been paid to hydrogen generation by hydrolysis of metals or metal hydrides for their high theoretical hydrogen yield. The main hydrogen generation metals include Al and Mg alloys, due to their light weight, abundance, and low cost and the byproduct being benign. The self-corrosion of hydrogen evolution reactions of aluminum has led to many studies focusing on hydrogen generation with aluminum and its alloys. However, aluminum has a strong affinity for oxygen, in the form of a dense oxide layer formed on the surface of aluminum and its alloys, making the corrosion potential shift nearly 1 V in the positive direction, thereby interrupting the corrosion reaction of aluminum [35]. This becomes the major challenge to realizing continuous hydrogen generation through aluminum corrosion. How to remove or prevent the formation of a dense aluminum oxide layer is the key issue in the hydrogen generation process by hydrolysis of aluminum alloys. Many methods have been reported to improve the hydrolysis properties of aluminum as, for example, changing the reaction solution (alkali solution, $NaAlO_2$ solution, Na_2SnO_3 solution, *etc.*) [36–38] and improving the reaction activity of aluminum alloys by ball milling and/or some special metal doping [39–41]. However, the corrosion of equipment in alkali or salt solution and the risk of corrosion of materials restrict its commercial utilization. Additional research is still needed to further improve the hydrogen generation performance with Al alloys. The hydrolysis reaction of Mg is rapidly interrupted because of the formation of a passive magnesium hydroxide layer.

Compared with pure metals or alloys, the metal hydrides can generate double the amount of hydrogen by hydrolysis, and the generated high purity hydrogen can be passed directly into the cells as fuel [42]. In addition, the fuel cell-generated water could be recycled and used for hydrolysis, thus achieving reduced weight of the system. Due to the hydrogen generation by hydrolysis having the excellent properties of widely raw material sources, high hydrogen generation yield, high hydrogen purity, and environmental friendliness, a large amount of metal hydrides have been studied as the mobile hydrogen source of fuel cell, such as LiH [43], CaH_2 [44], MgH_2 [45–48], $NaBH_4$ [49–51], $LiBH_4$ [52,53], $LiAlH_4$ [42,52]. Among

207

them, LiH, CaH$_2$, and LiAlH$_4$ can react with water violently, so the uncontrollable reaction leads to poor practicability. Among these methods, production of hydrogen using NaBH$_4$ is safety and controllability, but the regeneration of NaBH$_4$ needs a great deal of energy leading to a higher cost. On the other hand, the efficiency and service life of the catalyst and the price of the whole system restrict the promotion and application of NaBH$_4$ hydrogen production technology. Therefore, looking for a cheap and practical technology to generate hydrogen always is a fervent focus of research all over the world. Comparing the hydrolysis of MgH$_2$ with that of NaBH$_4$, it is not necessary for a highly basic solution to stabilize the NaBH$_4$, whereas MgH$_2$ can generate hydrogen in the absence of a catalyst at room temperature. Furthermore, as the existence of Mg(OH)$_2$, the hydrolysis product could be easily recycled. In comparison to ammonia borane (AB), there is no requirement for a catalyst to improve the hydrolysis rate, only an adjustment to the particle size or solution, which influence hydrolysis rate and hydrogen yield.

MgH$_2$-based materials have higher theoretical hydrogen content (15.2%) and Mg element is abundant in the earth's crust (2.4%). They are low cost, have a high hydrogen yield, and are gentle on the environment, making them a potential candidate for high-quality hydrogen generation material. The progress of reaction can be concluding as in Equation (1):

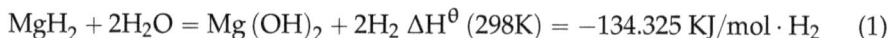

$$MgH_2 + 2H_2O = Mg\,(OH)_2 + 2H_2\;\Delta H^\theta\,(298K) = -134.325\;KJ/mol \cdot H_2 \quad (1)$$

The hydrolysis production of MgH$_2$-based hydride (Mg(OH)$_2$) is harmless to the environment and easily reused and recycled [54]. Fortunately, the hydrolysis reaction of MgH$_2$ occurred immediately when it had contact with water at room temperature. However, the hydrolysis reaction of MgH$_2$ is rapidly interrupted because of the formation of a passive magnesium hydroxide layer, which prevents the diffusion of the water to the particle inside, so that the hydrolysis reaction after the early high-speed reaction stage is quickly stagnated.

In order to improve the hydrolysis efficiency and the reaction rate, the main methods such as ball milling, alloying, changing hydrolysis solution composition, catalyst introduction, *etc.* were adopted. Research has shown that although adding acid to form Mg^{2+} ion is an effective method, it may pollute the environment, introduce an additional hazard to the equipment, and lower overall safety [55,56]. The addition of CaH$_2$ appears to be very promising, but this reagent is too reactive and needs an extended milling time of 10 h that is not propitious for industrial production in terms of feasibility and safety [57]. In addition, hydrolyses of high-energy ball-milled MgH$_2$ composites in aqueous NaCl solution [58–60], aqueous KCl solution [61], or in different alcoholic solutions [62] have been used to produce hydrogen, but the state-of-the-art hydrolysis performance is still far from commercial utilization. Even when Ni was added to MgH_2 in KCl solution, the hydrolysis reaction did not show any significant reactivity improvement [61]. Recently, ultrasonication has been shown to enhance hydrogen generation during the hydrolysis process of magnesium hydride, but it has the disadvantages of requiring extra ultrasonic equipment and the associated extra energy consumption [63]. Therefore, the key issue relating to MgH_2 hydrolysis remains the circumvention of the barrier of oxide/hydroxide layer formation during the hydrolysis process so as to realize complete hydrolysis [47,64]. In this paper, we review our recent research results about hydrolysis of MgH_2-based hydrides and confirm the effect of the ball-milling and catalyst-introducing methods to break the magnesium hydroxide passivation layer to enhance the hydrolysis rate and hydrogen generation yield. The hydrolysis rate and yield in the pure water of *in situ* synthetic MgH_2-based composites could be further adjusted by controlling the particle size via ball milling. The ammonium chloride solution can destroy the $Mg(OH)_2$ passivation layer and accelerate the hydrolysis process of MgH_2, which can produce 1711.2 mL/g hydrogen in 10 min at 298 K with a hydrolytic conversion rate of 99.5% in the 27.1% ammonium chloride solution. Moreover, the optimization of the $MgCl_2$ aqueous solution was also investigated. It not only eliminates the introduction of impure gas and byproduct, but also simplifies the regeneration process. Such a result opens a promising route for improving hydrolysis properties for commercial hydrogen production.

2. Experimental

2.1. Sample Preparation

The reagent of MgH_2 was purchased from Sigma-Aldrich Inc. The rare earth ingot was broken into small particles and filtered through a 400-mesh sieve, then hydrogenated at 573 K for 4 h with the hydrogen pressure of 4 MPa by using an AMC gas reaction controller (Advance Materials Corporation, Pittsburgh, PA, USA) to prepare REH_3 hydride. The MgH_2 was ball milled under hydrogen atmosphere for 3 h using QM-3SP4 planetary ball mill (Nanjing NanDa Instrument Plant, Nanjing, China) with a ball-to-powder mass ratio of 20:1 at rotational speed of 500 rpm. The MgH_2/REH_3 composite was obtained by ball milling LaH_3 and MgH_2 with different atomic ratios of 3:1, 5:1, 8.5:1 under hydrogen atmosphere for 4 h. The ball milling was performed on a QM-3SP4 planetary ball mill rotating at 500 rpm with the ball-to-powder mass ratio of 20:1. To prevent samples and raw materials from oxidation and/or hydroxide formation, all samples were stored and handled in an Ar filled glove box. The solution of $MgCl_2$ and NH_4Cl was prepared to be used as hydrolysis solution: 2.38 g $MgCl_2$ was dissolved into 500 mL deionized water to form a solution of 0.5 mol/L $MgCl_2$. 4.7 g, 9.3 g, 18.6 g and 37.2 g NH_4Cl were dissolved into 100 mL of deionized water to form solutions with the mass ratio of 4.5%, 8.5%, 15.7%, and 27.1%, respectively.

2.2. Hydrolysis Experiment

The hydrolysis reactions of the MgH_2 or MgH_2/LaH_3 composite were carried out in a 250 mL Pyrex glass reactor with three openings, one for water addition, one for hydrogen exhausting and one for inserting the thermometer. All the experiments were carried out at room temperature (298 K), without external heating. Figure 1 shows the schematic hydrolysis reaction equipment which was used to quantify the hydrogen production rate and yield. Samples of 0.1–0.2 g were added into the Pyrex glass, before 20 mL water or $MgCl_2$ solution were added to react with the alloys. Hydrogen production reaction started when the alloys contacted with pure water or solution. The generated hydrogen was exhausted through a Tygon tube, then passed through a Monteggia washing bottle filled with water at room temperature in order to condense the water vapor, and collected hydrogen by extracting water in a beaker which was put on an electronic scales to recorded the weight changes over time, in order to measure the quantity of hydrogen. The hydrogen generation rate and yield can be calculated from the reaction time and hydrogen volume, which was measured and analyzed by the computer. Each experiment test was repeated at least two times in order to confirm its reproducibility.

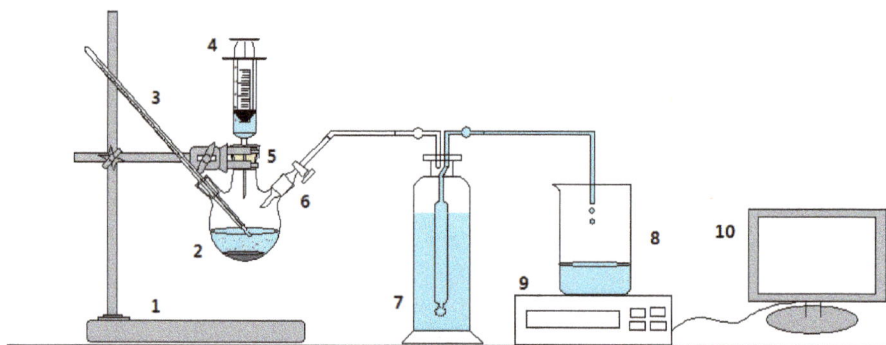

Figure 1. The schematic of equipment used to quantify hydrogen production rates and yields: (1) iron support; (2) Pyrex glass reactor; (3) thermometer; (4) injector; (5) double gum plug; (6) piston joint; (7) Monteggia washing bottle filled with water at room temperature; (8) beaker to collect the extracted water; (9) electronic scales; (10) computer.

2.3. Sample Characterization

The morphological characteristics of fabricated structures were studied by field emission scanning electron microscopy (FE-SEM, JEOL XL-30) (JEOL, Tokyo,

Japan) using secondary electrons with the acceleration voltage of 10 kV, and their structural and chemical characteristics were investigated using transmission electron microscopy (TEM, JEOL-2100 (JEOL, Tokyo, Japan) equipped with an energy dispersive spectroscopy (EDS) system with an operating voltage of 200 kV).

3. Results and Discussion

3.1. Enhancement of the Hydrolysis Properties of MgH_2-Based Hydrides by Formation of a Composite Structure

The hydrolysis reaction of MgH_2 is interrupted by the formation of a magnesium hydroxide passivation layer. To improve the hydrolysis performance of MgH_2-based materials, a composite structure of MgH_2 with a different hydride was synthesized through ball milling. By adjusting the atomic ratio, MgH_2/LaH_3 composites were designed to adjust the hydrolysis performance of the MgH_2-based composite. Figure 2 shows the hydrogen production curves by hydrolysis of the MgH_2/LaH_3 composites with the different atomic ratios of 3:1, 5:1, and 8.5:1 obtained after milling for 4 h. As shown in Figure 2, the overall hydrolysis rate of the composites improved; the hydrolysis rate was the same for all the composites during the first 5 min, and then increased with the LaH_3 content. The hydrolysis of LaH_3 can produce conductive ions on the surface of the composite, which is beneficial for the work of the micro galvanic cells and results in a synergetic effect leading to a higher hydrolysis rate [65]. The MgH_2/LaH_3 composites with the MgH_2/LaH_3 atomic ratio of 3:1 and 8.5:1 can generate 706.7 mL/g of H_2 in 40 min and 473.0 mL/g of H_2 in 60 min, respectively. The highest observed hydrolysis yield is equal to 1195 mL/g of H_2 in 80 min and belongs to the MgH_2/LaH_3 composite with the MgH_2/LaH_3 atomic ratio of 5:1. Notably, by decreasing the LaH_3 content, the hydrolysis synergetic effect was enhanced and consequently the hydrogen yield and rate increased. However, with the increase of the LaH_3 content, the theoretical hydrogen generation yield of the composites decreased because of the relatively low hydrogen yield of LaH_3.

The ball-milling method used to synthesize the composite structure of MgH_2 with other hydrides is relatively expensive, as LaH_3 and MgH_2 are obtained upon hydrogenation of Mg and La at relative high temperature. To decrease the costs of the MgH_2-based composites, an Mg-based alloy hydrogenated at room temperature with low pressure was designed. The Mg_3La and $Mg_3LaNi_{0.1}$ alloys prepared by induction melting with pure Mg and La could be hydrogenated *in situ* at room temperature and 3.8 MPa of hydrogen pressure to form the LaH_3/MgH_2 (hereafter referred to as H-Mg_3La) and $LaH_3/MgH_2/Ni$ (hereafter referred to as H-$Mg_3LaNi_{0.1}$) composites [66]. To further study the influence of the light rare earth (RE) elements on the hydrolysis performance of the composite formed *in situ*, hydrogenated Mg_3Mm (H-Mg_3Mm, Mm denotes the mischmetal), Mg_3La (H-Mg_3La), Mg_3Ce (H-Mg_3Ce),

211

Mg$_3$Pr (H-Mg$_3$Pr), and Mg$_3$Nd (H-Mg$_3$Nd) composites were also prepared by hydrogenating the Mg$_3$Mm, Mg$_3$La, Mg$_3$Ce, Mg$_3$Pr, and Mg$_3$Nd (H-Mg$_3$Nd) alloys at 298 K [67]. The hydrolysis performances of the H-Mg$_3$Mm, H-Mg$_3$La, H-Mg$_3$Ce, H-Mg$_3$Pr, and H-Mg$_3$Nd composites were investigated at 298 K, as shown in Figure 3. The H-Mg$_3$Mm, H-Mg$_3$La, H-Mg$_3$Ce, H-Mg$_3$Pr, and H-Mg$_3$Nd composites generated 1097 mL/g, 949 mL/g, 1025 mL/g, 905 mL/g, and 657 mL/g of hydrogen by hydrolysis, with total hydrogen yields of 9.79 wt.%, 8.47 wt.%, 9.15 wt.%, 8.08 wt.%, and 5.86 wt.%, respectively.

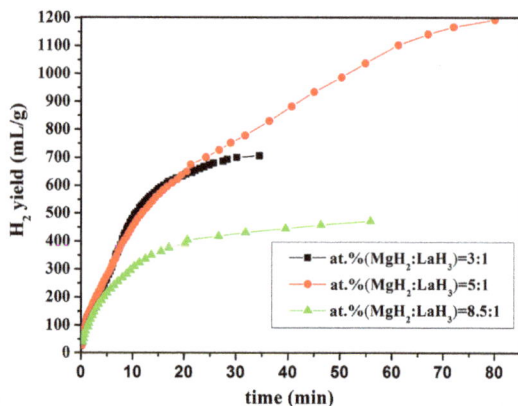

Figure 2. The hydrogen generation curves of ball-milled MgH$_2$ and LaH$_3$ with different atomic ratio.

Notably, H-Mg$_3$Mm exhibited the best hydrolysis performance (Figure 3, magenta line), showing the fastest hydrolysis rate and producing 695 mL/g of H$_2$ in 5 min, 784 mL/g of H$_2$ in 10 min, 828 mL/g of H$_2$ in 15 min, and 1097 mL/g of H$_2$ in 36 h. H-Mg$_3$La generated 463 mL/g of H$_2$ in 5 min and 653 mL/g of H$_2$ in 10 min (Figure 3, black line). H-Mg$_3$Ce produced 542 mL/g of H$_2$ in 5 min and 705 mL/g of H$_2$ in 10 min (Figure 3, red line). The blue line in Figure 3 indicates the hydrolysis of the H-Mg$_3$Pr sample; the conversion yields were equal to 545 mL/g of H$_2$ in 5 min and 699 mL/g of H$_2$ in 10 min. For the hydrolysis of the H-Mg$_3$Nd sample (Figure 3, green line), the hydrolysis yield was relatively low and the sample only generated 434 mL/g of H$_2$ in 5 min and 581 mL/g of H$_2$ in 10 min. All the H-Mg$_3$RE samples exhibited a faster hydrolysis rate than Mg or MgH$_2$, indicating that the rare-earth hydrides may significantly improve the hydrolysis performance of MgH$_2$. For commercial use, pure rare-earth metal La was replaced by the mischmetal to reduce the costs. An inexpensive and efficient hydrolysis material, namely, as-hydrogenated Mg$_3$Mm (abbreviated as H-Mg$_3$Mm, where Mm denotes the La-rich and Ce-rich mischmetal), was identified as an excellent hydrolysis material. Pure REH$_3$ can fully react with water. Conversely, the hydrolysis of MgH$_2$ can be rapidly interrupted by the formation of a magnesium hydroxide passivation layer onto the reactive materials. During the H-Mg$_3$RE hydrolysis, REH$_3$ could continually hydrolyze and produce a reaction tunnel for the further hydrolysis of MgH$_2$, leading to the break of the magnesium hydroxide passivation layer. Thus, REH$_3$ can accelerate the hydrolysis rate of MgH$_2$ and lead to the completion of the hydrolysis process.

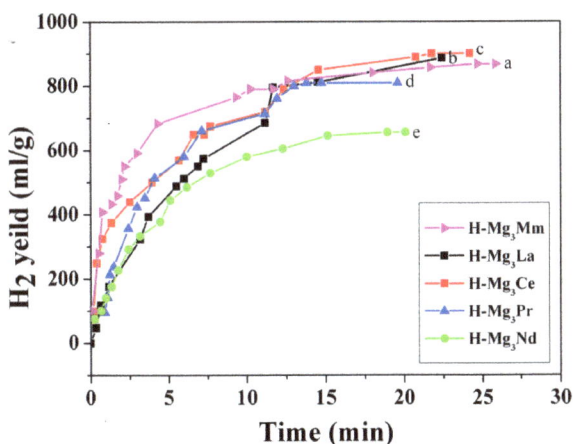

Figure 3. The hydrogen evolution curves for the hydrolysis of (a) H-Mg$_3$Mm; (b) H-Mg$_3$La; (c) H-Mg$_3$Ce; (d) H-Mg$_3$Pr and (e) H-Mg$_3$Nd alloys at 298 K [67]. Reproduced with permission from [67], copyright 2013 Elsevier.

Comparing the hydrogen evolution curves for the hydrolysis of the LaH$_3$/MgH$_2$ and LaH$_3$/MgH$_2$/Ni composites at room temperature (298 K), the hydrogen generation rate of LaH$_3$/MgH$_2$/Ni is higher than that of LaH$_3$/MgH$_2$ during the first 3 min [66]. The addition of Ni plays an active role in enhancing the reaction rate. However, to study the hydrolysis performance of *in situ* formed MgH$_2$/LaH$_3$ composites affected by the different atomic ratio of MgH$_2$ to LaH$_3$, MgH$_2$/LaH$_3$ composites hydrogenated from Mg$_3$La (MgH$_2$:LaH$_3$ = 3:1, hereafter referred to as H-Mg$_3$La) and La$_2$Mg$_{17}$ (MgH$_2$:LaH$_3$ = 17:2, hereafter referred to as H-La$_2$Mg$_{17}$) were synthesized [47]. The results show that the hydrolysis rate of H-Mg$_3$La was higher than that of H-La$_2$Mg$_{17}$ (918.4 mL/g and 851.7 mL/g for H-Mg$_3$La and H-La$_2$Mg$_{17}$, respectively, in 21 min), while the hydrogen yield decreased from 1224.5 mL/g to 983.7 mL/g in 4.4 h for H-La$_2$Mg$_{17}$ and H-Mg$_3$La, respectively. By increasing the content of LaH$_3$ from H-La$_2$Mg$_{17}$ to H-Mg$_3$La, the interface density between the LaH$_3$ and MgH$_2$ phases significantly increased, leading a change in the rate controlling steps from a one-dimensional diffusion process to a three-dimensional interface reaction process. As the hydrolytic rate depends on the content of LaH$_3$, the hydrolytic rate of MgH$_2$ is lower than that of the hydrides in the hydrogenated La-Mg system. As a result, the kinetic properties of the LaH$_3$/MgH$_2$ composites hydrogenated from Mg_3La and La_2Mg_{17} alloys are significantly superior to those of pure MgH_2.

To further examine the composite structure, we measured the microstructure by analyzing the TEM images. As shown in Figure 4a, the bright field image of H-Mg_3Ce demonstrates the presence of MgH_2 and $CeH_{2.73}$. Figure 4b shows the selected area electron diffraction (SAED) patterns, which was obtained by selecting the diffraction spots of a circular area. The MgH_2 (PDF card 012-0697) phase with a tetragonal structure and $CeH_{2.73}$ (PDF card 089-3694) phase with an FCC structure are shown in the SAED image. A very fine plate-like or lamellar mixture of $CeH_{2.73}$ (dark) and MgH_2 (white) is observed, as identified by the SAED patterns. In the bright field image in Figure 4a, the $CeH_{2.73}$ platelet has an average width of 20 nm and a planar length of about 150 nm, whereas the MgH_2 matrix is nanocrystalline with an average grain size of approximately 20 nm. The $CeH_{2.73}$ plate is embedded in the MgH_2 phases, uniformly. The hydrolysis of $CeH_{2.73}$ can produce conductive ions on the surface of the composite results in a synergetic effect leading to a higher hydrolysis rate.

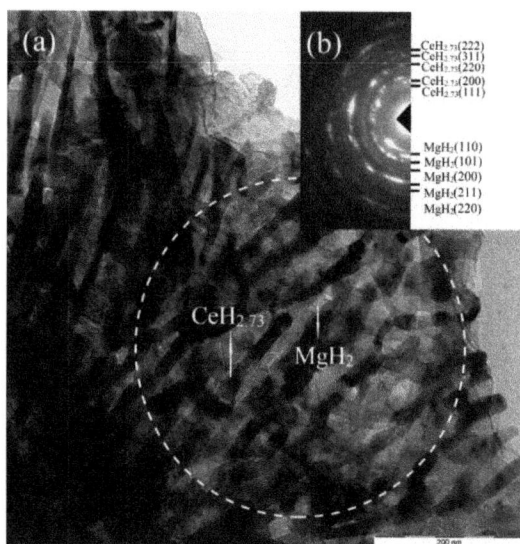

Figure 4. The bright field image of H-Mg_3Ce composite structure (**a**) and selected area diffraction patterns (**b**).

3.2. Acceleration of the Hydrolysis Reaction of MgH_2 by Controlling the Particle Size via Ball Milling

The composite structure formed via ball milling could significantly improve the hydrolysis properties. Consequently, the ball milling effect should also be

214

investigated. Figure 5 shows the hydrogen evolution curves by hydrolysis of the as-received (untreated) and 3 h ball-milled MgH_2 in pure water. As shown in Figure 5a, the hydrolysis rate of untreated MgH_2 in pure water is very low, and only 27.4 mL/g of H_2 could be collected in the Monteggia washing bottle after 20 min. In contrast, after 3 h ball milling, the hydrolysis rate of MgH_2 in pure water increased significantly, as shown in Figure 5b, and produced 394.6 mL/g of H_2 in 5 min. Unfortunately, the hydrolysis reaction is interrupted and there is no hydrogen produced after 5 min because of the formation of the magnesium hydroxide passivation layer. According to the hydrogen evolution curves of untreated MgH_2 and ball-milled MgH_2 in pure water by hydrolysis shown in Figure 5, the ball-milling method can effectively improve the reaction rate and yield of hydrogen generation by hydrolysis. This phenomenon could be attributed to the fact that the ball milling process refines the particle and grain size, and increases the specific surface area. In addition, the introduction of structural defects, phase change, and nanocrystalline structures is also beneficial for the MgH_2 hydrolysis. Although ball milling could improve the hydrolysis properties of MgH_2-based materials, the hydrolysis reaction of MgH_2 was still interrupted because of the formation of a magnesium hydroxide passivation layer.

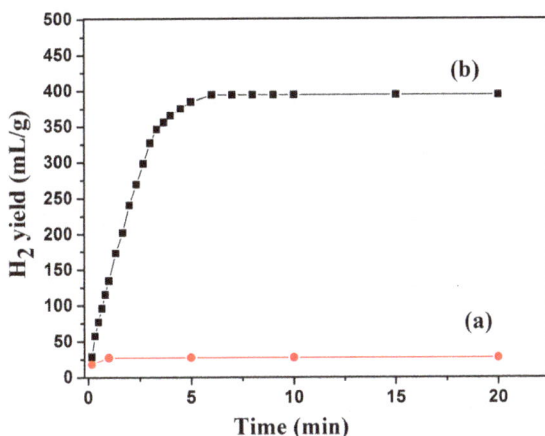

Figure 5. The hydrogen evolution curves by hydrolysis of as-received MgH_2 (a) and 3 h ball-milled MgH_2 (b) in pure water.

As a consequence, the ball-milling method has only a limited effect on the hydrolysis properties for pure MgH_2 and the MgH_2 particle size is difficult to control. To understand the effects of the particle size and ball-milling effect on the hydrolysis properties of the MgH_2-based hydrides, H-Mg_3La with different particle sizes but fixed grain sizes of LaH_3 (~17 nm) and MgH_2 (~33 nm) was prepared *in situ* by hydrogenating Mg_3La and controlling the hydrogenation conditions [68].

215

Figure 6 shows the hydrogen evolution curves of H-Mg$_3$La with different particle sizes. Clearly, the hydrolysis rate and hydrogen yield were significantly affected by the particle size. The hydrogenated Mg$_3$La with the smallest particle size (<12 μm) exhibited a higher hydrolysis yield of 863 mL/g (7.70 wt.%) of H$_2$. The final hydrogen hydrolysis yield for the samples with particle sizes of >38 μm, 38–23 μm, and 23–18 μm were 1.67 wt.%, 5.06 wt.%, and 6.26 wt.%, respectively. The smaller the particle size, the higher the observed hydrolysis rate. The above results reveal that ball milling is an effective and simple method to improve the hydrolysis properties of MgH$_2$-based materials.

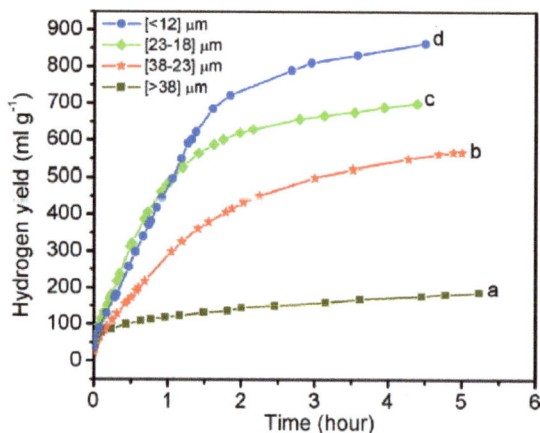

Figure 6. The hydrogen evolution curves of H-Mg$_3$La with different particle sizes [68]. Reproduced with permission from [68], copyright 2014 Elsevier.

Analyzing the hydrolysis curves and the hydrolysis products, we found that, as the particle size decreased, the hydrolysis rate accelerated and the hydrogen yield increased, promoting the hydrolysis of H-Mg$_3$La. Figure 7a,b shows the SEM images of H-Mg$_3$La with particle sizes >38 μm and <12 μm. The H-Mg$_3$La sample with particle sizes of <12 μm exhibited larger surface areas and more defects than the sample with particle size >38 μm. As the particle size decreased, the surface area and defects of the sample increased, providing larger surfaces and available sites for the water. As more contact area with water was available, the reaction proceeded more violently, preventing the magnesium hydroxide layer from covering the sample surface. The hydrolysis reaction can spontaneously continue until fully completed. Therefore, reducing the particle size is an effective and simple method to improve the hydrolysis properties of the MgH$_2$-based materials.

Figure 7. SEM images of H-Mg₃La with particle sizes of (**a**) [>38] μm and (**b**) [<12] μm [68]. Reproduced with permission from [68], copyright 2014 Elsevier.

3.3. Enhancement of the Hydrolysis Properties by Optimization of the Solution Composition

The hydrolysis reaction of MgH_2 can be accelerated both by controlling the particles size via ball milling and by forming composite structures. It would be significant to develop a simple method to get the high hydrolysis reaction rate of MgH_2. To break the $Mg(OH)_2$ passivation layer and accelerate the hydrolysis process, acids were used as effective hydrolytic media to enhance the hydrolysis properties. However, acids may cause corrosion damage to the reaction equipment and increase the cost of the process. The hydrolysis of MgH_2 at different concentrations of NH_4Cl aqueous solution was investigated in this work. The hydrolysis kinetics of MgH_2 at different concentrations of ammonium chloride solution is shown in Figure 8. The hydrolytic rate and hydrogen yield increased with the increase of the NH_4Cl concentration. The hydrogen evolution curves also show that the hydrogen yield kinetics of MgH_2 improved with the increase of the ammonium chloride concentration. This result confirms that the hydrolysis speed can be controlled by adjusting the solution concentration. The hydrolysis of MgH_2 produced 1711.2 mL/g of H_2 in 10 min at 298 K in the 27.1% ammonium chloride solution, and the hydrolytic conversion rate reached the value of 99.5%.

The introduction of a high concentration NH_4Cl solution leads to a high hydrolysis rate, but it also causes some issues, such as resource waste and equipment corrosion damage. Besides, the hydrolysis reaction in NH_4Cl solution introduces impurities and byproducts, which make the regeneration process and hydrolysis gas more complex. Thus, adding a stable catalyst to improve the hydrolysis rate may be a simple way to solve the key issue of hydrogen generation. The 0.05 mol/L $MgCl_2$ aqueous solution has been optimized for the hydrolysis of MgH_2. The hydrogen evolution curves by hydrolysis of the ball-milled MgH_2 in pure water and in a 0.05 mol/L $MgCl_2$ solution are shown in Figure 9. The hydrolysis of ball-milled

MgH$_2$ produced 394.6 mL/g of H$_2$ in 5 min at 298 K, either in the MgCl$_2$ solution or in pure water, indicating that the hydrolysis rate and yield are similar in the initial stage. The hydrolysis of ball-milled MgH$_2$ produced 1137.4 mL/g in 90 min at 298 K in the 0.05 mol/L MgCl$_2$ solution, whereas the hydrolysis in pure water was interrupted. This result suggests that the presence of MgCl$_2$ played a key role for the hydrolysis of MgH$_2$. The Mg^{2+} and Cl$^-$ ions can actively affect the formation of a relatively loose magnesium hydroxide layer to further accelerate the hydrolysis of MgH$_2$ in the MgCl$_2$ solution. Furthermore, the byproduct is the MgO and MgCl$_2$ composite and no separation process is necessary for the regeneration. A simple method involves the reaction of the MgO and MgCl$_2$ composite with chlorine at 900 °C to produce MgCl$_2$; Mg can then be industrially produced by electrolysis of fused MgCl$_2$. This strategy opens a novel and effective route to modulate the hydrogen yield and hydrogen generation rate in the hydrolysis process.

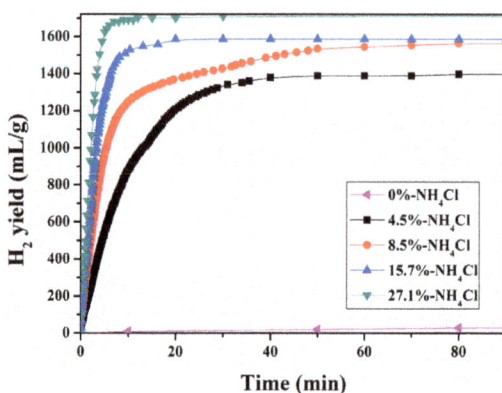

Figure 8. The hydrolysis kinetics of MgH$_2$ in different concentrations of ammonium chloride solution.

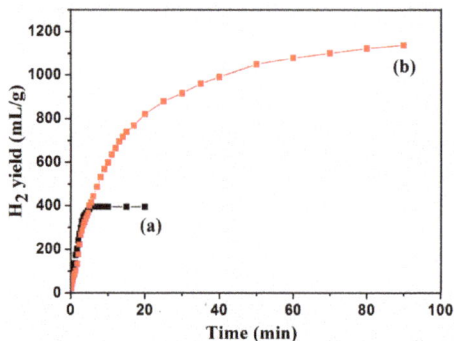

Figure 9. The hydrogen evolution curves by hydrolysis of 3 h ball-milled MgH$_2$ in pure water (a) and in 0.05M MgCl$_2$ solution (b).

4. Conclusions

MgH_2/LaH_3 composites were prepared by ball milling of MgH_2 and LaH_3 and showed a synergetic effect during the hydrolysis process. A low-cost method for the synthesis of MgH_2/REH_3-based composites was developed by *in situ* hydrogenation of Mg_3RE-based alloys at room temperature, and showed an enhanced hydrolysis rate. The H-Mg_3Mm, H-Mg_3La, H-Mg_3Ce, H-Mg_3Pr, and H-Mg_3Nd composites generated 1097 mL/g, 949 mL/g, 1025 mL/g, 905 mL/g, and 657 mL/g of hydrogen by hydrolysis, with the total hydrogen yields of 9.79 wt.%, 8.47 wt.%, 9.15 wt.%, 8.08 wt.%, and 5.86 wt.%, respectively. The hydrolysis properties of the MgH_2/REH_3 composites could be further adjusted by controlling the particle size via ball milling. The ammonium chloride solution, which can destroy the $Mg(OH)_2$ passivation layer and accelerate the hydrolysis process, with the MgH_2 produced 1,711.2 mL/g of H_2 in 10 min at 298 K with a hydrolytic conversion rate of 99.5%. To eliminate the impure gas and simplify the regeneration process, the 0.05 mol/L $MgCl_2$ aqueous solution was optimized for the hydrolysis of MgH_2 and the hydrogen yield was 1137.4 mL/g in 90 min at 298 K. In conclusion, MgH_2-based hydride composites, ball milling, and tuning of the solution composition are all effective methods to enhance the hydrolysis reaction rate and hydrogen generation yield of MgH_2.

Acknowledgments: This work was supported by the National Natural Science Foundation of China Projects (Nos. 51431001, U1201241 and 51271078), by Guangdong Natural Science Foundation (2014A030311004) and by International Science and Technology Cooperation Program of China (2015DFA51750). The Project Supported by Guangdong Province Universities and Colleges Pearl River Scholar Funded Scheme (2014) is also acknowledged.

Author Contributions: Miaolian Ma, Minghong Huang, and Ruoming Duan did the experiment and collected the data; Liuzhang Ouyang and Miaolian Ma write the manuscript; Min Zhu, Hui Wang and Lixian Sun discussed the results and modified the draft.

Conflicts of Interest: The authors declare no conflict of interest.

References

1. Kikkinides, E.S. Design and optimization of hydrogen storage units using advanced solid materials: General mathematical framework and recent developments. *Comput. Chem. Eng.* **2011**, *35*, 1923–1936.
2. Bauen, A. Future energy sources and systems—Acting on climate change and energy security. *J. Power Sources* **2006**, *157*, 893–901.
3. Zhu, M.; Peng, C.H.; Ouyang, L.Z.; Tong, Y.Q. The effect of nanocrystalline formation on the hydrogen storage properties of AB_3-base Ml–Mg–Ni multi-phase alloys. *J. Alloys Compd.* **2006**, *426*, 316–321.
4. Shi, Q.; Hu, R.; Ouyang, L.; Zeng, M.; Zhu, M. High-capacity LiV_3O_8 thin-film cathode with a mixed amorphous-nanocrystalline microstructure prepared by RF magnetron sputtering. *Electrochem. Commun.* **2009**, *11*, 2169–2172.

5. Jain, I.P.; Lal, C.; Jain, A. Hydrogen storage in Mg: A most promising material. *Int. J. Hydrog. Energy* **2010**, *35*, 5133–5144.

6. Muir, S.S.; Yao, X. Progress in sodium borohydride as a hydrogen storage material: Development of hydrolysis catalysts and reaction systems. *Int. J. Hydrog. Energy* **2011**, *36*, 5983–5997.

7. Yumurtaci, Z.; Bilgen, E. Hydrogen production from excess power in small hydroelectric installations. *Int. J. Hydrog. Energy* **2004**, *29*, 687–693.

8. Gratzel, M. Perspectives for dye-sensitized nanocrystalline solar cells. *Prog. Photovolt.* **2000**, *8*, 171–185.

9. Turner, J.; Sverdrup, G.; Mann, M.K.; Maness, P.C.; Kroposki, B.; Ghirardi, M.; Evans, R.J.; Blake, D. Renewable hydrogen production. *Int. J. Energ. Res.* **2008**, *32*, 379–407.

10. Funk, J.E. Thermochemical hydrogen production: past and present. *Int. J. Hydrog. Energy* **2001**, *26*, 185–190.

11. Urbaniec, K.; Friedl, A.; Huisingh, D.; Claassen, P. Hydrogen for a sustainable global economy. *J. Clean. Prod.* **2010**, *18*, S1–S3.

12. Midilli, A.; Ay, M.; Dincer, I.; Rosen, M.A. On hydrogen and hydrogen energy strategies: I: Current status and needs. *Renew. Sustain. Energy Rev.* **2005**, *9*, 255–271.

13. Demirci, U.B.; Miele, P. Overview of the relative greenness of the main hydrogen production processes. *J. Clean. Prod.* **2013**, *52*, 1–10.

14. Ersoz, A.; Olgun, H.; Ozdogan, S. Reforming options for hydrogen production from fossil fuels for PEM fuel cells. *J. Power Sources* **2006**, *154*, 67–73.

15. Das, D.; Veziroglu, T.N. Advances in biological hydrogen production processes. *Int. J. Hydrog. Energy* **2008**, *33*, 6046–6057.

16. Balachandran, U.; Dorris, S.E.; Bose, A.C.; Stiegel, G.J.; Lee, T.H. Method of generating hydrogen by catalytic decomposition of water. U.S. Patent US6468499 B1, 22 November 2002.

17. Hiraki, T.; Hiroi, S.; Akashi, T.; Okinaka, N.; Akiyama, T. Chemical equilibrium analysis for hydrolysis of magnesium hydride to generate hydrogen. *Int. J. Hydrog. Energy* **2012**, *37*, 12114–12119.

18. Ni, M.; Leung, D.Y.C.; Leung, M.K.H.; Sumathy, K. An overview of hydrogen production from biomass. *Fuel Process. Technol.* **2006**, *87*, 461–472.

19. Holladay, J.D.; Hu, J.; King, D.L.; Wang, Y. An overview of hydrogen production technologies. *Catal. Today* **2009**, *139*, 244–260.

20. Manish, S.; Banerjee, R. Comparison of biohydrogen production processes. *Int. J. Hydrog. Energy* **2008**, *33*, 279–286.

21. Antal, T.K.; Krendeleva, T.E.; Rubin, A.B. Acclimation of green algae to sulfur deficiency: underlying mechanisms and application for hydrogen production. *Appl. Microbiol. Biotechnol.* **2011**, *89*, 3–15.

22. Basak, N.; Das, D. The Prospect of Purple Non-Sulfur (PNS) photosynthetic bacteria for hydrogen production: The present state of the art. *World J. Microbiol. Biotechnol.* **2007**, *23*, 31–42.

23. Huesemann, M.H.; Hausmann, T.S.; Carter, B.M.; Gerschler, J.J.; Benemann, J.R. Hydrogen generation through indirect biophotolysis in batch cultures of the nonheterocystous nitrogen-fixing cyanobacterium plectonema boryanum. *Appl. Biochem. Biotechnol.* **2010**, *162*, 208–220.

24. Xie, G.J.; Liu, B.F.; Ding, J.; Ren, H.Y.; Xing, D.F.; Ren, N.Q. Hydrogen production by photo-fermentative bacteria immobilized on fluidized bio-carrier. *Fuel Energy Abstr.* **2011**, *36*, 13991–13996.

25. Kovács, K.L.; Maróti, G.; Rákhely, G. A novel approach for biohydrogen production. *Int. J. Hydrog. Energy* **2006**, *31*, 1460–1468.

26. Laurinavichene, T.; Kosourov, S.; Ghirardi, M.; Seibert, M.; Tsygankov, A. Prolongation of H_2 photoproduction by immobilized, sulfur-limited Chlamydomonas reinhardtii cultures. *J. Biotechnol.* **2008**, *134*, 275–277.

27. Hydrogen, Fuel Cells and Infrastructure Technologies Program Multi-Year Research, Development and Demonstration Plan. Available online: http://www.nrel.gov/docs/fy08osti/39146.pdf (accessed on 5 May 2015).

28. Levin, D.B.; Pitt, L.; Love, M. Biohydrogen production: Prospects and limitations to practical application. *Int. J. Hydrog. Energy* **2004**, *29*, 173–185.

29. Bhandari, R.; Trudewind, C.A.; Zapp, P. Life cycle assessment of hydrogen production via electrolysis—A review. *J. Clean. Prod.* **2014**, *85*, 151–163.

30. Janssen, H.; Bringmann, J.C.; Emonts, B.; Schroeder, V. Safety-related studies on hydrogen production in high-pressure electrolysers. *Int. J. Hydrog. Energy* **2004**, *29*, 759–770.

31. Koroneos, C.; Dompros, A.; Roumbas, G.; Moussiopoulos, N. Life cycle assessment of hydrogen fuel production processes. *Int. J. Hydrog. Energy* **2004**, *29*, 1443–1450.

32. Fujishima, A.; Honda, K. Electrochemical photolysis of water at a semiconductor electrode. *Nature* **1972**, *238*, 37–38.

33. Liu, G.; Shi, J.; Zhang, F.; Chen, Z.; Han, J.; Ding, C.; Chen, S.; Wang, Z.; Han, H.; Li, C. A tantalum nitride photoanode modified with a hole-storage layer for highly stable solar water splitting. *Angew. Chem.* **2014**, *53*, 7295–7299.

34. Aroutiounian, V.M.; Arakelyan, V.M.; Shahnazaryan, G.E. Metal oxide photoelectrodes for hydrogen generation using solar radiation-driven water splitting. *Sol. Energy* **2005**, *78*, 581–592.

35. Wang, H.Z.; Leung, D.Y.C.; Leung, M.K.H.; Ni, M. A review on hydrogen production using aluminum and aluminum alloys. *Renew. Sustain. Energ. Rev.* **2009**, *13*, 845–853.

36. Aleksandrov, Y.A.; Tsyganova, E.I.; Pisarev, A.L. Reaction of Aluminum with Dilute Aqueous NaOH Solutions. *Russ. J. Gen. Chem.* **2003**, *73*, 689–694.

37. Soler, L.; Candela, A.M.; Macanás, J.; Muñoz, M.; Casado, J. *In situ* generation of hydrogen from water by aluminum corrosion in solutions of sodium aluminate. *J. Power Sources* **2009**, *192*, 21–26.

38. Soler, L.; Candela, A.M.; Macanás, J.; Muñoz, M.; Casado, J. Hydrogen generation from water and aluminum promoted by sodium stannate. *Int. J. Hydrog. Energy* **2010**, *35*, 1038–1048.

39. Alinejad, B.; Mahmoodi, K. A novel method for generating hydrogen by hydrolysis of highly activated aluminum nanoparticles in pure water. *Int. J. Hydrog. Energy* **2009**, *34*, 7934–7938.

40. Deng, Z.Y.; Liu, Y.F.; Tanaka, Y.; Ye, J.; Sakka, Y. Modification of Al Particle Surfaces by γ-Al_2O_3 and Its Effect on the Corrosion Behavior of Al. *J. Am. Ceram. Soc.* **2005**, *88*, 977–979.

41. Huang, T.; Gao, Q.; Liu, D.; Xu, S.; Guo, C.; Zou, J.; Wei, C. Preparation of Al-Ga-In-Sn-Bi quinary alloy and its hydrogen production via water splitting. *Int. J. Hydrog. Energy* **2015**, *40*, 2354–2362.

42. Kong, V.C.Y.; Foulkes, F.R.; Kirk, D.W.; Hinatsu, J.T. Development of hydrogen storage for fuel cellgenerators. I: Hydrogen generation using hydrolysishydrides. *Int. J. Hydrog. Energy* **1999**, *24*, 665–675.

43. Beattie, S.D.; Langmi, H.W.; McGrady, G.S. *In situ* thermal desorption of H_2 from $LiNH_2$–2LiH monitored by environmental SEM. *Int. J. Hydrog. Energy* **2009**, *34*, 376–379.

44. Ward, C.A.; Stanga, D.; Pataki, L.; Venter, R.D. Design for the cold start-up of a man-portable fuel cell and hydrogen storage system. *J. Power Sources* **1993**, *41*, 335–352.

45. Grosjean, M.H.; Roué, L. Hydrolysis of Mg-salt and MgH_2-salt mixtures prepared by ball milling for hydrogen production. *J. Alloys Compd.* **2006**, *416*, 296–302.

46. Lukashev, R.V.; Yakovleva, N.A.; Klyamkin, S.N.; Tarasov, B.P. Effect of mechanical activation on the reaction of magnesium hydride with water. *Russ. J. Inorg. Chem.* **2008**, *53*, 343–349.

47. Ouyang, L.Z.; Xu, Y.J.; Dong, H.W.; Sun, L.X.; Zhu, M. Production of hydrogen via hydrolysis of hydrides in Mg–La system. *Int. J. Hydrog. Energy* **2009**, *34*, 9671–9676.

48. Zhao, Z.; Zhu, Y.; Li, L. Efficient catalysis by $MgCl_2$ in hydrogen generation via hydrolysis of Mg-based hydride prepared by hydriding combustion synthesis. *Chem. Commun.* **2012**, *48*, 5509–5511.

49. Kojima, Y.; Haga, T. Recycling process of sodium metaborate to sodium borohydride. *Int. J. Hydrog. Energy* **2003**, *28*, 989–993.

50. Demirci, U.B.; Akdim, O.; Miele, P. Ten-year efforts and a no-go recommendation for sodium borohydride for on-board automotive hydrogen storage. *Int. J. Hydrog. Energy* **2009**, *34*, 2638–2645.

51. Stearns, J.E.; Matthews, M.A.; Reger, D.L.; Collins, J.E. The thermal characterization of novel complex hydrides. *Int. J. Hydrog. Energy* **1998**, *23*, 469–474.

52. Aiello, R.; Sharp, J.H.; Matthews, M.A. Production of hydrogen from chemical hydrides via hydrolysis with steam. *Int. J. Hydrog. Energy* **1999**, *24*, 1123–1130.

53. Kojima, Y. Hydrogen generation by hydrolysis reaction of lithium borohydride. *Int. J. Hydrog. Energy* **2004**, *29*, 1213–1217.

54. Kato, Y.; Sasaki, Y.; Yoshizawa, Y. Magnesium oxide/water chemical heat pump to enhance energy utilization of a cogeneration system. *Energy* **2005**, *30*, 2144–2155.

55. Uan, J.Y.; Yu, S.H.; Lin, M.C.; Chen, L.F.; Lin, H.I. Evolution of hydrogen from magnesium alloy scraps in citric acid-added seawater without catalyst. *Int. J. Hydrog. Energy* **2009**, *34*, 6137–6142.

56. Huot, J.; Liang, G.; Schulz, R. Magnesium-based nanocomposites chemical hydrides. *J. Alloys Compd.* **2003**, *353*, L12–L15.

57. Cho, C.Y.; Wang, K.H.; Uan, J.Y. Evaluation of a new hydrogen generating system: Ni-Rich magnesium alloy catalyzed by platinum wire in sodium chloride solution. *Mater. Trans.* **2005**, *46*, 2704–2708.

58. Uan, J.Y.; Cho, C.Y.; Liu, K.T. Generation of hydrogen from magnesium alloy scraps catalyzed by platinum-coated titanium net in NaCl aqueous solution. *Int. J. Hydrog. Energy* **2007**, *32*, 2337–2343.

59. Uan, J.Y.; Lin, M.C.; Cho, C.Y.; Liu, K.T.; Lin, H.I. Producing hydrogen in an aqueous NaCl solution by the hydrolysis of metallic couples of low-grade magnesium scrap and noble metal net. *Int. J. Hydrog. Energy* **2009**, *34*, 1677–1687.

60. Grosjean, M.H.; Zidoune, M.; Huot, J.Y.; Roué, L. Hydrogen generation via alcoholysis reaction using ball-milled Mg-based materials. *Int. J. Hydrog. Energy* **2006**, *31*, 1159–1163.

61. Grosjean, M.H.; Zidoune, M.; Roué, L.; Huot, J.Y. Hydrogen production via hydrolysis reaction from ball-milled Mg-based materials. *Int. J. Hydrog. Energy* **2006**, *31*, 109–119.

62. Hiroi, S.; Hosokai, S.; Akiyama, T. Ultrasonic irradiation on hydrolysis of magnesium hydride to enhance hydrogen generation. *Int. J. Hydrog. Energy* **2011**, *36*, 1442–1447.

63. Tessier, J.P.; Palau, P.; Huot, J.; Schulz, R.; Guay, D. Hydrogen production and crystal structure of ball-milled MgH_2–Ca and MgH_2–CaH_2 mixtures. *J. Alloys Compd.* **2004**, *376*, 180–185.

64. Izumi, F.; Ikeda, T. A rietveld-analysis programm RIETAN-98 and its applications to zeolites. *Mater. Sci. Forum.* **2000**, *321–324*, 198–205.

65. Fan, M.Q.; Xu, F.; Sun, L.X.; Zhao, J.N.; Jiang, T.; Li, W.X. Hydrolysis of ball milling Al–Bi–hydride and Al–Bi–salt mixture for hydrogen generation. *J. Alloys Compd.* **2008**, *460*, 125–129.

66. Ouyang, L.Z.; Wen, Y.J.; Xu, Y.J.; Yang, X.S.; Sun, L.X.; Zhu, M. The effect of Ni and Al addition on hydrogen generation of Mg_3La hydrides via hydrolysis. *Int. J. Hydrog. Energy* **2010**, *35*, 8161–8165.

67. Ouyang, L.Z.; Huang, J.M.; Wang, H.; Wen, Y.J.; Zhang, Q.A.; Sun, D.L.; Zhu, M. Excellent hydrolysis performances of Mg_3RE hydrides. *Int. J. Hydrog. Energy* **2013**, *38*, 2973–2978.

68. Huang, J.M.; Duan, R.M.; Ouyang, L.Z.; Wen, Y.J.; Wang, H.; Zhu, M. The effect of particle size on hydrolysis properties of Mg_3La hydrides. *Int. J. Hydrog. Energy* **2014**, *39*, 13564–13568.

Effect of Magnesium Fluoride on Hydrogenation Properties of Magnesium Hydride

Pragya Jain, Viney Dixit, Ankur Jain, Onkar N. Srivastava and Jacques Huot

Abstract: A cost effective catalyst is of great importance for consideration of MgH_2 as potential hydrogen storage material. In this regard, we investigated the catalytic role of alkaline metal fluoride on the hydrogen storage behavior of MgH_2. Samples were synthesized by admixing 5 mol % MgF_2 into MgH_2 powder using planetary ball mill. Hydrogenation measurements made at 335 °C showed that in comparison to only 70% absorption by pure MgH_2, catalyzed material absorbed 92% of theoretical capacity in less than 20 min and desorbed completely in almost the same time. Sorption studies done at lower temperatures revealed that complete absorption at temperature as low as 145 °C is possible. This is due to uniform distribution of MgF_2 nano particles within the MgH_2 powder. X-ray diffraction patterns also showed the presence of stable MgF_2 phase that does not decompose upon hydrogen absorption-desorption. Cyclic measurements done at 310 °C showed negligible loss in the overall storage capacity with cycling. These results reveal that the presence of the chemically inert and stable MgF_2 phase is responsible for good reversible characteristic and improved kinetics.

Reprinted from *Energies*. Cite as: Jain, P.; Dixit, V.; Jain, A.; Srivastava, O.N.; Huot, J. Effect of Magnesium Fluoride on Hydrogenation Properties of Magnesium Hydride. *Energies* **2015**, *8*, 12546–12556.

1. Introduction

Magnesium hydride is a potential candidate for hydrogen storage because of its high gravimetric and volumetric capacities. Pure magnesium's low environmental impact and abundant availability makes it very attractive for hydrogen storage application. However, high working temperature and slow kinetics limit its potential as hydrogen storage material for practical applications. Therefore, research is required to circumvent these difficulties and make MgH_2 a viable hydrogen storage material. Nano-structuring of MgH_2 is one of the most adopted methods to improve the hydrogenation performance [1]. However, this method has a limitation to achieve the nanocrystalline size (<5 nm) required for destabilization of MgH_2 [2].

Further improvements in sorption behaviour have been achieved by adding a wide variety of pure transition metals [3,4], their oxides [5] and halides [6–8]. The remarkable catalytic effect of transition metal oxide, Nb_2O_5 has been well reported.

However, during cycling at elevated temperatures, reduction of Nb_2O_5 occurs with augmentation of MgO content [5]. Later, it was found that some transition metal halides, such as FeF_3, $CrCl_3$, NiF_2, $NbCl_5$ and $TiCl_3$ possess better catalytic activity than pure metals or their oxides [6–8]. In the case of halides, Malka *et al.* [9] showed that fluorides are better catalysts than chlorides for MgH_2. Addition of transition metal fluorides during milling helps to lower the hydrogen release temperature and increase the rate of hydrogen uptake by MgH_2. It has been shown by different groups [7–9] that during milling of MgH_2 with transition metal fluorides, the formed MgF_2 phase replaces the original oxide layer and provides a reactive and protective fluorinated surface for hydrogen uptake. This compound possesses high affinity with hydrogen because of the F-anion, which weakens the Mg-H bonding and improves the sorption properties [9].

However, not much work has been done on direct use of MgF_2 as an additive for MgH_2. Ivanov *et al.* [10] reported that addition of 5 wt % MgF_2 to pure Mg during milling leads to 5 wt % hydrogen absorption in over 20 h but has an insignificant effect on dehydrogenation kinetics of MgH_2. Loss in absorption capacity from the second cycle onwards was also observed. Recently, Ma *et al.* [11] investigated the catalytic effects of MgF_2 and TiH_2 to understand the kinetic improvements obtained when MgH_2 was ball milled with 4 mol % TiF_3. They reported that sole addition of 6 mol % MgF_2 has negligible catalytic effect on MgH_2 at an operating temperature of 150 °C.

The limited and inconsistent results attained on catalytic effect of MgF_2 on MgH_2 shows that more work needs to be done to understand this system both from hydrogenation and material perspective. The present work is aimed to investigate the microstructural, morphological and hydrogenation behaviour of MgH_2 when MgF_2 is used as additive.

2. Results

2.1. Comparison of Undoped and 5 mol % MgF₂ Doped MgH₂ at 335 °C

The X-ray diffraction (XRD) patterns of MgH_2 without and with 5 mol % MgF_2 prepared by 1 h ball milling are shown in Figure 1. For the undoped sample, the diffraction pattern peaks are associated with main phase of β-MgH_2 and some unreacted Mg. There is no evidence of the metastable γ-MgH_2 phase. This is due to the short milling time and low milling intensity. A broad peak centered at 43° is attributed to MgO. The crystallite size of β-MgH_2 is evaluated from Rietveld refinement to be 23.4 ± 0.3 nm. Milling with MgF_2 additive was even more effective for reduction of crystallite size of MgH_2 that was evaluated as 10.9 ± 0.3 nm. During milling there is physical interaction between the different species during the repeating collisions. Therefore, the MgF_2 has also some mechanical effect on

MgH$_2$. Unfortunately, the alloying of brittle-brittle system is poorly understood [12]. However, from the present experiment it seems that addition of small amount of MgF$_2$ improves crystallite size reduction but the exact mechanism is still unclear.

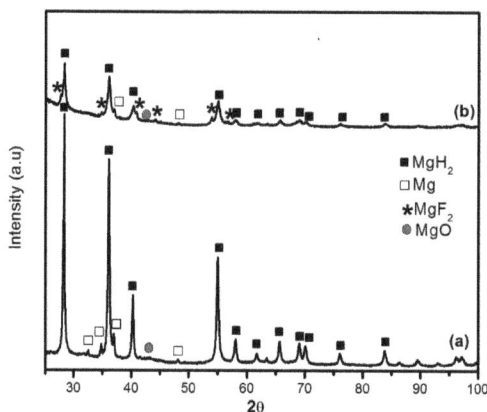

Figure 1. XRD patterns of ball milled samples milled 1 h: (a) pure MgH$_2$ and (b) MgH$_2$ + 5 mol % MgF$_2$.

In practical applications, desorption will be performed under a pressure of at least 100 kPa of hydrogen. However, in order to study the behaviour of MgH$_2$–MgF$_2$ system, we decided to fully dehydride the samples after ball milling. Therefore, after milling the samples were completely desorbed at 335 °C under dynamic vacuum before investigating their hydrogenation properties. Representative hydrogenation and dehydrogenation characteristics are shown in Figure 2.

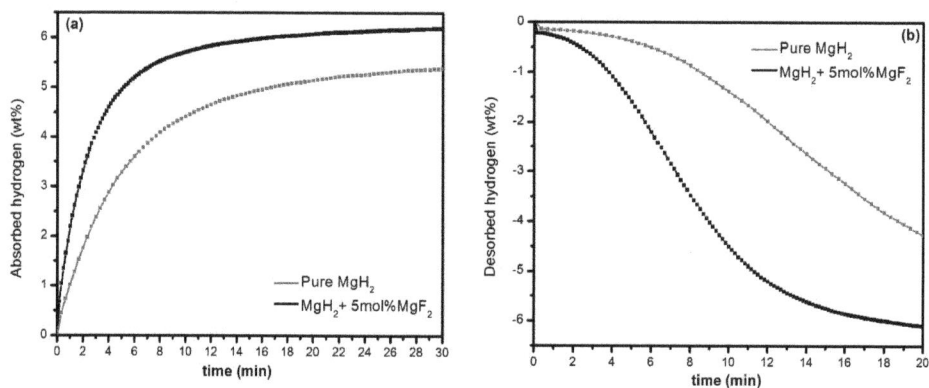

Figure 2. Hydrogen sorption kinetics at 335 °C of 1 h milled MgH$_2$ without and with 5 mol % MgF$_2$. (a) First absorption under 1000 kPa H$_2$; (b) desorption under 100 kPa H$_2$.

It is observed that at 335 °C under 1000 kPa H_2 pressure MgH_2 + 5 mol % MgF_2 system absorbs 6.2 wt % hydrogen in 30 min in comparison to only 5.3 wt % absorption by pure MgH_2. This shows a large improvement in absorption capacity is achieved, yielding 92% of the theoretical capacity in comparison to 70% for the pure MgH_2. In addition, significant improvement in desorption kinetics is achieved with complete desorption of the hydride phase in less than 20 min in presence of MgF_2. Thus, the beneficial effect of MgF_2 is clearly evident on the hydriding/dehydriding aspect of MgH_2.

Figure 3 shows the diffraction patterns of the doped sample in its desorbed and reabsorbed states. The desorbed pattern shows a small amount of un-desorbed MgH_2. The interesting fact is that MgF_2 is still present in the sample. This could be expected because it is known that for MgH_2-transition metal (TM) fluoride systems, milling or dehydrogenation induces the formation of MgF_2 and TM hydride [4,11]. Thus, MgF_2 is a stable compound and does not react to form MgH_2.

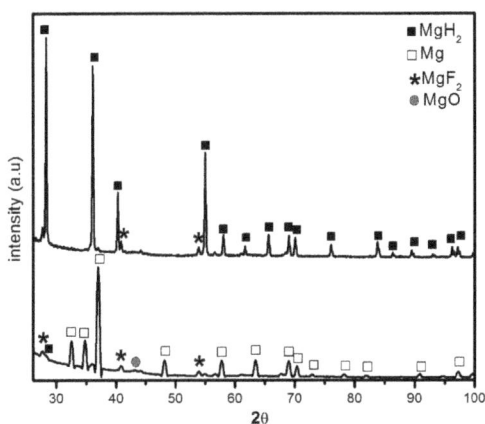

Figure 3. XRD patterns of MgH_2 + 5 mol % MgF_2 (**a**) after desorption at 335 °C under 100 kPa H_2 and (**b**) after re-hydrogenation at 335 °C under 1000 kPa H_2.

This is confirmed by the diffraction pattern of fully hydrided sample. The phases present are MgH_2 and MgF_2 along with small amount of unreacted Mg. Compared to the patterns of Figure 1 we see that the peaks of patterns of Figure 3 are not as broad, implying that the crystallite size increased. From Rietveld analysis we found that the crystallite size of Mg in the dehydrided pattern is 49.1 ± 0.8 nm while the crystallite size of MgH_2 in the reabsorbed pattern is 64 ± 2 nm. This shows that there is grain growth compared to the as-milled sample. This may be due to the high temperature of hydrogenation and also because of desorption/absorption itself.

Figure 4 shows the SEM images of MgH_2 + 5 mol % MgF_2 composite in (Figure 4a) desorbed state and (Figure 4b) after re-hydrogenation at 335 °C in comparison with pure MgH$_2$ (Figure 4c). The images show that ball milling with additive leads to effective decrease in particle size. In addition, energy dispersive X-ray (EDX) mapping done at higher magnification (Figure 5) shows that agglomerates consist of smaller MgH$_2$ particles and additive.

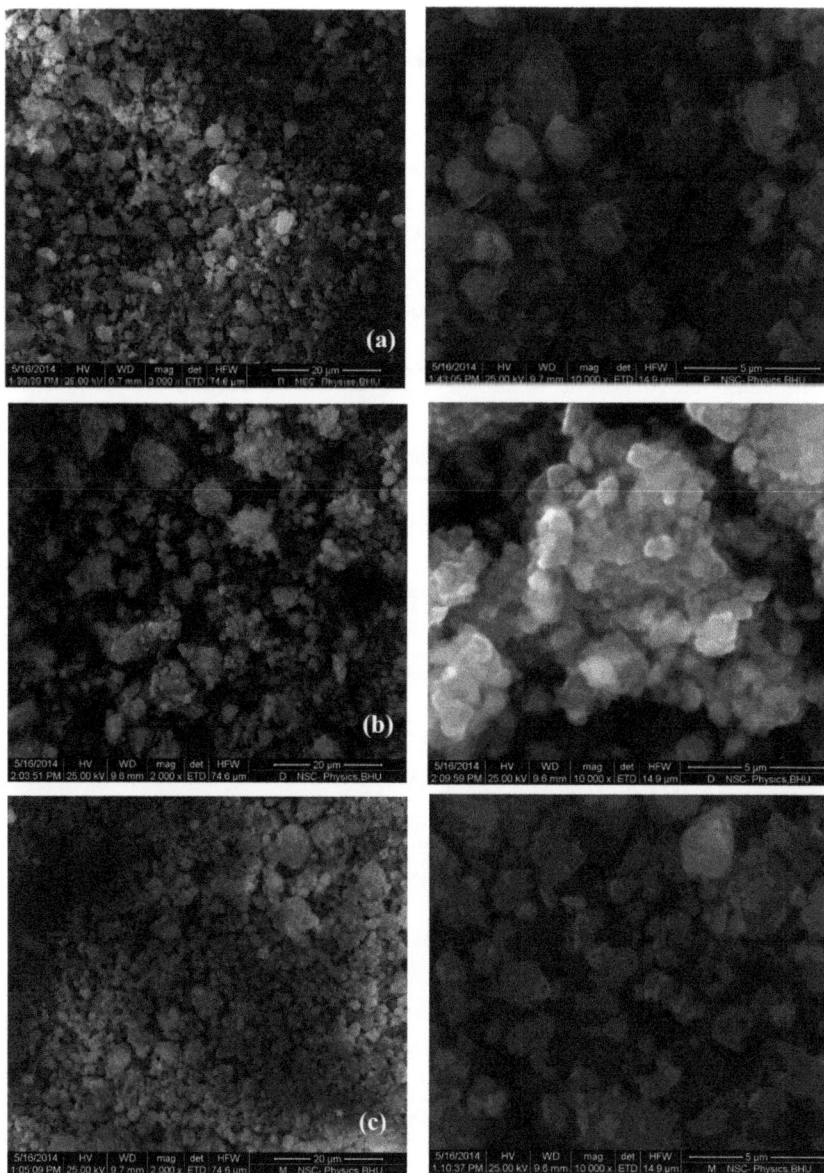

Figure 4. SEM images for MgH$_2$ + 5 mol % MgF$_2$ in (a) desorbed state and (b) re-hydrogenated state in comparison to (c) pure MgH$_2$.

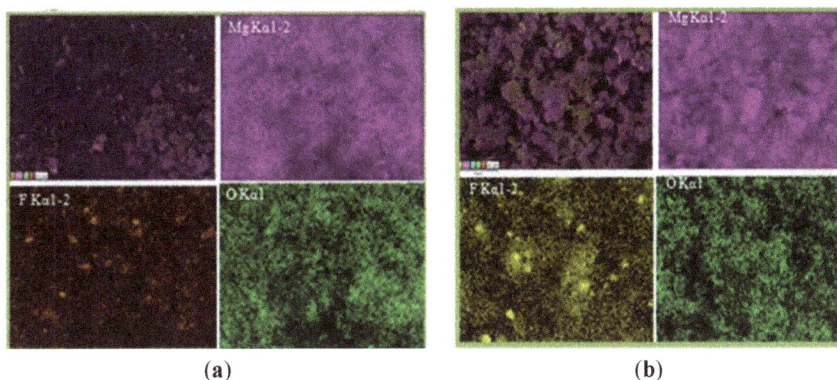

Figure 5. Elemental mapping showing particle morphology and distribution of 1 h milled MgH_2 + 5 mol % MgF_2: (**a**) after desorption and (**b**) after re-hydrogenation at 335 °C.

Elemental mapping made on MgH_2 + 5 mol % MgF_2 in both the desorbed state (Figure 5a) and re-hydrogenated state (Figure 5b) confirms homogenous distribution of MgF_2.

High energy milling leads to uniform dispersion of MgF_2 phase in MgH_2 matrix which may act as a catalytic layer and contributes in improving sorption properties. Chemical analysis performed by EDX spectroscopy during transmission electron microscopy (TEM) investigation of the desorbed sample gave the average atomic composition of different elements as 9.8% O, 14.7% F and 75.7% Mg which is very close to the nominal composition (86% Mg and 14% F). The presence of oxygen in EDX pattern in comparison to its small trace in XRD pattern could be due to small crystallite size of MgO making it peak difficult to distinguish from the background. A similar EDX investigation was performed on a sample that has been submitted to five dehydrogenation/hydrogenation cycles. Because abundances vary from point to point, we average over four different localisations. We found that, after cycling, the atomic composition of different elements was 15% $\pm$ 4% O, 11% $\pm$ 4% F and 74% $\pm$ 6% Mg. Within experimental error, these values are similar to the ones before cycling. However, this may be an indication that cycling induces a loss of MgF_2 and increase of MgO. Typical TEM micrographs presented in Figure 6 shows the morphology of the desorbed sample.

The image shows presence of large number of particles agglomerated together with no visibility clear particle boundaries. These observations are quite similar to those reported recently by Grzech *et al.* [13]. High resolution pictures taken over region-1 in Figure 6 and its corresponding selected area electron diffraction (SAED) patterns shows reflections at *d*-spacing 2.45, 1.90, 1.60, 1.36 and 1.22 Å which are characteristic of Mg (101), (102), (110), (112) and (202) planes respectively along with reflections at d values 2.10 and 1.48 Å, corresponding to MgO (200) and (220) planes. Thus, the surface consists of small crystallites of MgO (forming well defined ring and represented by red rings) surrounding the large crystallites of Mg (seen as discontinuous spots and represented by white rings). While the multiple SAED patterns acquired from the region-2 were well indexed as a mixture of large crystallite Mg and small crystallite of MgF_2 (seen as well-defined rings and colored yellow). The absence of oxide in region-2 is evidence that presence of fluoride limits MgO only to the surface. Both structural and morphological studies support the presence of MgF_2 phase even after complete hydrogen absorption/desorption cycle at 335 °C.

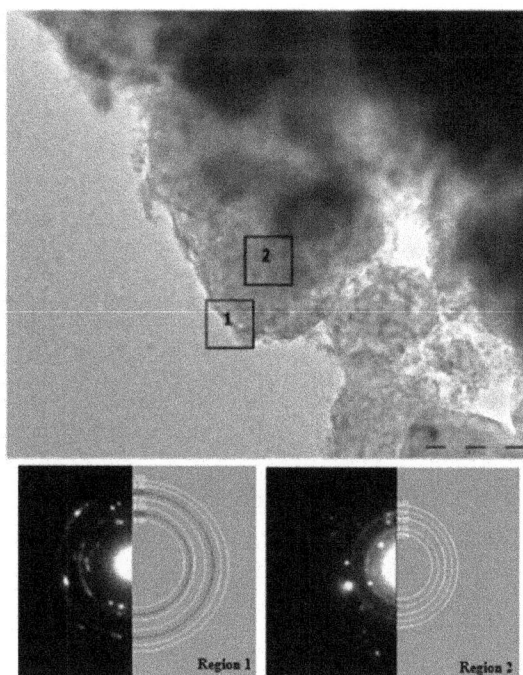

Figure 6. Transmission electron microscopy (TEM) micrograph of MgH_2 + 5 mol % MgF_2 sample after desorption at 335 °C with selected area electron diffraction (SAED) patterns and simulations. Region 1 is composed of Mg (**white rings**) covered with MgO layer (**red rings**) in simulated data while Region 2 shows diffraction rings corresponding to Mg (**white rings**) and MgF_2 (**yellow rings**).

2.2. Hydrogenation Characteristics of MgH_2 + 5 mol % MgF_2 at Lower Temperatures

The catalytic effect of 5 mol % MgF_2 on hydrogen sorption properties of MgH_2 was further investigated at lower temperatures. Figure 7 shows the absorption kinetics at 335, 310, 285 and 145 °C under 1000 kPa of hydrogen.

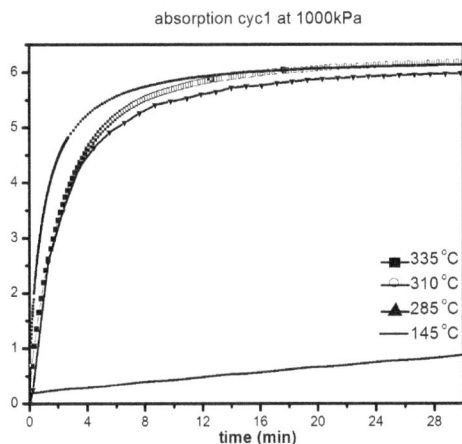

Figure 7. First absorption under 1000 kPa H$_2$ at different temperatures of 1 h milled MgH$_2$ + 5 mol % MgF$_2$. The insert is a compete absorption curve at 145 °C.

It should be pointed out that the samples were initially desorbed at 335 °C in order to ensure that full desorption was achieved before all absorption measurements. We notice only a slight loss in absorption capacity with reduction of temperature from 335 °C (6.2 wt % H$_2$) to 285 °C (5.8 wt % H$_2$). As seen in Figure 7, there was slight loss in kinetics and capacity in the temperature range 335–285 °C with the material reaching its complete capacity in less than 30 min. However, at 145 °C, the kinetics are much slower, but after 20 h, a capacity of 5.5 wt % is reached, as shown in Figure 7b. Desorption kinetic under 100 kPa H$_2$ at 285, 310 and 335 °C are shown in Figure 8.

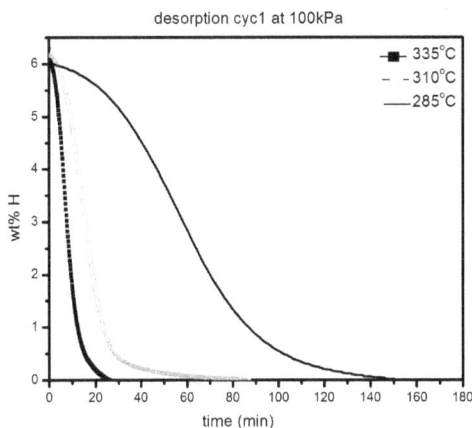

Figure 8. Desorption under 100 kPa H$_2$ at different temperatures of 1 h milled MgH$_2$ + 5 mol % MgF$_2$.

231

As expected, the kinetics are getting slower as temperature decreases but is still relatively fast even at 285 °C were complete desorption takes place in less than 3 h. These results reveal that even by sole addition of alkaline metal fluorides, improvements in hydrogenation characteristics of magnesium hydride can be achieved.

2.3. Cyclic Stability of MgH_2 + 5 mol % MgF_2 at T = 310 °C

Micro structural results have confirmed that MgF_2 phase does not decompose and no new phase formation occurs during hydrogen absorption/desorption measurements for MgH_2 + 5 mol % MgF_2 system. Therefore, cyclic performance of catalyzed magnesium hydride was examined at moderate operating temperature of 310 °C at pressure of 1000 kPa (for absorption) and 10 kPa (for desorption) to evaluate performance stability. Prior to the measurements the sample was completely desorbed at 335 °C. Figure 9 shows that the absorption capacity goes down from 5.9 wt % in first cycle to 5.6 wt % in the 10th cycle.

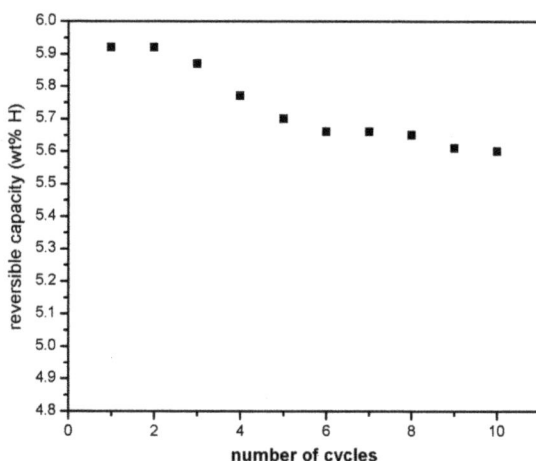

Figure 9. Hydrogen absorption kinetics at 310 °C under 1000 kPa of hydrogen of ball milled MgH_2 + 5 mol % MgF_2 for 10 cycles.

The observed loss of 0.3 wt % in capacity was reached within the first five cycles and thereafter the maximum capacity achieved by the system is more or less stabilized. These results show that magnesium hydride exhibits good hydrogen storage capacity and cyclic stability when magnesium fluoride is used as catalyst in comparison to the use of transition metal fluoride like NbF_5 or ZrF_4 where sharp decline in storage capacity was observed by Malka *et al.* [14] in the first 10 hydrogenation cycles recorded at 325 °C. X-ray diffraction patterns of the sample taken after 1st and 10th desorption cycle are presented in Figure 10. It shows that the β-MgH$_2$ phase and the catalytic material remain intact while small increase in content of MgO occurs. Thus, the growth of the MgO layer is mostly responsible for an observed loss in capacity.

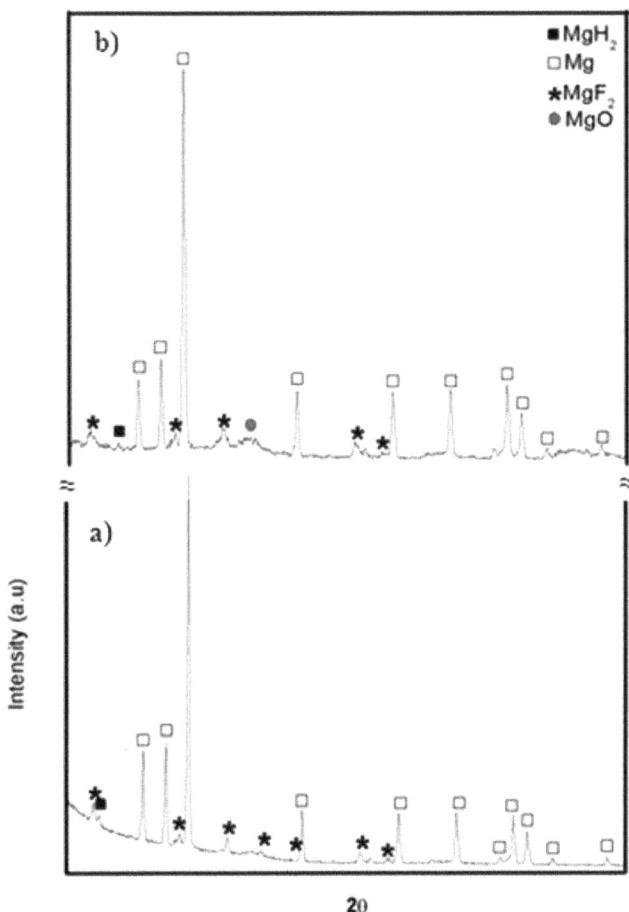

Figure 10. X-ray diffraction patterns of ball milled MgH$_2$ + 5 mol % MgF$_2$ after (**a**) one desorption and (**b**) 10 desorption cycles at 310 °C.

3. Discussion

The structural and hydrogenation results suggest that hydrogen absorption/desorption kinetics of MgF$_2$ doped MgH$_2$ is relatively slower than that attained with transition metal fluorides (TmF = TiF$_3$, ZrF$_4$, NbF$_5$, TaF$_5$). This could be explained by the presence of only one catalytically active phase in the present case (MgF$_2$) while two active phases are present when transition metal fluorides are used as additives. More explicitly, upon dehydrogenation the following reaction takes place in the present case.

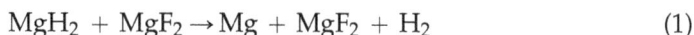

$$MgH_2 + MgF_2 \rightarrow Mg + MgF_2 + H_2 \tag{1}$$

While, as reported by Ma *et al.* [11], when transition metal fluoride is added the reaction taking place is:

$$3MgH_2 + 2TiF_3 \rightarrow 3\,MgF_2 + 2TiH_2 + H_2 \tag{2}$$

As TiH_2 possess more negative enthalpy formation (-136 kJ/mol) than MgH_2 (-75 kJ/mol) it will remain as a stable phase during desorption of MgH_2 in later cycles [13]. Furthermore, presence and concentration of TiH_2 phase would increase on multiple absorption/desorption cycling, which results in reduction of overall storage capacity. In addition, the transition metal fluoride is very sensitive to atmospheric conditions. Ball milled MgH_2 + TM-fluoride samples require oxygen and moisture level to be less than 0.1 ppm for obtaining good hydrogenation results [7,9,11,15]. Additionally, even in the desorbed state the material is pyrophoric, which makes it difficult to handle. Moreover, transition metals are much heavier than magnesium thereby increasing the mass of entire system. It thus seems more practical to use MgF_2 as a doping agent to increase the hydrogenation/dehydrogenation kinetics than transition metal fluorides.

4. Experimental Section

The starting materials MgH_2 (99.8% purity) and MgF_2 (99.9% purity) purchased from Alfa Aesar (Ward Hill, MA, USA) were vacuum annealed for few hours at 80 °C before using them for experiments. Afterwards, MgH_2 powder with 2, 5, and 10 mol % MgF_2, was milled under Ar atmosphere using Fritsch P4 planetary mill (Idar-Oberstein, Germany) with ball to powder ratio of 50:1 at a crucible rotation speed of 220 rpm. Milling was done for 60 min with 15 min rest after every 15 min of milling. The final milled products were handled in a glove box with oxygen and moisture level below 0.1 ppm. Initial hydrogen desorption curves taken at 335 °C under 100 kPa H_2 pressure showed that with 2 mol % catalytic material the kinetics was too slow which could be improved by increasing the additive content to 5 mol %. Further increase in concentration of catalytic material to 10 mol % didn't cause any significant change in kinetics. Therefore, MgF_2 concentration was restricted to 5 mol % for further investigation.

The hydrogenation characteristics were measured on homemade Sievert-type apparatus and the cyclic studies were made on an automated-four channel apparatus called Multi Channel Hydride Evaluation System from Advanced Materials Corporation, Petersburg, VA, USA. Approximately 400 mg of powder was placed in a sample cell and completely desorbed under dynamic vacuum at 335 °C prior to any measurement. Thereafter all measurements were made under 1000 kPa H_2 pressure for absorption and 100 kPa H_2 pressure for desorption at temperatures ranging from 335 °C to 145 °C. X-ray diffraction was performed using Bruker D8 Focus X-Ray apparatus (Bruker, Madison, WI, USA) with $CuK\alpha$ radiation. Phase abundances were evaluated from Rietveld method using Topas software [16]. Small quantity of milled MgH_2 + 5 mol % MgF_2 (a) after desorption at 325 °C and (b) after rehydrogenation under 1000 kPa H_2 was characterized for morphological studies with chemical analysis using JEOL JSM-5500 scanning electron microscope (JEOL, Tokyo, Japan). The sample was filled in air tight bottles and taken to SEM-EDX lab were they were slightly exposed to air for loading in SEM chamber. TEM analysis was performed on FEI: Technai $20G^2$ electron microscope (FEI, Hillsboro, OR, USA), operating at 200 kV accelerating voltage. TEM samples were prepared by dry dispersion of the powder onto a carbon substrate supported by copper TEM grid. This was done in an argon glove box before the TEM session, and the prepared sample was sealed by covering with parafilm tape to be carried to TEM lab. The sample was exposed to air for short duration during loading onto the TEM holder. Thus, partially transformed samples were characterised using scanning and transmission electron microscopy.

5. Conclusions

This investigation showed that magnesium fluoride could significantly influence the hydrogen sorption properties of magnesium hydride. It has been shown that MgF_2 additive acts as a catalyst for MgH_2, thereby improving its hydrogenation/dehydrogenation kinetics. These kinetic improvements are due to the presence of chemically stable MgF_2 powder well mixed in MgH_2 matrix and MgO layer being limited only to the surface. Cyclic stability reveals that 5 mol % MgF_2 helps to accelerate the reversible kinetics of MgH_2 with higher capacity in comparison to other transition metal fluoride catalysts. This is probably due to the persistence of MgF_2 phase during hydrogen cycling. These results suggest that owing to its fast sorption properties, low sensitivity to atmospheric conditions and easy handling ability, this material can be used in applications where operation at relatively high temperature is not considered a significant issue.

Acknowledgments: Pragya Jain would like to thank University Grant Commission (UGC) for the Dr. D.S. Kothari post-doctoral fellowship to carry out this work.

Author Contributions: Study conception and design: Pragya Jain, Jacques Huot; Acquisition of data: Pragya Jain, Viney Dixit; Analysis and interpretation of data: Pragya Jain, Jacques Huot, Onkar N. Srivastava; Drafting of manuscript: Pragya Jain, Ankur Jain, Jacques Huot; Critical revision: Pragya Jain, Jacques Huot.

Conflicts of Interest: The authors declare no conflict of interest.

References

1. Jain, I.P.; Lal, C.; Jain, A. Hydrogen storage in Mg: A most promising material. *Int. J. Hydrog. Energy* **2010**, *35*, 5133–5144.

2. Wu, Z.; Allendorf, M.D.; Grossman, J.C. Quantum Monte Carlo Simulation of Nanoscale MgH_2 Cluster Thermodynamics. *J. Am. Chem. Soc.* **2009**, *131*, 13918–13919.

3. Dornheim, M.; Eigen, N.; Barkhordarian, G.; Klassen, T.; Bormann, R. Tailoring Hydrogen Storage Materials Towards Application. *Adv. Eng. Mater.* **2006**, *8*, 377–385.

4. Hanada, N.; Ichikawa, T.; Fujii, H. Catalytic effect of nanoparticle 3d-transition metals on hydrogen storage properties in magnesium hydride MgH_2 prepared by mechanical milling. *J. Phys. Chem. B* **2005**, *109*, 7188–7194.

5. Aguey-Zinsou, K.F.; Ares Fernandez, J.R.; Klassen, T.; Bormann, R. Effect of Nb_2O_5 on MgH_2 properties during mechanical milling. *Int. J. Hydrog. Energy* **2007**, *32*, 2400–2407.

6. Ma, L.P.; Wang, P.; Cheng, H.M. Improving hydrogen sorption kinetics of MgH_2 by mechanical milling with TiF_3. *J. Alloys Compd.* **2007**, *432*, L1–L4.

7. Yavari, A.R.; LeMoulec, A.; de Castro, F.R.; Deledda, S.; Friedrichs, O.; Botta, W.J.; Vaughan, G.; Klassen, T.; Fernandez, A.; Kvick, Å. Improvement in H-sorption kinetics of MgH_2 powders by using Fe nanoparticles generated by reactive FeF_3 addition. *Scr. Mater.* **2005**, *52*, 719–724.

8. Deledda, S.; Borissova, A.; Poinsignon, C.; Botta, W.J.; Dornheim, M.; Klassen, T. H-sorption in MgH_2 nanocomposites containing Fe or Ni with fluorine. *J. Alloys Compd.* **2005**.

9. Malka, I.E.; Pisarek, M.; Czujko, T.; Bystrzycki, J. A study of the ZrF_4, NbF_5, TaF_5, and $TiCl_3$ influences on the MgH_2 sorption properties. *Int. J. Hydrog. Energy* **2011**, *36*, 12909–12917.

10. Ivanov, E.; Konstanchuk, I.; Bokhonov, B.; Boldyrev, V. Hydrogen interaction with mechanically alloyed magnesium-salt composite materials. *J. Alloys Compd.* **2003**, *359*, 320–325.

11. Ma, L.-P.; Wang, P.; Cheng, H.-M. Hydrogen sorption kinetics of MgH_2 catalyzed with titanium compounds. *Int. J. Hydrog. Energy* **2010**, *35*, 3046–3050.

12. Soni, P.R. *Mechanical Alloying: Fundamentals and Applications*; Cambridge International Science Publishing: Cambridge, UK, 2000; p. 160.

13. Grzech, A.; Lafont, U.; Magusin, P.; Mulder, F.M. Microscopic Study of TiF_3 as Hydrogen Storage Catalyst for MgH_2. *J. Phys. Chem. C* **2012**, *116*, 26027–26035.

14. Malka, I.E.; Bystrzycki, J.; Płociński, T.; Czujko, T. Microstructure and hydrogen storage capacity of magnesium hydride with zirconium and niobium fluoride additives after cyclic loading. *J. Alloys Compd.* **2011**, *509*, S616–S620.

15. Jin, S.-A.; Shim, J.-H.; Cho, Y.W.; Yi, K.-W. Dehydrogenation and hydrogenation characteristics of MgH_2 with transition metal fluorides. *J. Power Sources* **2007**, *172*, 859–862.

16. *Topas V4: General Profile and Structure Analysis Software for Powder Diffraction Data*; Bruker_AXS: Karlsruhe, Germany, 2008.

Hydrogen Storage in Pristine and d10-Block Metal-Anchored Activated Carbon Made from Local Wastes

Mohamed F. Aly Aboud, Zeid A. ALOthman, Mohamed A. Habila,
Claudia Zlotea, Michel Latroche and Fermin Cuevas

Abstract: Activated carbon has been synthesized from local palm shell, cardboard and plastics municipal waste in the Kingdom of Saudi Arabia. It exhibits a surface area of 930 m^2/g and total pore volume of 0.42 cm^3/g. This pristine activated carbon has been further anchored with nickel, palladium and platinum metal particles by ultrasound-assisted impregnation. Deposition of nanosized Pt particles as small as 3 nm has been achieved, while for Ni and Pd their size reaches 100 nm. The solid-gas hydrogenation properties of the pristine and metal-anchored activated carbon have been determined. The pristine material exhibits a reversible hydrogen storage capacity of 2.3 wt% at 77 K and 3 MPa which is higher than for the doped ones. In these materials, the spillover effect due to metal doping is of minor importance in enhancing the hydrogen uptake compared with the counter-effect of the additional mass of the metal particles and pore blocking on the carbon surface.

Reprinted from *Energies*. Cite as: Aboud, M.F.A.; ALOthman, Z.A.; Habila, M.A.; Zlotea, C.; Latroche, M.; Cuevas, F. Hydrogen Storage in Pristine and d10-Block Metal-Anchored Activated Carbon Made from Local Wastes. *Energies* **2015**, *8*, 3578–3590.

1. Introduction

Hydrogen has become a promising substitute for fossil fuels in automotive applications due to its abundance and its higher chemical energy. Hydrogen holds three times the specific energy of gasoline [1]. Besides, it is environmentally friendly since its only oxidation product is water [2]. Owing to its extremely low density, a lot of efforts are in progress to reversibly store hydrogen by different methods. The main avenues can be summarized in two groups. The first avenue is chemical storage where chemical bonds occur between hydrogen atoms and receptors to form hydride phases [3,4]. Some of these materials are heavy, expensive and difficult to regenerate on-board. Additionally the thermodynamic stability of these compounds is usually high and a lot of energy is needed for releasing the hydrogen back during the desorption process. The second avenue is physical storage where hydrogen is kept in its molecular state. Physical storage can be performed in three ways:

i) Liquefaction of hydrogen, which requires cryogenic temperatures and efficient insulation. This wastes around 30% of the available energy from burning the hydrogen which makes this storage method a costly process [5].

ii) Compression of hydrogen gas in high-pressure tanks raises safety concern issues mainly as results of hydrogen embrittlement effects. Additionally the low gravimetric storage should also be considered if we include the mass of the tank [6].

iii) Physisorption in porous materials, which attracts scientific interests due to its simplicity. The large surface areas of porous materials, especially of nanostructures, allow the adsorption of large quantities of hydrogen.

Activated carbon (AC) is among the potential candidates for the last route because of its abundance, low cost, non-toxicity, good chemical stability, low density, good recycling characteristics and wide range of pore structures and forms [7–12]. The main factor that limits the storage of hydrogen on the carbon surface is the low energy of adsorption, in the range of 4–8 kJ/mol H_2 for most porous carbons.

Metal atom decoration on graphitic skeletons can increase the hydrogen uptake by introducing strong hydrogen interaction sites on the carbonaceous surfaces. Besides, these metallic centers facilitate the dissociation of hydrogen molecules and eventually increase the hydrogen content over the carbon receptor by a spillover mechanism [13].

In this work, high surface area activated carbon was produced from local wastes and decorated with some d^{10}-block transition elements (Ni, Pd and Pt). Park *et al.* reported that the amount of hydrogen storage capacity for Multi Walled Carbon Nanotubes (MWCNTs) increased in proportion to the Pt content with an optimum loading value of 4 wt% [14]. Zubizarreta *et al.* reported that the optimum nickel content for improving the hydrogen storage in carbon nanospheres was 5 wt% [15], while, Kim *et al.* found that hydrogen storage increased up to 1.23 wt% after loading with 8 wt% of nickel nanoparticles due to the spillover effect [16]. For Pd doping, Lueking *et al.* reported that the optimum loading for hydrogen storage on activated carbon is 14 wt%, while it is 5.7 wt% in single wall carbon nanotubes [17]. Mu *et al.* used 20 wt% of palladium for enhancing the hydrogen storage capacity in etched carbon nanotubes [18].

After performing detailed chemical, structural and thermal characterizations, the hydrogen storage properties of the metal-anchored activated carbons were determined and compared to those of the pristine material.

2. Results and Discussion

2.1. Structural, Chemical and Thermal Characterizations

Figure 1 displays the X-Ray Diffraction (XRD) patterns of all studied samples. For the pristine activated carbon almost no diffraction peaks are detected due to its poor crystallinity. However, a tiny peak around $2\theta° = 26.5°$ might correspond to the (002) plane of graphite [19]. After Pt-anchoring, an additional and very broad peak is detected at $2\theta = 39.7°$. It is assigned to the diffraction plane (111) of the *fcc* structure of Pt according to JCPDS card No. 03-065-2868. This strong peak broadening is attributed to the ultra-small size of platinum particles, which will be later supported by High-Resolution Transmission Electron Microscope (HRTEM) analysis. For the Ni-anchored sample, well-defined diffraction peaks are detected at 44.5° (111), 52° (200) and 76.5° (220) in 2θ. They are characteristic of *fcc* nickel in agreement with the JCPDS card No. 03-065-2865. Finally, for the Pd-anchored sample sharp diffraction peaks are detected at $2\theta°$ of 39.88° (111), 46.45° (200), 67.96° (220), 82° (311) and 86° (222) showing the occurrence of *fcc* Palladium (JCPDS card No. 03-065-2867).

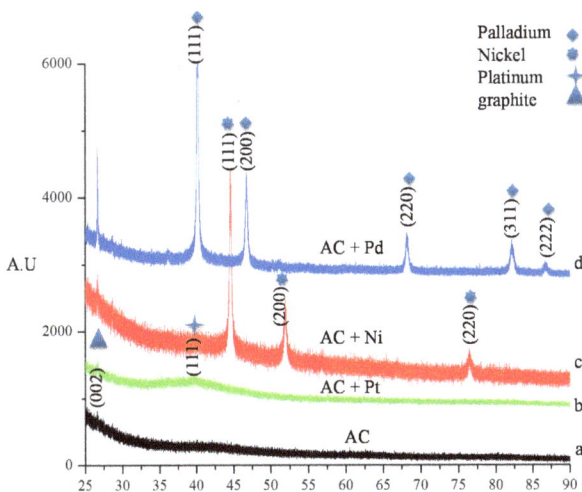

Figure 1. XRD diffraction patterns of (**a**) activated carbon (AC), (**b**) Pt-anchored AC, (**c**) Ni-anchored AC and (**d**) Pd-anchored AC.

Backscattered SEM images for pristine activated carbon AC, Pt-, Ni- and Pd-anchored AC are shown in Figure 2a to Figure 2d, respectively. The pristine AC exhibits a fibrous layer structure with particle dimensions of ~30 μm in length and ~4 μm in thickness. Cavities and porosity are observed as result of the evaporation of the activating agent. Both Ni- and Pd-anchored particles have spherical morphology with particle size ~100 nm. Nickel particles are uniformly dispersed over the AC, while Pd particles tend to agglomerate. SEM images failed to detect the presence of platinum particles on the surface of the activated carbon (Figure 2b) due to its low spatial resolution. Therefore, HRTEM investigations were used for Pt-AC as shown in Figure 3. The platinum particles are homogeneously distributed with ultra-small particle size of ~3 nm. The formation of such ultra-small particles of platinum compared to other anchored metals is consistent with the reported observation by Okhlopkova *et al.* that referred to the aggregation effect of palladium compared to platinum [20].

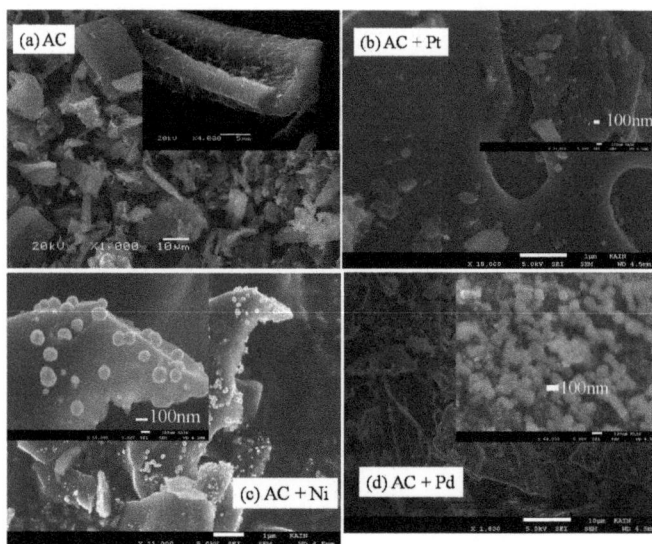

Figure 2. SEM images of (**a**) activated carbon (AC), (**b**) Pt-anchored AC, (**c**) Ni-anchored AC and (**d**) Pd-anchored AC.

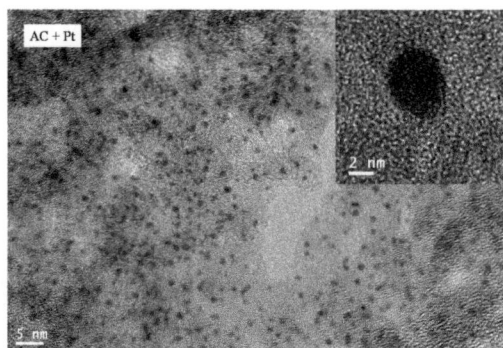

Figure 3. HRTEM images of Pt-anchored activated carbon.

The FTIR spectra recorded for all samples in the 400–4,000 cm^{-1} range are displayed in Figure 4. Almost no difference is seen between the activated carbon and the metal anchored activated carbons. The main groups are both alkene and aromatic C=C functions in addition to hydroxyl (OH$^-$) groups.

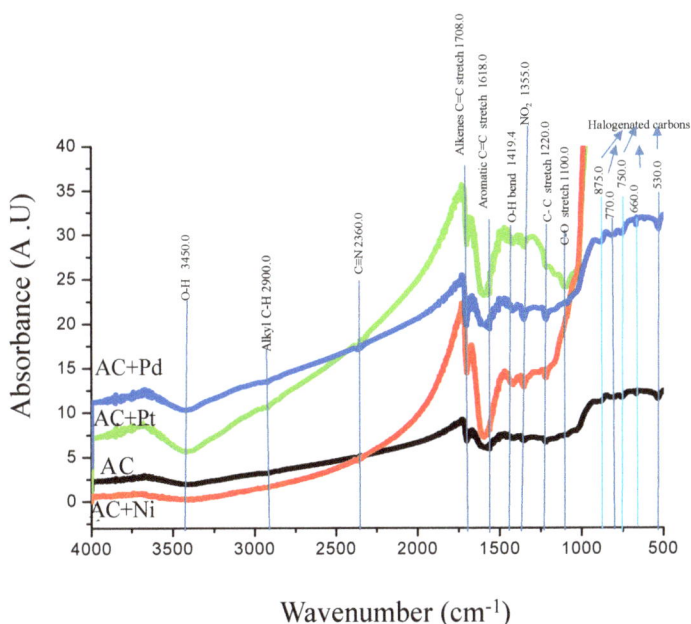

Figure 4. FTIR spectra for activated carbon (AC) and metal-anchored AC + Ni, AC + Pd and AC + Pt.

Figure 5 shows the monitored Differential Scanning Calorimetry (DSC) (Figure 5a) and Thermal Gravimetric Analysis (TGA) (Figure 5b) traces under air flow. For all samples, exothermic reactions due to carbon oxidation are observed. DSC peak temperatures are observed at ~820 K for both the AC and Ni-anchored materials, while they occur at ~770 K and 650 K for Pd- and Pt-anchored ones, respectively. The DSC onset reaction temperatures, as determined from the tangential method, are observed at ~675 K for AC, Ni- and Pd-anchored samples. Interestingly, the onset temperature for Pt-anchored AC is as low as 600 K. The lower oxidation temperature for the Pt containing sample is attributed to the ultra-fine Pt particles which enhance the oxidation reaction.

Figure 5b shows the TGA traces for all samples. Weight losses below 400 K are likely related to desorption of residual adsorbed moisture and volatile matter. Next, in agreement with the DSC analyses, release of CO_2 due to carbon oxidation takes place in the 600–873 K range leading to severe mass loss. Above this temperature range, no significant mass changes occur. The reported TGA curves allow estimating the actual metal doping of the different samples. It can be evaluated from the final weighted TGA mass considering that the amount of carbon ashes is constant for all samples (evaluated as *ca.* 9 wt% for pristine AC) and that the anchored metals form oxide species NiO, PdO, and PtO_2 by reaction with air. The thus-estimated metal doping is 6.5 wt% Ni, 16 wt% Pd, and 5 wt% Pt for nickel-, palladium- and platinum-anchored activated carbon samples, respectively.

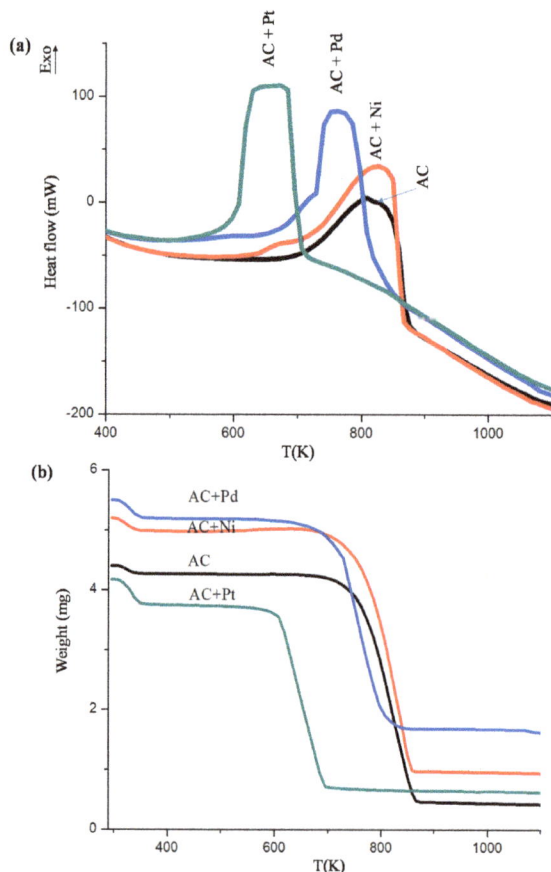

Figure 5. (a) DSC analysis under air flow of activated carbon (AC) and metal-anchored AC + Ni, AC + Pd and AC + Pt, and (b) TGA analysis under air flow of AC and metal-anchored AC + Ni, AC + Pd and AC + Pt.

2.2. Textural Characterization

Figure 6 shows the nitrogen adsorption isotherms at 77 K of pristine AC and Ni-, Pt-, Pd-anchored activated carbon. The isotherms are Type I, typical of microporous materials with a narrow pore size distribution. The textural properties are summarized in Table 1. The pristine AC as well as both Ni- and Pt-anchored samples have similar textural properties with Brunauer-Emmett-Teller (BET) surface area $A_s \sim 900$ m^2/g, total pore volume $V_{tot} = 0.42$ cm^3/g and micropore volume $V_\mu \sim 0.35$ cm^3/g. The Pd-anchored sample exhibits much lower textural properties.

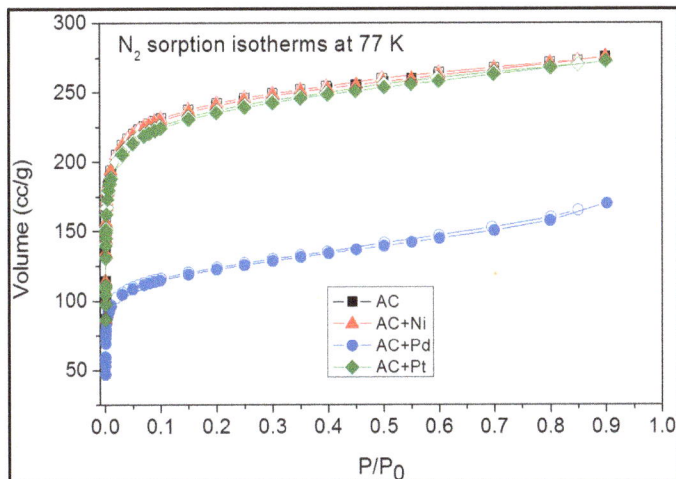

Figure 6. Nitrogen adsorption isotherms at 77 K of pristine and Ni-, Pd- and Pt-anchored activated carbon.

Table 1. Textural properties of pristine and Ni-, Pt-, Pd-anchored activated carbon.

Sample	Skeletal Density g/cm^3	A_s (BET) m^2/g	V_{tot} (P/P_0 = 0.9) cm^3/g	V_μ (DR) cm^3/g
Activated carbon	1.7 ± 0.1	931 ± 20	0.42 ± 0.1	0.36 ± 0.1
AC + Ni	1.9 ± 0.1	910 ± 20	0.42 ± 0.1	0.35 ± 0.1
AC + Pt	1.9 ± 0.1	889 ± 20	0.42 ± 0.1	0.34 ± 0.1
AC + Pd	2.0 ± 0.1	455 ± 20	0.26 ± 0.1	0.17 ± 0.1

Although the textural properties of activated carbons can vary depending on the processing techniques and precursors, the values reported here are almost identical to activated carbon made from Malaysian palm shells that were physically activated with steam ($A_s = 941$ m^2/g, $V_{tot} = 0.52$ cm^3/g) [21]. Moreover, they are much better than for H$_3$PO$_4$ chemically activated Malaysian palm ($A_s = 615$ m^2/g, $V_{tot} = 0.28$ cm^3/g) [22]. Also, the textural properties of our AC is superior to the activated carbon made from palm shells that underwent both chemical and physical treatments ($A_s = 642$ m^2/g, $V_{tot} = 0.28$ cm^3/g) [22].

The slight decrease for the surface area of the Ni- and Pt-anchored ACs compared with the pristine one is consistent with the reported data for Pt doped super-activated carbons [23]. This decrease was attributed to metal particles filling or blocking pores in the activated carbon. The drastic decrease for Pd-AC, with BET surface area about half that of Ni- and Pt-anchored AC, likely results from the high agglomeration of Pd-particles at the AC surface (Figure 2d).

The increase in the density of the metal anchored AC compared with the pristine one is due to the higher density of metal particles as compared to that of the adsorbent. Such an effect is consistent with the higher reported density of doped Pd-activated carbon fibers compared to non-doped ones [24]. The low surface area and pore volumes for palladium supported activated carbon due to agglomeration result in more packed and dense material compared to other doped activated carbons due its higher loading as was confirmed by the TGA measurements (Figure 5b).

2.3. Hydrogen Storage Measurements

Pressure Composition Isotherm (PCI) curves at 77 K and 298 K for all samples are shown in Figure 7 and the hydrogen uptake capacities are summarized in Table 2. In all cases, no hysteresis (below the experimental error) is observed, indicating that the adsorbed hydrogen can be desorbed reversibly when the pressure is released.

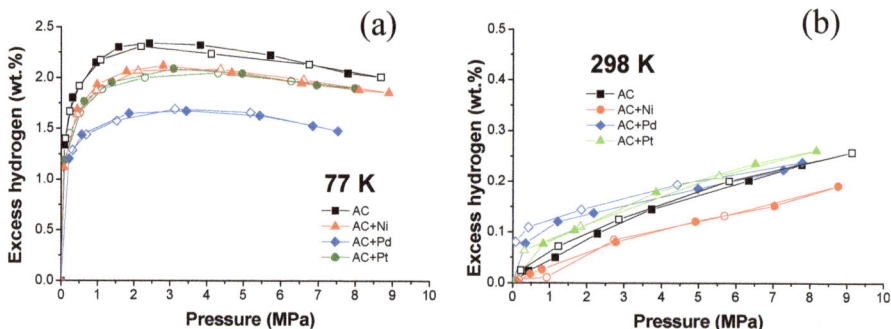

Figure 7. Pressure composition isotherm curves at (**a**) 77 K and (**b**) 298 K for the activated carbon (AC), Ni- , Pd- and Pt-anchored AC. Full and empty symbols stand for adsorption and desorption measurements, respectively.

All low temperature (77 K) the PCI curves exhibit Type I isotherms. This indicates monolayer adsorption of hydrogen molecules, which extent depends on the micropore volume and its filling. The hydrogen uptake increases up to 3 MPa and then reaches a maximum due to the completion of micropore filling. The amount of adsorbed hydrogen in the pristine activated carbon is consistent with the maximum theoretical capacity established by Ströbel *et al.* for carbonaceous materials [25]. These authors showed that the hydrogen uptake on carbon surface could be derived as:

$$H\,(\mathrm{wt\%}) \;=\; 2.27 \times 10^{-3} \cdot A_s\,(\mathrm{m^2/g}) \tag{1}$$

where A_s is the surface area of the carbonaceous material. Using Equation (1) and substituting for the observed surface area $A_s = 931$ m^2/g, we obtain a theoretical hydrogen uptake of 2.1 wt% in close agreement with the reported value of 2.3 wt%. Moreover, these capacities are consistent with the reported data for wide range of carbon-based materials with wide textural properties [26–29]. The density of the adsorbed hydrogen at 77 K can be calculated from the hydrogen uptake (2.3 wt%) and the volume occupied by hydrogen in the micropore volume (0.36 cm^3/g). The calculated density equals 0.063 g/cm^3, lower than the reported density for liquid hydrogen at 20 K (0.071 g/cm^3) [30].

Table 2. Hydrogen uptake for pristine activated carbon and Ni-, Pt- and Pd-anchored activated carbon at 77 and 298 K.

Sample	H$_2$ Excess Capacity (wt%)	
	77 K at 3 MPa	298 K at 8 MPa
Activated carbon	2.3 ± 0.1	0.24 ± 0.01
AC + Ni	2.1 ± 0.1	0.17 ± 0.01
AC + Pt	2.1 ± 0.1	0.26 ± 0.01
AC + Pd	1.7 ± 0.1	0.25 ± 0.01

At 298 K and high-pressure (Figure 7b), both adsorption and desorption PCI isotherms exhibit linear behavior with low hydrogen uptake (≤0.26 wt%). Such a low capacity is attributed to the low enthalpy of hydrogen adsorption [31] and concurs with earlier reports on platinum doping in super activated carbon [18] and palladium in mesoporous carbon [32]. Similar capacities have been reported for hydrogen storage in activated carbon of even higher surface area $A_s = 3{,}500$ g/cm^3 [23]. At high pressure, pure AC, Pd- and Pt-anchored activated carbons exhibit rather similar hydrogen uptake, whereas Ni-anchored activated carbon has lower capacity. At room temperature and low pressure, Pd-anchored activated carbon has higher capacity than all other materials. This confirms the formation of palladium hydride at room temperature and low pressure, as demonstrated previously [28,29,32]. However, at high pressure, the favorable effect of palladium hydride formation is counteracted by the weight of the dopant, the filling of adsorbent micropores and pore blocking. These latter drawbacks could be counterbalanced if metal deposition would induce a significant hydrogen uptake by spillover effect. Unfortunately, the low hydrogen uptake at high pressure suggests that the spillover effect, if it exists, turns out to be of minor importance in the studied systems.

3. Experimental Section

Palm shell, cardboard and plastics collected from a municipal solid waste station in Riyadh, Saudi Arabia were used as raw material precursors in the present study. The mixing ratio was 1:1:0.33 by weight. The choice of such raw materials was given due to their abundance in the kingdom which will be in great advantage for a green environment. The elemental analysis for the raw materials and their activation process are given in [33,34]. The used metal precursors were nickel acetate tetrahydrate ($Ni(OCOCH_3)_2 \cdot 4H_2O$, 99%, E. Merck, darmstadt, Germany), palladium acetate ($Pd(OCOCH_3)_2$, 99%, Sigma Aldrich, St. Louis, MO, USA) and hexachloroplatenic acid ($H_2PtCl_6 \cdot XH_2O$, assay 40.0% Loba Chemie, Mumbai, India).

The ultrasound-assisted impregnation method was used to anchor metal particles on activated carbon surface because of its simplicity and its large scale [35–38]. Deionized water was used as dissolving media for both nickel acetate and platinum chloride while ultra-high purity benzene was used for dissolving palladium acetate. In the same dissolving media, activated carbon was dispersed using an ultrasonic bath and mixed with the dissolved metal salt to nominally form a 10 wt% of anchored metal activated carbon. The mixture was left several days under the hood with repetitive stirring to achieve complete evaporation of the solvents. The resulted impregnated powders were grounded in ceramic mortar for homogenizing the impregnated materials. Same solvents were added to the resulted impregnated powders extra two times and followed by same procedure to ensure well dispersion. The resulted metal supported activated carbons were annealed at 523 K for 2 h under a continuous flow of hydrogen atmosphere to reduce the supported metal salts.

All samples underwent structural analysis by X-ray diffraction (XRD) using a X'Pert PRO Philips diffractometer (Eindhoven, The Netherlands), operated at 40 mA and 40 kV with CuK_α radiation and a nickel filter. Measurements were performed over the 2-theta range from 2° to 100° by steps of 0.02°, with a sampling time of one second per step. Sample morphology was investigated using a Field-Emission Scanning Electron Microscope (FE-SEM, model FEI-200NNL, Hillsboro, OR, USA) and a High-Resolution Transmission Electron Microscope (HRTEM, model JEM-2100F, JEOL, Tokyo, Japan). Surface chemistry was studied by Fourier Transform Infrared (FT-IR) transmission spectroscopy using a Nicolet 6700 FT-IR spectrophotometer (Thermo Scientific, Waltham, MA, USA). To this aim, samples were mixed with 98% wt% of KBr and then finely ground to make pellets. Thermal stability was investigated using a Differential Scanning Calorimetry–Thermal Gravimetric Analysis (TGA/DSC1 Star^e) system (Mettler Toledo, Columbus, OH, USA). TGA/DSC runs were performed under air within the temperature range 298–1,100 K with a heating rate of 5 K/min.

The textural properties were determined from nitrogen adsorption/desorption isotherms using an Autosorb-iQ Automated Gas Sorption Analyzer from Quantachrome Instruments (Hartley Wintney, UK). All isotherms have been measured by volumetric means. Samples were previously degassed overnight at 473 K. The specific surface area was obtained by the Brunauer-Emmett-Teller (BET) method. The total pore volume was computed from the amount of gas adsorbed at $P/P_0 = 0.9$, and the micropore volume was calculated using the Dubinin-Radushkevich (DR) equation.

Hydrogen sorption properties of samples were determined by measuring the Pressure-Composition-Isotherms (PCI) at 77 and 298 K up to 9 MPa hydrogen pressure. The PCI curves were recorded using a manual volumetric device (Sievert's method) equipped with calibrated and thermostated volumes and pressure gauges. The samples were enclosed in a stainless steel sample holder closed with a metal seal. Before any sorption measurements, the samples were degassed under secondary vacuum at 473 K for 12 h. Weight losses were measured as 3.1, 3.5, 6.8 and 5.6 wt% for non-doped, Ni-doped, Pt-doped and Pd-doped samples, respectively. The sample holder is immersed in a liquid nitrogen Dewar at 77 K or a thermostated water bath maintained at 298 K, and high purity hydrogen (6N) is introduced step by step up to 9 MPa. The pressure variations due to both gas expansion and hydrogen sorption are measured after reaching thermodynamic equilibrium, usually in the range of minutes. Real equation of state for hydrogen gas is used from the program GASPAK V3.32. The PCI curves were measured twice (*i.e.*, two full adsorption–desorption cycles) in order to check the hysteresis effect and the measurement repeatability. Good repeatability has been obtained for all samples. All capacities reported are excess hydrogen sorption quantities and refer to the sample dry mass (*i.e.*, degassed mass). Sample volume correction is derived from skeletal density measurements and data correspond to the excess values obtained by He pycnometry (Utrapyc 1200 Quantachrome, Hartley Wintney, UK).

4. Conclusions

Activated carbon was successfully synthesized from local palm shell and other wastes. The pristine activated carbon was impregnated with nickel, palladium and platinum. Coarse metallic nanoparticles of typical size 100 nm are formed for Ni- and Pd-anchored samples, whereas ultra-small particles size of 3 nm were obtained for the Pt case.

The hydrogen adsorption mechanism for all the synthesized samples occurs via micropore filling. No significant hysteresis was observed for any of the samples, indicating that adsorbed hydrogen can be desorbed reversibly when the pressure is released. At 298 K and 8 MPa both adsorption and desorption showed linear behavior with a low amount of hydrogen uptake. Low hydrogen uptake at ambient temperature is attributed to the low enthalpy of hydrogen adsorption on carbon surface.

At 77 K and 3 MPa, the activated carbon showed hydrogen storage of 2.3 wt% which is consistent with the BET surface area of ~931 m^2/g. Under the same conditions, metal-anchored activated carbons exhibited lower hydrogen uptakes ranging between 1.7 and 2.1 wt%. For the materials used in this study, possible spillover effects induced by the presence of metallic dopants are counterbalanced in terms of hydrogen storage by the concomitant addition of metal mass and reduction of carbon porosity.

Acknowledgments: The authors gratefully thank Sustainable Energy Technologies (SET) Center at King Saud University for financing this work.

Author Contributions: Mohamed F. Aly Aboud conceived and designed the experiments, performed the doping experiments and all data analysis; Mohamed F. Aly Aboud wrote the paper; Claudia Zlotea performed the textural characterization and hydrogen storage measurements; Claudia Zlotea, Michel Latroche and Fermin Cuevas participated in data analysis, paper writing and corrections; Zeid A. ALOthman and Mohamed A. Habila provided the virgin activated carbons and participated in FTIR analysis and provided important suggestions. All authors examined and approved the final manuscript.

Conflicts of Interest: The authors declare no conflict of interest.

References

1. Schlapbach, L.; Züttel, A. Hydrogen-storage materials for mobile applications. *Nature* **2001**, *414*, 353–358.

2. Rostrup-Nielsen, J.R.; Rostrup-Nielsen, T. Large-scale hydrogen production. *Cattech* **2002**, *6*, 150–159.

3. Krasae-In, S.; Stang, J.H.; Neksa, P. Simulation on a proposed large-scale liquid hydrogen plant using a multi-component refrigerant refrigeration system. *Int. J. Hydrog. Energy* **2010**, *35*, 4524–4533.

4. Barthélémy, H. Effects of pressure and purity on the hydrogen embrittlement of steels. *Int. J. Hydrog. Energy* **2011**, *36*, 2750–2758.

5. Kleperis, J.; Wójcik, G.; Czerwinski, A.; Skowronski, J.; Kopczyk, M.; Beltowska-Brzezinska, M. Electrochemical behavior of metal hydrides. *J. Solid State Electrochem.* **2001**, *5*, 229–249.

6. Ovshinsky, S.R.; Fetcenko, M.A.; Ross, J.A. Nickel metal hydride battery for electric vehicles. *Science* **1993**, *260*, 176–181.

7. Ebbesen, T.W.; Lezec, H.J.; Hiura, H.; Bennett, J.W.; Ghaemi, H.F.; Thio, T. Electrical conductivity of individual carbon nanotubes. *Nature* **1996**, *382*, 54–56.

8. Wong, E.W.; Sheehan, P.E.; Lieber, C.M. Nanobeam mechanics: Elasticity, strength, and toughness of nanorods and nanotubes. *Science* **1997**, *277*, 1971–1975.

9. Collins, P.G.; Bradley, K.; Ishigami, M.; Zettl, A. Extreme oxygen sensitivity of electronic properties of carbon nanotubes. *Science* **2000**, *287*, 1801–1804.

10. Liang, W.; Yokojima, S.; Ng, M.F.; Chen, G.; He, G. Optical properties of single-walled 4 Å carbon nanotubes. *J. Am. Chem. Soc.* **2001**, *123*, 9830–9836.

11. Minot, E.D.; Yaish, Y.; Sazonova, V.; McEuen, P.L. Determination of electron orbital magnetic moments in carbon nanotubes. *Nature* **2004**, *428*, 536–539.

12. Motta, M.; Li, Y.L.; Kinloch, I.; Windle, A. Mechanical properties of continuously spun fibers of carbon nano tubes. *Nano Lett.* **2005**, *5*, 1529–1533.

13. Psofogiannakis, G.M.; Froudakis, G.E. Study of hydrogen spill over mechanism on Pt doped graphite. *J. Phys. Chem. C* **2009**, *113*, 14908–14915.

14. Park, S.J.; Lee, S.Y. Hydrogen storage behaviors of platinum-supported multi-walled carbon nanotubes. *Int. J. Hydrog. Energy* **2010**, *35*, 13048–13054.

15. Zubizarreta, L.; Menéndez, J.A.; Pis, J.J.; Arenillas, A. Improving hydrogen storage in Ni-doped carbon nanospheres. *Int. J. Hydrog. Energy* **2009**, *34*, 3070–3076.

16. Kim, J.H.; Han, K.S. Ni nanoparticles-hollow carbon spheres hybrids for their enhanced room temperature hydrogen storage performance. *Trans. Korean Hydrog. New Energy Soc.* **2013**, *24*, 550–557.

17. Lueking, A.D.; Yang, R.T. Hydrogen spillover to enhance hydrogen storage—Study of the effect of carbon physicochemical properties. *Appl. Catal. A Gen.* **2004**, *265*, 259–268.

18. Mu, S.C.; Tang, H.L.; Qian, S.H.; Mu, P.; Yuan, R.Z. Hydrogen storage in carbon nanotubes modified by microwave plasma etching and Pd decoration. *Carbon* **2006**, *44*, 762–767.

19. Li, Z.Q.; Lu, C.J.; Xia, Z.P.; Zhou, Y.; Luo, Z. X-ray diffraction patterns of graphite and turbostratic carbon. *Carbon* **2007**, *45*, 1686–1695.

20. Okhlopkova, L.B.; Lisitsyn, A.S.; Likholobov, V.A.; Gurrath, M.; Boehm, H.P. Properties of Pt/C and Pd/C catalysts prepared by reduction with hydrogen of adsorbed metal chlorides: Influence of pore structure of the support. *Appl. Catal. A Gen.* **2000**, *204*, 229–240.

21. Arouaa, M.K.; Daud, W.M.A.W.; Yin, C.Y.; Adinata, D. Adsorption capacities of carbon dioxide, oxygen, nitrogen and methane on carbon molecular basket derived from polyethyleneimine impregnation on microporous palm shell activated carbon. *Sep. Purif. Technol.* **2008**, *62*, 609–613.

22. Arami-Niya, A.; Daud, W.M.A.W.; Mjalli, S.F.; Abnisa, F.; Shafeeyan, M.S. Production of microporous palm shell based activated carbon for methane adsorption: Modeling and optimization using response surface methodology. *Chem. Eng. Res. Des.* **2012**, *90*, 776–784.

23. Stadie, N.P.; Purewal, J.J.; Ahn, C.C.; Fultz, B. Measurements of hydrogen spillover in platinum doped super activated carbon. *Langmuir* **2010**, *26*, 15481–15485.

24. Contescu, C.I.; Benthem, K.V.; Sa, L.; Bonifacio, C.S.; Pennycook, S.J.; Jena, P.; Gallego, N.C. Single Pd atoms in activated carbon fibers and their contribution to hydrogen storage. *Carbon* **2011**, *49*, 4050–4058.

25. Ströbel, R.; Garche, J.; Moseley, P.T.; Jörissen, L.; Wolf, G. Hydrogen storage by carbon materials. *J. Power Sources* **2006**, *159*, 781–801.

26. Zubizarreta, L.; Gomez, E.I.; Arenillas, A.; Ania, C.O.; Parra, J.B.; Pis, J.J. Hydrogen storage in carbon materials. *Adsorption* **2008**, *14*, 557–566.

27. Yürüm, Y.; Taralp, A.; Veziroglu, N.T. Storage of hydrogen in nano structured carbon materials. *Int. J. Hydrog. Energy* **2009**, *34*, 3784–3798.

28. Dibandjo, P.; Zlotea, C.; Gadiou, R.; Ghimbeu, C.; Cuevas, F.; Latroche, M.; Leroy, E.; Vix-Guterl, C. Hydrogen storage in hybrid nanostructured carbon/palladium materials: Influence of particle size and surface chemistry. *Int. J. Hydrog. Energy* **2013**, *38*, 952–965.

29. Zhao, W.; Fierro, V.; Zlotea, C.; Chevalier-César, C.; Izquierdo, M.T.; Latroche, M.; Celzard, A. Carbons doped with Pd nanoparticles for hydrogen storage. *Int. J. Hydrog. Energy* **2012**, *27*, 5072–5080.

30. Midilli, A.; Ay, M.; Dincer, I.; Rosen, M.A. On hydrogen and hydrogen energy strategies: I: Current status and needs. *Renew. Sustain. Energy Rev.* **2005**, *9*, 255–271.

31. Thomas, M.K. Hydrogen adsorption and storage on porous materials. *Catal. Today* **2007**, *120*, 389–398.

32. Zlotea, C.; Cuevas, F.; Boncour, V.P.; Leroy, E.; Dibandjo, P.; Gadiou, R.; Guterl, C.V.; Latroche, M. Size-dependent hydrogen sorption in ultrasmall Pd clusters embedded in mesoporous carbon template. *J. Am. Chem. Soc.* **2010**, *132*, 7720–7729.

33. AlOthman, Z.A.; Habila, M.A.; Ali, R. Preparation of activated carbon using the copyrolysis of agricultural and municipal solid wastes at a low carbonization temperature. *Int. Conf. Biol. Environ. Chem.* **2011**, *24*, 67–72.

34. Habila, M.; Yilmaz, E.; ALOthman, Z.A.; Soylak, M. Flame atomic absorption spectrometric determination of Cd, Pb, and Cu in food samples after pre-concentration using 4-(2-thiazolylazo) resorcinol-modified activated carbon. *J. Ind. Eng. Chem.* **2014**, *20*, 3989–3993.

35. Zha, Q.F.; Hu, X.H.; Guo, Y.S.; Wu, M.B.; Li, Z.F.; Zhang, Y.Z. Improved antioxidative ability of porous carbons by boron-doping. *New Carbon Mater.* **2008**, *23*, 356–360.

36. Li, Y.; Yang, R.T. Hydrogen storage on platinum nanoparticles doped on superactivated carbon. *J. Phys. Chem. C* **2007**, *111*, 11086–11094.

37. Matsumoto, T.; Komatsu, T.; Nakanoa, H.; Arai, K.; Nagashima, Y.; Yooa, E. Efficient usage of highly dispersed Pt on carbon nano tubes for electrode catalysts of polymer electrolyte fuel cells. *Catal. Today* **2004**, *90*, 277–281.

38. Şayan, E. Ultrasound assisted preparation of activated carbon from alkaline impregnated hazelnut shell: An optimization study on removal of copper ion from aqueous solution. *Chem. Eng. J.* **2006**, *115*, 213–218.

MDPI AG

St. Alban-Anlage 66

4052 Basel, Switzerland

Tel. +41 61 683 77 34

Fax +41 61 302 89 18

http://www.mdpi.com

Energies Editorial Office

E-mail: energies@mdpi.com

http://www.mdpi.com/journal/energies

www.ingramcontent.com/pod-product-compliance
Lightning Source LLC
Chambersburg PA
CBHW080133240326
41458CB00128B/6379

* 9 7 8 3 0 3 8 4 2 2 0 8 2 *